Decision Engineering

Series editor

Rajkumar Roy, Cranfield, Bedfordshire, UK

More information about this series at http://www.springer.com/series/5112

Louis Redding · Rajkumar Roy
Editors

Through-life Engineering Services

Motivation, Theory, and Practice

Springer

Editors
Louis Redding
EPSRC Centre for Innovative Manufacture in Through-life Engineering Services
Cranfield University
Cranfield, Bedfordshire
UK

Rajkumar Roy
EPSRC Centre for Innovative Manufacture in Through-life Engineering Services
Cranfield University
Cranfield, Bedfordshire
UK

ISSN 1619-5736 ISSN 2197-6589 (electronic)
Decision Engineering
ISBN 978-3-319-12110-9 ISBN 978-3-319-12111-6 (eBook)
DOI 10.1007/978-3-319-12111-6

Library of Congress Control Number: 2014957710

Springer Cham Heidelberg New York Dordrecht London

Printed on acid-free paper

Springer International Publishing AG Switzerland is part of Springer Science+Business Media (www.springer.com)

Some Food for Thought!

The major difference between a thing that might go wrong and a thing that cannot possibly go wrong is that when a thing that cannot possibly go wrong goes wrong, it usually turns out to be impossible to get at or repair

Douglas Adams

Strive for perfection in everything you do. Take the best that exists and make it better. Where it does not exist, design it

Sir Henry Royce

A good engineer thinks in reverse and asks himself about the …… consequences of the components and systems he proposes

Helmut Jahn

Knowledge has to be improved, challenged, and increased constantly or it will disappear

Peter F. Drucker

A superficial knowledge is not enough. It must be ... knowledge capable of analyzing a situation quickly and making an immediate decision

Cavert Robert

Acquire new knowledge whilst thinking over the old, and you may become a teacher of others

Confucius

Foreword

History is a mirror to the future. We can learn a great deal from it. A product generates wealthy data during its life cycle, and we can maximize its value if we understand how customers have used this data.

As engineers, we often focus on designing the product's functions to solve the defined known problems, and analyzing them using the scientific knowledge and our experiences. But often we fail to address its meaning to customers due to lack of understanding of the value during its life cycle. The total ownership value of a product must be well understood before we can optimize the value.

Generally speaking, product life cycle issues can be mapped into two spaces: visible and invisible. Some examples of visible issues include machine failure, product defects, poor cycle times, long time delays, drop in overall equipment effectiveness (OEE), etc. On the other hand, invisible issues may occur as machine degradation and component wear. In each of the spaces, issues are treated at both deterministic and uncertain levels.

Traditional practices depict problem solving for well-defined problems, such as quality, productivity, and costs issues, etc., through continuous improvement, best practices, and standard work. Competitive companies use new methods and techniques to work with their suppliers and partners to integrate design and manufacturing for problem avoidance. World-class companies develop innovation and business models to provide value-added solutions for their customers. Examples include the power-by-the hour aviation engine to reduce maintenance costs and John Deere's Agri Service to provide farmers with crop yield management beyond farming equipment. Such evolution requires the utilization of advance knowledge and tools so that data can be systematically processed into information that can explain the uncertainties and thereby make more "informed" decisions.

Through-life Engineering Services (TES) is a transformative approach that addresses the value of the product life cycle system, wherein data obtained through health monitoring, maintenance, repair, and overhaul service delivery systems is further used to impact design and manufacturing for improved performance, quality, reliability, and sustainability.

For the past 10 years, I have witnessed the transformation of Condition-based Maintenance to Prognostics and Health Management (PHM), and then Smart Asset Management through the advances of intelligent data instrumentation, analytics, and decision supporting tools. TES is a critical research domain that leverages these fields to not only develop insights into future equipment performance and estimate the time to failure, but also reduce the impact of these uncertainties and give users the opportunity to proactively implement adjustment solutions to prevent performance loss of the engineering system.

This book provides systematic knowledge of TES including service design, maintenance data analytics, autonomous maintenance, and diagnostics/prognostics based on decision processes informed by a deep understanding of a component's performance, degradation mechanism, and usage profile. These well-selected topics will offer readers insight into this existing field as well as the relevant practices for improved quality, productivity, and performance.

Jay Lee, D. Sc.
Founding Director of NSF Industry/University
Cooperative Research Center on Intelligent
Maintenance Systems (IMS)
University of Cincinnati
Cincinnati, OH
USA

Preface

Manufacturers of complex engineering systems are increasingly offering through-life engineering services (TES) in addition to their product provision. The industrial trend toward servitization has encouraged new research in novel and applied technologies and engineering solutions to support TES. This book provides a foundation on different aspects of TES. It illustrates example case studies to aid understanding of the challenges in implementing TES solutions. There are 25 chapters in the book that are categorized into seven parts. The chapters also demonstrate the latest thinking in the area of TES. The task of editing this book was considered as a part of the EPSRC Centre in Through-life Engineering Services activities in order to support the academic and industrial community. The book structure is grounded by the views from practitioners and researchers in TES from a think tank meeting. The authors are from different countries around the world and also represent multiple manufacturing sectors: aerospace, railway, and automotive. The TES Book is the first book in this area. The book also helps in defining a boundary for the TES.

TES involves the application of "service knowledge" across the life cycle of a product in order to mitigate degradation and reduce maintenance through design and manufacturing. The services are necessary to support recently developed "performance-based contracts". The initial taxonomy for TES has highlighted the key technical applications involved. The definition and the taxonomy are the bases to scope the TES activities against the traditional maintenance initiatives. TES provides a technology-enabled service delivery system. The NedTrain example presented in the book outlines major challenges in implementing TES across a large enterprise and has highlighted the need to better understand the technical states of all relevant systems in real-time to improve the maintenance planning and execution. The emphasis on data, diagnostics, and prognostics is continued in the next part of the book with technical and people issues discussed relative to dealing with no-fault found faults and the monitoring of mechanical and control systems. The study of component and system degradation in use and over time is a major theme for TES. The book presents an overview of non-destructive testing techniques application to assess the degradation and automate the process. The latest developments in maintenance-repair-overhaul (MRO) are presented with emphasis on cleaning technologies, repair and overhaul approaches and planning, and digital assistance. The impact of these technologies on sustainable enterprises is outlined. These parts of the book also present how the degradation information could be used for maintenance planning and scheduling. Changing design of legacy systems based on service feedback is not trivial; the book proposes a "use of functional blueprints" to manage the design change. Providing TES over the entire operational life of a product involves significant additional uncertainty and risks. One such risk is in the area of obsolescence management and in our ability to design for obsolescence. The book highlights several types of risks across the supply chain, these uncertainties and associated risks also have major impact on the whole life cost of the product. New and effective standards to support the obsolescence management are essential to reduce the cost. Autonomous maintenance is setting the agenda for future developments in TES. Two main directions are highlighted, one to develop self-healing technologies to reduce the maintenance activities and therefore reduce the cost and the other to apply robots for maintenance to bring automation and thus reduce human effort in maintenance. The book concludes with a discussion relative to the future challenges and opportunities in TES. The research and business trends are determined through a number of structured interactions with industry and academic studies. Although the interactions are not comprehensive in terms of the industry sectors included, the conclusions provide direction for researchers and practitioners alike. This book will help to support the TES community with better views on the foundation and technology trends with real-life examples.

Louis Redding
Rajkumar Roy

Acknowledgments

The editors would like to acknowledge the guidance and support of all who contributed to the writing of this text. Initial guidance with regard to the scope and focus matter resulted from the outputs and notes taken during a "Think Tank" event held at the EPSRC Centre for Through-life Engineering Services at Cranfield University, which was facilitated by Mr. Andy Shaw (Centre Manager). The event was attended by representatives from academia, research, industry, consultants, and the UK-MOD. While there are far too many to mention, the editors would like to thank all who attended and made the event a success. Your insights and opinions were the impetus for the writing of this book.

Secondly, the editors would like to thank the "unknown" peer reviewers who, upon receipt of the book proposal, gave their strong support and constructive comments as to how the book should be structured and the subject areas for inclusion.

The editors would also like to express their sincere gratitude to all the authors of the chapters, who took time from their busy schedules to contribute the results of their work and offer informed insight and guidance within this compendium of chapters. Without the contribution and support of such leading scholars and practitioners there could not have been this book.

Special recognition should also be made of the research staff within the EPSRC Centre for Through-life Engineering Services, who had to accommodate our (eds.) constant requests for information, resource, and support as the book was being assembled.

Finally, the editors wish to thank the publishers (Springer) for their proactive support throughout the project.

Contents

Chapter 1
Introduction

Louis Redding and Andy Shaw

Abstract Through-life Engineering Services (TES) are emerging as a key facilitator to technology enabled service delivery systems in support of Product Service System (PSS) generic business models. This chapter briefly introduces the concept as an enabler and risk mitigation initiative to business revenue streams, to the successful delivery of PSS, and 'availability contracts'. The chapter gives insight to those who may benefit from the contents of this compendium of contributions from eminent scholars and practitioners within the field. It gives clarity to the reader relative to how the scope and content of this manuscript was developed in order to illustrate alignment to the current requirements of academic, research, and practitioner requirements.

1.1 Background and Rationale for This Book

As manufacturing organisations adopt 'Availability Contracting' [1, 2] facilitated by complex service delivery systems the availability of the product for use becomes ever-more important. This has resulted in an increasing focus upon the technologies and infrastructures required to support the product throughout its life. Through-life Engineering Services (TES) [3, 4] supported by 'in-use' condition based management [5, 6] and bespoke maintenance, repair, and overhaul (MRO) activities are central to the effective delivery of availability contracts.

Through-life Engineering Services relate to the means by which data obtained through health monitoring systems and maintenance, repair and overhaul service delivery systems are used, or have the potential to be used, to inform both the design and manufacturing functions within manufacturing organisations and system integrators [7]. They are seen as being distinct from 'asset management' [8]

L. Redding (✉) · A. Shaw
EPSRC Centre for Through-Life Engineering Services, Cranfield University, Cranfield, Bedfordshire MK43 0AL, UK
e-mail: louis.redding@btinternet.com

L. Redding and R. Roy (eds.), *Through-life Engineering Services*,
Decision Engineering, DOI 10.1007/978-3-319-12111-6_1

and Product life-cycle Management (PLM) [9]. The literature relating to asset management tends to be related to large infrastructure assets and constructed entities, whilst PLM tends to be more focused on software and system architecture based solutions for the whole life support of the asset or product.

The particular focus of this book is upon the technologies, structures, and processes that underpin the service delivery system in providing whole life service support for the complex manufactured engineering product. It will be seen that the ability to deliver such service support requires the application of many technical innovations which offer both real time 'dynamic' diagnosis and prognostics, and design for service which is informed by the current operating experience and effective use of historical service and maintenance knowledge which is both explicit and tacit in nature [10, 11]. The ability to deliver TES based solutions is enabled by an aligned hybrid mix of all of these generic initiatives ranging from condition based applications [typically Integrated Vehicle Health Management (IVHM)] [12] supported by an in-depth understanding and knowledge of component degradation mechanisms, and the ability to undertake accurate diagnosis of degradation and the calculation of Remaining Useful Life (RUL) through the application of novel algorithms. The TES solution described within this book includes service led design, maintenance data capture, data trending and application, autonomous maintenance, and diagnostics/prognostics based upon decision processes informed by a deep understanding of a component's performance, degradation mechanisms and usage profile.

This book is primarily aimed at academics, postgraduate students, researchers, and industrial practitioners who have an interest in the following fields:

- Maintenance, Repair, and Overhaul (MRO) activities in support of Product Service Systems (PSS)
- Life-cycle Engineering (LCE)
- Asset Management (AM)
- Product Life-cycle Management (PLM)
- Design for Service (D4S)
- Original Equipment Manufacturers (OEM's) offering Product Service Solutions
- Maintenance, Repair and Overhaul (MRO) Organisations working through the supply chain in support of Product Service Solutions (PSS) and Availability Contracting
- Consultants operating in the field of Product Service Systems and the maintenance related arena

In addition the text could appeal as a supplementary text for students studying for Postgraduate qualification in the following generic areas:

- MSc Through-life Systems Sustainment
- MSc Integrated Vehicle Health Management
- MSc/PGCert Cost Engineering
- MSc Maintenance Engineering and Asset Management
- MSc Maintenance Engineering

- MSc Maintenance Management
- MSc Aircraft Management Systems
- Any MSc Course relating to Maintenance Repair, and Overhaul
- Any MSc Course relating to Product Service Systems

This book is supported by the research and work that is being undertaken by the UK Engineering and Physical Sciences Research Council (EPSRC) for Innovative Manufacture in Through-life Engineering Services and continues to build on the EPSRC—IMRC research findings relative to Product Service Systems (PSS), 'servitization', and on-going case work with major industrial organisations. The work is enriched by contributions from eminent academics, researchers, and practitioners within the field who have both research and industrial interest within the subject area.

1.2 Derivation of the Book Structure

In undertaking the task of drafting this edited book which focuses upon the motivation for the adoption of Through-life Engineering Services, the associated theories and examples of application from industry several questions naturally arise. Typically the following questions were asked by the editors:

i. What areas of TES should be discussed in the text?
ii. How is the decision made as to what to include and what to exclude?
iii. How is the selection of authors to be made?
iv. How does the book align to the current research interest and the requirements of practitioners?
v. Who are the potential readers of the text and how can the work meet their need?
vi. What level of science should the book address?

Initially guidance was sought from four sources when seeking to address these questions and formulate the structure of this book. Firstly a study of the literature relative to TES was undertaken. The findings showed that content and structure of the literature was fragmented when the idea for the book was first floated (2012) with little contribution existing which addressed the concept of TES directly [3]. Secondly, an information source informing the generation of the above questions by the author included a survey of UK based practitioners who were competing by offering TES generic solutions, or had the potential to do so, in 2012 [3]. The survey conducted by Redding et al. [3] sought to identify the existing levels of adoption, and report upon the challenges and opportunities which existed as identified by the practitioners responding and the findings of the study were reported in 2014. Thirdly the survey also sought to identify a stratified sample of industrial practitioners from those responding who were invited to an industry day where the findings of the work were reported and a discussion forum was facilitated during the event.

Finally a 'Think Tank' workshop was hosted by the EPSRC Centre for Through-life Engineering Services during October 2012 [14] which was attended by invited leading academics and practitioners who had been identified has being either stakeholders or having an interest within the TES concept during the first three stages. The purpose of the workshop was to inform the strategic direction of the research to be undertaken by the EPSRC Centre for Through-life Engineering Services so that it aligned to the interests and needs of the stakeholders (i.e. academics and industry) in order to ensure "*that the investment for further projects is focused*" [14]. To achieve this, a road mapping process was conducted with guidance sought from the work of Farrukh et al. [13]. The process followed sought to identify the 'Trends and Drivers', 'Markets', 'Opportunities', 'Technologies and Capabilities', and 'Enablers' relative to TES whilst seeking to identify the position against three time stages (Fig. 1.1).

Taking this general guidance the process was further decomposed to inform the five lenses identified. This decomposition is seen as '*indicative*' and not '*prescriptive*' in nature and reports a consensus view of those attending the workshop (Fig. 1.2). It can be seen that the time series has five dimensions (i) Past, (ii) Short Term, (iii) Medium Term, (iv) Long Term, and (v) Vision. The exercise seeks to answer four main questions, they being:

- What external trends, drivers, and market needs will influence developments in the TES landscape?
- What are the opportunities (Products and Services etc.) which will be delivered by the changing TES landscape?

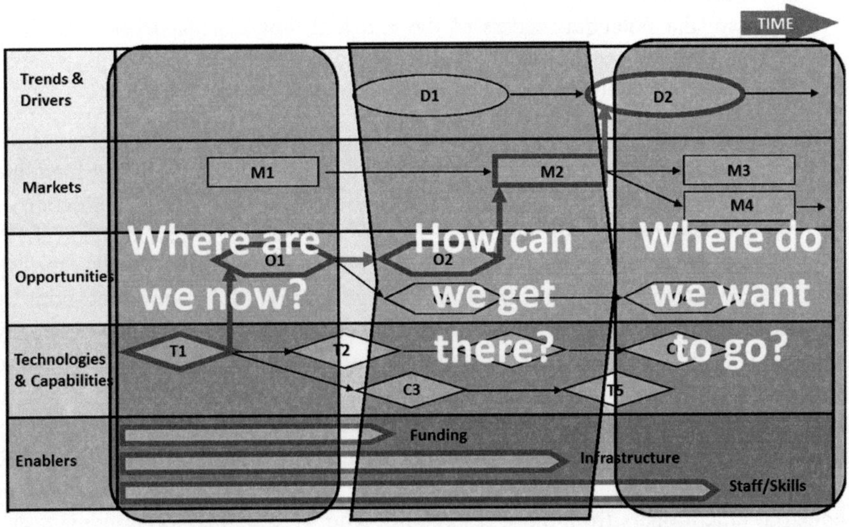

Fig. 1.1 Road mapping—Linking future to the present [14]

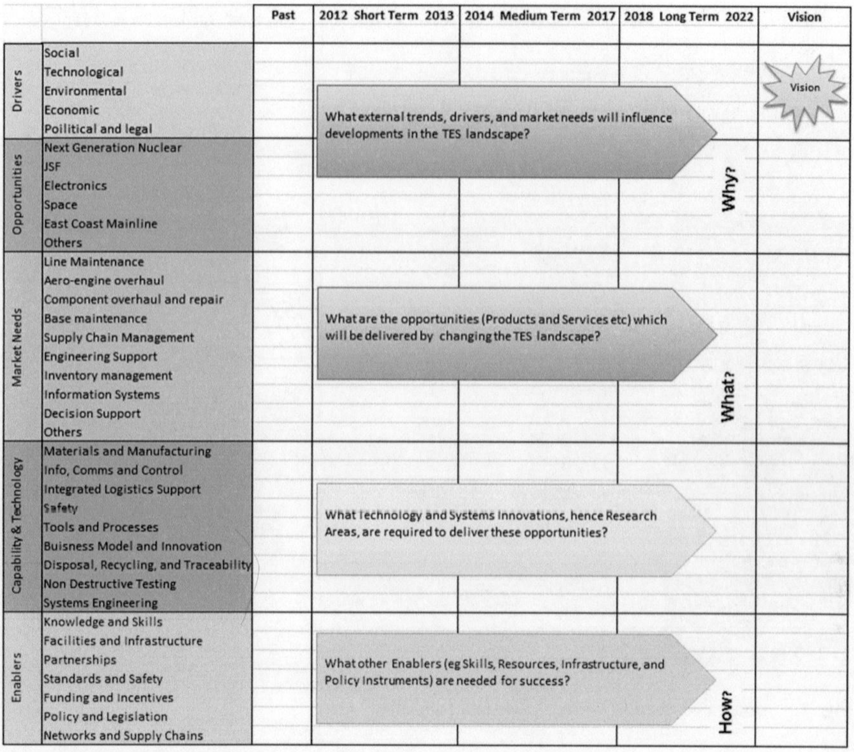

Fig. 1.2 Decomposition of road mapping focus [14]

- What Technology and Systems innovations, and hence research areas are required to deliver these opportunities?
- What other Enablers (e.g. Skills, resources, Infrastructure, and Policy Instruments) are needed for success?

In seeking to scan the landscape (relative to TES) and identify potential links, focus, and offer a priority, the cohort were given 2 min to select the FIVE most important issues and then post these on the road map. In so doing each participant was requested to cluster similar initiatives identified. If delegates were aware of potential initiatives and/or opportunities that were related to other posts, the focus of which had been missed, they were encouraged to add these to the road map.

Having completed this task by the whole cohort, the facilitators of the workshop then clustered all similar posts together by way of a 'rough cut' of the data which resulted in '*islands*' or '*rough trends*' emerging.

Finally each delegate was given 'ten' sticky dots to use in order to highlight their order of importance to the successful implementation of TES from their own perspectives. This resulted in the following 'Raw data Road Map' (Fig. 1.3).

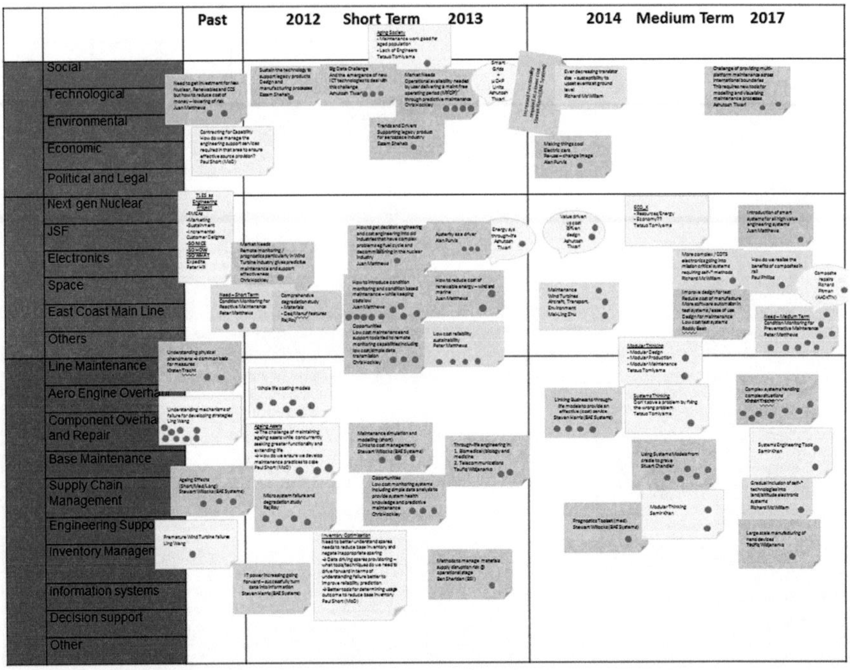

Fig. 1.3 Illustrated section of the 'raw data road map' [15]

In seeking to process the road map the 'post it' notes were then re-colored by the facilitating team in order to match the 'six' themes emerging.

Having re-colour coded the 'post it' notes as illustrated in Fig. 1.4 and re-clustered the 'post it' notes in line with the contents detailed upon them a ranked trend and time-line position emerges (Fig. 1.5). The number of votes for each of the 'points of interest' is also recorded (Fig. 1.4).

Finally all the classified 'post it' notes were removed from the road map and the unclassified ones reviewed to see if the workshop had missed any other trends. After careful consideration two additional high scoring areas were identified as being:

- Future resource scarcity
- Prognostics and diagnostics

This road map identifies 'six' main research areas that are seen to be of significance for the cohort of researchers and practitioners participating in the workshop. Further work was conducted upon the data in order to inform the strategic direction of the research to be conducted but is not discussed within this book although elements of the findings will be used to identify future directions of interest (Chap. 25). Whilst the process followed is not academically rigorous the findings do offer insight as to the main areas which should be investigated. Upon

Standards – 15	
Systems – 25	
Cost and Opportunities – 39	
Degradation – 20	
Data – 50	
Autonomy - 53	

Fig. 1.4 'Post it' notes colour key (Adapted from Shaw (2012) [14]

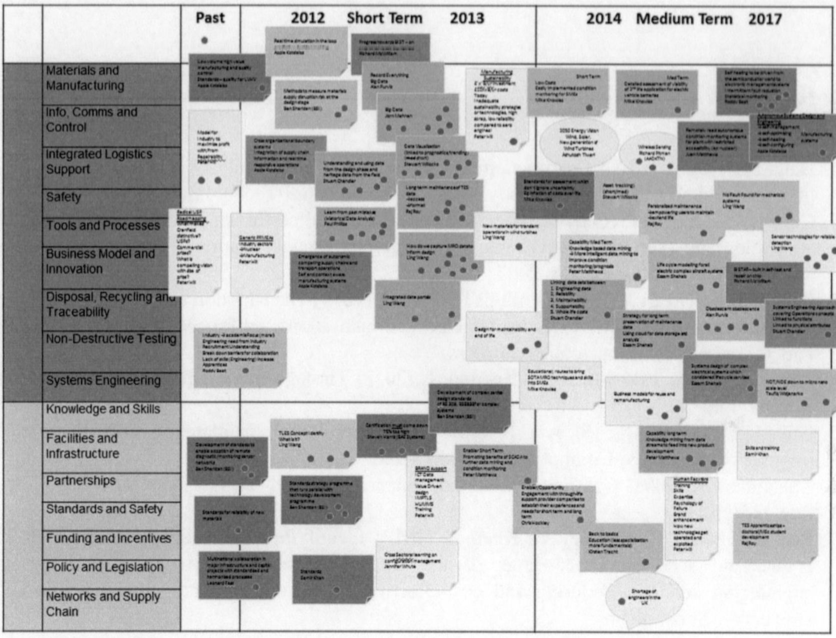

Fig. 1.5 Illustrated section of the 'post processed' road map [15]

review by the authors the data was used to inform the sections of this book in which the chapter contributions are found.

The book is structured therefore thus:

- Part A: Introduction to Through-life Engineering Services (TES)
- Part B: Data, Diagnostics and Prognostics
- Part C: Component Degradation and Design

- Part D: Systems Degradation and Design
- Part E: Cost, Uncertainty, Risk, and Standards in TES
- Part F: Autonomous Maintenance
- Part G: Future Challenges and Opportunities in TES Conclusions/Summary

Each section of this book has contributions from eminent academics, researchers, practitioners and leaders in the field of Through-life Engineering Services and related technologies, applications and strategies. Whilst the editors cannot include everyone it is hoped that this first book on the subject of Through-life Engineering Services which deals with the subject explicitly gives the reader valuable insight into the concept, scope, content, and context of this applied application of technology in support of the service delivery system enabling greater adoption of Product Service Systems through the process of '*servitization*'. It is hoped that this work will trigger further development in this facilitator of change in the operating model of the manufacturing organisation.

References

1. Datta PP, Roy R (2010) Cost modelling techniques for availability type support contracts: a literature review and empirical study. CRIP J Manufact Sci Technol 3(2):142–157
2. Cohen MA (2007) Power by the hour: can paying only for performance redefine how products are sold and serviced? Available at: http://knowledge.wharton.upenn.edu/article.cfm?articleid=1665. Accessed Nov 2007
3. Redding LE, Tiwari A, Roy R, Phillips P, Shaw A (2014) The adoption and use of through-life engineering services within UK manufacturing organisations. In: Proceedings of the IMechE, Part B, Journal of Manufacture. (in press)
4. Roy R, Shaw A, Erkoyuncu JA, Redding L (2013) Through-life engineering services. Meas Control 46(6):172–175
5. Gulledge T, Hiroshige S, Iyer R (2010) Condition-based maintenance and the product improvement process. Comput Ind 61(9):813–832
6. Chatterjee SK (2008) Condition-based maintenance management: enhances reliability. Chem Eng 115(12):46
7. Roy R, Shehab E, Hockley C, Khan, S (2012) Identifying the challenges in through-life engineering services. Proceedings of the 1st international conference on through-life engineering services: enduring and cost-effective engineering support solutions. Cranfield University, Shrivenham
8. Lee J, Lapira S, Yang S, Kao A, Yen N (2012) New PHM techniques and trends for asset lifecycle management. Proceedings of the 1st international conference on through-life engineering services: enduring and cost-effective engineering support solutions. Cranfield University, Shrivenham, pp 13–22
9. Savant P (2009) Developing a product data structure to implement PLM in a company. PhD thesis, Cranfield University, Cranfield, Bedfordshire
10. Mountney S (2009) Acquisition and shaping of innovative manufacturing knowledge for preliminary design. PhD thesis, Cranfield University, Cranfield, Bedfordshire
11. Doultsinou A (2010). Service knowledge capture and re-use to support product design. PhD thesis, Cranfield University, Cranfield, Bedfordshire
12. Jennions IK (2011) Integrated vehicle health management: perspectives on an emerging field. SAE International, Warrendale, PA, USA

13. Farrukh C, Phaal R, Probert D (2004) Technology road mapping: linking technology resources to business planning. Int J Technol Manage 26(1):2–19
14. Shaw A (2012) Think tap research roadmap—Workshop 3rd October [Unpublished slide pack]
15. Shaw A (2012) Through-life engineering services: think tank report (Presentation Slide Pack), [Unpublished]

Part I
Introduction to Through-life Engineering Services

Chapter 2
Through-Life Engineering Services: Definition and Scope: A Perspective from the Literature

Louis Redding

Abstract Through-life Engineering Services (TES) provide product support throughout each stage of the product-lifecycle; from conception, through design, manufacture and operational life, to end of life disposal. They are seen as a natural stage in the evolution of product support and maintenance, repair and overhaul strategy. They are the sum of many diverse product support strategies which use emerging and traditional technologies, processes, and applications. Whilst there are increasing numbers of contributions to be found within the literature defining the content, scope, purpose and application of the supporting technologies one sees no definition for TES emerging. This chapter offers a definition for Through-life Engineering Services which states what the concept **is**. It gives dimension, application, and purpose for TES in its role as a facilitator of Technology Enabled Service Delivery Systems which support manufacturing organisations wishing to compete through the adoption of Product Service Systems. An initial taxonomy is also presented.

2.1 Introduction to, and Definition of, Through-Life Engineering Services (TES)

As manufacturing organisations seek '*whole-life*' revenue streams the role of service in support of their products becomes ever more important. Driven by the requirement to offer sustainable solutions due to the realisation that material resource is limited, and the need to be ever more competitive in the face of increasing competition from both the emerging industrial (BRIC) economies and the global market, one sees Product-Service Systems (PSS) [1–3] and *servitization* [4, 5] emerge. The traditional contract of design, manufacture, sell, and maybe offer

L. Redding (✉)
EPSRC Centre for Through-Life Engineering Services, Cranfield University, Cranfield, Bedfordshire MK43 0AL, UK
e-mail: l.e.redding@cranfield.ac.uk; louis.redding@btinternet.com

L. Redding and R. Roy (eds.), *Through-life Engineering Services*,
Decision Engineering, DOI 10.1007/978-3-319-12111-6_2

a limited service by way of spares and repair, is no longer seen as the sole order qualifiers for those manufacturers who provide complex engineering products (typically aeroplanes, ships, trains, machine tools, etc.) [6]. Organisations who operate/use such products are choosing to only pay for the use (or the availability to use) these products resulting in manufacturing organisations offering whole-life service support for their offerings. One such often quoted and leading initiative is Rolls Royce's 'Power by the Hour' [7]. Here the operators of Rolls Royce engines pay for the '*availability for use*' of the asset to the manufacturer with the engine typically being owned by the leasing company or bank. This structure is also observed with other transport systems such as railway rolling stock. In the UK Rail Sector trains are designed and produced by the manufacturer who then sells them to Rolling Stock Operating Companies (ROSCO's) who are fundamentally leasing companies. These in turn offer franchises to the Train Operating Companies (TOC's) under fixed term franchise agreements. The revenue stream back to the manufacturer is therefore heavily reliant upon the train (or other such product) being available for use.

Underpinning all of these initiatives is the role of the engineering service support function. Such support can manifest itself as either/or a hybrid solution of dynamic 'real time' support (the monitoring and resulting decision systems) and maintenance, repair, and overhaul function. Through-life Engineering Services as discussed in this chapter (and in the following chapters) will be seen as the sum of both these dimensions which, when applied within the service delivery system, not only increases the product's availability for use, but offers the potential to 'engineer out' inherent degradation mechanisms (or at least minimise their occurrence) by using whole life product related knowledge to inform design. This thus has the potential to greatly reduce whole life cost for the stakeholders who obtain revenue through manufacture, service support, and operation of the products function.

2.1.1 Evolution of Maintenance in Support of Manufactured Products

Pinjala et al. [8] and Waeyenburgh et al. [9] suggest that there are few management disciplines that have undergone so many changes as those observed in the field of maintenance and product support. Initially the maintenance repair and overhaul function (MRO) was nothing more than an essential activity which was called upon to remedy a failed or degraded component, system, or product. It was seen as a function which was secondary to manufacturing, a necessary 'evil' and, if no more than an irritant, a constant drain upon the organisations balance sheet that was to be minimised at all cost. Typically service support and maintenance at this level consisted of the provision of spares, repair call outs which were treated as a separate standalone contract, routine inspection of equipment covered by statutory requirement (i.e. insurance driven), and in some cases engineering upgrades of

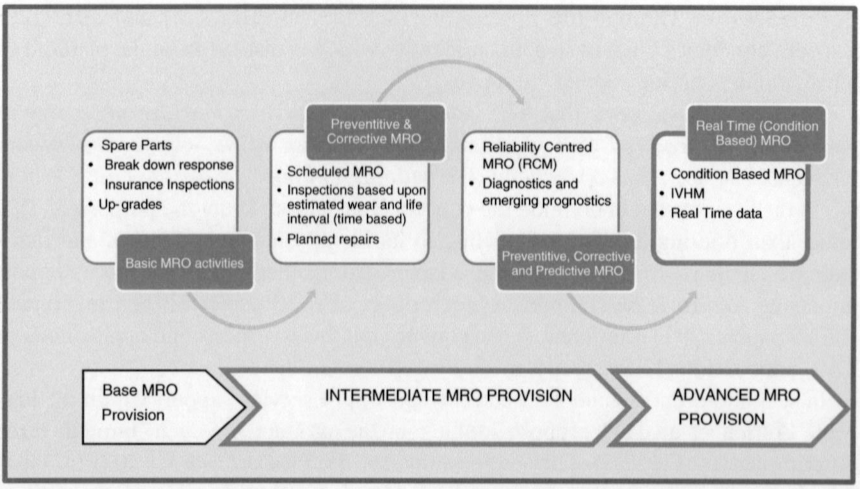

Fig. 2.1 The Evolution of maintenance repair and overhaul (MRO) strategies

controlled products. In categorising service levels the author takes guidance from Baines et al. [3] who suggest that services fall into three categories, they being '*base*', '*intermediate*', and '*advanced*'.

Base level services were sufficient in a period where assets and products were simple stand-alone items offering little by way of revenue to the manufacturer post of point of sale. However as products become more complex their reliability and availability become more important to the user and the commercial performance of the manufacturing organisation. The drivers of technological advancement, the emergence of modular design, reduction of redundancy within a given design, and more bespoke designs based upon aligned customer requirements and specifications, call for products that are ever more reliable and available for use. This sees a move into the 'intermediate' service provision. Here scheduled service activities which are time based evolve with estimates for service intervals based upon wear and service (as used) life emerging. This evolution is illustrated in Fig. 2.1.

2.2 Through-Life Engineering Services—a Definition

Whilst the literature offers both explicit and implicit insights [10–14] into the role of services in support of servitization and PSS it offers little by way of a definition for Through-life Engineering Services (TES). The concept finds its introduction based in the literature and associated research relating to Condition Based Maintenance (CBM) [15–17], Integrated Vehicle Health Management (IVHM) [18–20], Technical Product Service Systems [10, 21] and Design for Service [22–24] which address service support from a 'whole-life' perspective driven by the changes in business paradigms. It is not until the emergence of the 1st and 2nd International

Conferences on Through-life Engineering Services in 2012 [25] and 2013 [26] respectively does TES start to gain an identity which is distinct from the plethora of other product service support initiatives.

Roy [25, 27] suggests that TES are "*technical services that are necessary to guarantee the required and predictable performance of an engineering system throughout its expected operational life with the optimum whole life cost*". Whilst this is the first attempt at defining the concept it says more about the purpose of TES rather than offering a definition relating to its identity, content, context, and rationale which helps to distinguish the concept from other product service support initiatives. Whilst these conference proceedings offer 53 papers (2012 Conference) and 79 papers (2013 Conference) respectively relative to aspects and applications of TES, none define TES directly.

In seeking to understand TES and the concept of service support within the life-cycle Aurich et al. [11] propose looking at the product life-cycle through three differing lenses (Fig. 2.2). They suggest that for the Product Life Cycle (PLC) lens the stages of evolution are '*engineering*', '*production*', '*usage*', and '*re-cycling/disposal*'. Whilst these are intuitive and well documented these stages are not simply sequential in all industrial sectors in the world of concurrent engineering of complex products. In the automotive sector, for example, significant overlap can be witnessed as OEM's launch advanced pilot build vehicles into the market to harvest usage performance prior to main volume build. During this process product processes are analysed, pilot vehicle performance harvested, degradation mechanisms identified, and service response defined, all of which are fed back into the design '*engineering*' activity before the final design stage gate is completed.

Aurich et al. [11] continue to propose that from the manufacturer's lens the 'product' consists of two attributes, the 'physical' (the physical product) and 'non-physical' (service support) elements of the product offering. From the manufacturing of the physical product perspective within the organisation the PLC is defined as 'product design', 'product manufacturing', 'spare parts manufacturing', and 'product re-manufacturing'. However, within manufacturing organisations producing complex engineering products there is also a service offering (the non-physical product) supplied by the company. From the manufacturer's service support department the life-cycle is different and is as illustrated in Fig. 2.2. It is seen that

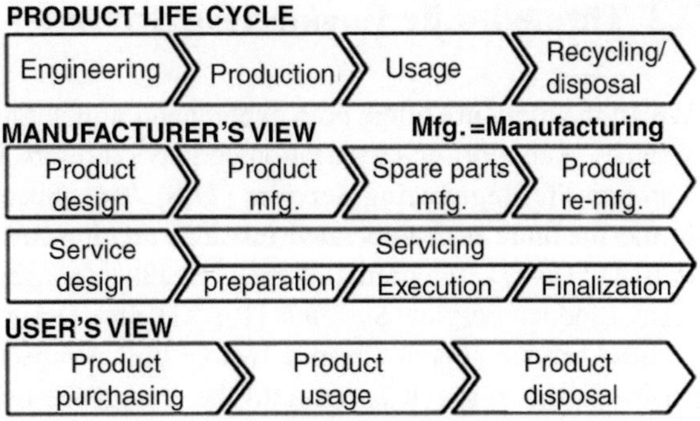

Fig. 2.2 Service in the product life-cycle (PLC) [11]

service design leads to the delivery of service through preparation, execution, and finalisation. The final lens defined by Aurich et al. [11] is that of the user and it is proposed that there are three elements within the lens as illustrated in Fig. 2.2.

Whilst Aurich et al's. [11] illustration (Fig. 2.2) serves to offer the reader a perspective of the PLC through the different lenses specified it does not offer insight as to the position and duration of each element against a time scale within the PLC. Traditionally, organisations designed and manufactured products to suit a specification relative to a defined function, with service activities being an afterthought based upon explicit (or tacit) knowledge gained by previous experience of the performance of generic parts. As maintenance, repair, and overhaul activities evolved (Fig. 2.1) so did the awareness of the benefits of putting service experience and knowledge into the design activity. In the world of increasing *servitization* the users of complex engineering products "expect the…(product)…suppliers to provide more efficient and reliable products and services, with lower and more predictable operating costs" [22].

As organisations evolve through the PSS continuum [28], from that of pure product provider to that of pure service provider through the process of *servitization* this, and the evolution in service delivery systems (maintenance, repair and overhaul), converge as both concepts become interdependent. This is no better illustrated than the initiatives being pursued within the aerospace sector (i.e. Rolls Royce—Total Care™) and its 'Power by the Hour' availability contracting solutions [7]. Harrison [22] identifies this convergence of evolution and interdependency when stating that "the real potential for quantum reductions …(in whole life cost)… comes when the product and service are designed in harmony. This requires a…shift… from '*offering a service around an existing product*' to '*designing a service and the product that supports it*'" [22].

Implicit in the discussion thus far is the role of service data and information from which knowledge and hopefully wisdom can be acquired relative to the product [29]. That is data which is relevant to the design and use of the product; the degradation and failure mechanisms experienced by the product; the means of detection of such degradation and failure; and the technologies and methodologies by which original design function can be restored and prolonged. Jagtap and Johnson [24] illustrate this in their study which proposes how 'in-service' information may be used by engineers when seeking to design the next generation of aero-engine [30]. In their paper they state that "in-service information related to … degradation mechanisms of similar aero-engines is utilised to avoid the same issues in future aero-engines …(and the)… flow of information from the service domain to designers is thus crucial for minimising in-service issues and can also reduce the cost of both planned and unplanned maintenance" [30].

In seeking to offer a definition which is grounded in the aforementioned literature the following questions are posed:

i. What do we mean by service knowledge in relation to engineering services?
ii. How do we define Through-life Engineering Services (TES) and how do they differ from other product service support systems?

Much is written within the literature relative to the distinctions between data, information and knowledge, and the *content*, *context*, and *structure* of each of these dimensions ranging from definitions for each dimension [31, 32], its structure and management [33–35], the comparison between semantic and syntactic knowledge structures relative to storage and retrieval systems [36], and studies as to how service knowledge can be used to benefit design [14, 37–39]. After review of all of these works cited the following definition for service knowledge is offered which is relevant to the focus of this work.

> Service knowledge is the ability to initiate a change of state in a product or asset facilitated by the awareness of the current condition of that product or asset, its historical usage, and means of restoring the 'as designed' functionality supported by explicit (codified) and tacit knowledge of the degradation and failure mechanisms of that component (Paper forthcoming by the author 2014)

Having provided a definition for service knowledge which is seen as the foundation from which TES operates and which is grounded in the literature, this definition becomes an essential component of any definition proposed for TES. In order to offer such a definition for TES the literature is reviewed to find similar service support concepts which can be applied to complex engineering products and their use in facilitating PSS through a process of servitization. Whilst not exhaustive, several concepts and definitions are presented in Table 2.1.

The literature relative to the definitions of engineering service support is very limited and of those offered they are all generic. Most definitions state purpose of the concept with few (if any) stating content. Of those service support concepts cited in Table 2.2 which support PSS several common factors emerge. These refer to through-life considerations and state that the service provision is the sum of element parts;

- Aurich et al. [11], physical and non-physical components and the need for stakeholder integration; (*Dimension*)
- Houschild et al. [40], application of scientific principle throughout life-cycle with consideration to sustainability (*Application*)
- Meir et al. [41], delivers value, integrated planning, development, provision, use; dynamic provision, enable availability (*Purpose*)
- Roy et al. [27], guarantee design performance and reduce whole-life cost (*Purpose*)

Whilst all of the definitions cited give value and insight into the identity of the concepts to which they refer (*Dimension, Application, Purpose*), none actually state what the concepts **are**. In consideration of these elements extracted from the definitions cited and the author's proposal that the underlying premise for such initiatives is that they are based upon the acquisition, storage, retrieval and application of service knowledge, the following definition of TES is offered.

> Through-life Engineering Services are a result of the application of explicit and tacit 'service knowledge' supported by the use of monitoring, diagnostic, prognostic technologies and decision support systems whilst the product is in use, and maintenance and repair

Table 2.1 Definitions and characterisation of differing service support concepts found in the literature (not exhaustive)

Author (ref)	Concept	Definition
Aurich et al. [10]	Technical product service systems (t-PSS)	"With respect to the understanding of technical product-service systems and their non-physical components… three constitutive characteristics can be identified, which distinguish technical services from physical products: • Technical services are mainly non-physical. Their realization can therefore often be performed at minimum consumption of resources, which is one of the decisive reasons for services being considered in the context of dematerialization. Furthermore, due to their non-physical character, services can neither be produced to stock nor distributed like physical products. Hence, the service provider must build up corresponding resources for 'on demand servicing' • Unlike physical products that are first manufactured and later consumed over a period of time, technical services are realized and consumed simultaneously. This principle is referred to as the 'uno acto principle' • The realization of technical services requires the integration of the customers in terms of providing the products, respectively, staff, to which a service (e.g. maintenance and user training) refers"
Hauschild et al. [40]	Life cycle engineering (LCE)	Engineering activities which include: "the application of technological and scientific principles to the design and manufacture of products, with the goal of protecting the environment and conserving resources, whilst encouraging economic progress, keeping in mind the need for sustainability, and at the same time optimizing the product life-cycle and minimising pollution and waste"

and overhaul (MRO) functions to mitigate degradation, restore 'as design' functionality, maximise product availability, thus reducing whole life operating cost.

This definition defines TES by stating its **content** and acknowledging that the underlying principle from which the concept is based is the acquisition, storage, retrieval, and application of **service knowledge** for the purpose of extending the life of the design function throughout the whole life-cycle. This increases the availability for use of the product by the owner/operator thus reducing the whole life cost of operation and/or ownership whilst mitigating risks to the revenue stream by component degradation and failure.

Table 2.2 Definitions and characterisation of differing service support concepts found in the literature (not exhaustive)

Author (ref)	Concept	Definition
Meir et al. [41]	Industrial product service systems (IPS^2)	"…is characterised by the integrated and mutually determined planning, development, provision, and use of product and service shares including its immanent software components in business to business applications and represents knowledge-intensive socio-technical system. This means in detail… • An IPS^2 is an integrated product and service offering that delivers value in industrial applications • IPS^2 is a new product understanding consisting of integrated product and service shares • IPS^2 comprises the integrated and mutually determined planning, development, provision and use • IPS^2 includes the dynamic adoption of changing customer demands and provider abilities • The partial substitution of product and service shares over the lifecycle is possible • This integrated understanding leads to new, customer-adjusted solutions • IPS^2 enable innovative function-, availability- or result-oriented business models"
Roy et al. [27]	Through-life engineering services (TES)	"Technical services that are necessary to guarantee the required and predictable performance of an engineering system throughout its expected operational life with optimum whole life cost"

2.3 The Scope of Through-Life Engineering Services

Having offered a definition for TES the next question which naturally follows relates to the scope of the concept. There are many product support initiatives discussed within the literature which cover many aspects of through-life product support. These include (i) knowledge of degradation mechanisms (their root causes and means of manifestation) which exist within a given product/system; (ii) knowledge of the means of restoring design functionality; (iii) knowledge of degradation/fault diagnostic methods and the means of assessing remaining useful life, (iv) knowledge of the mode of use which can be either in 'real time' and/or historic in nature; and (v) knowledge of the design function and manufacturing methods used to produce the product. These inputs either collectively or individually facilitate TES enabled solutions which in turn offer numerous benefits to the organisation when seeking to compete through PSS generic solutions (Fig. 2.3).

When reviewing the inputs to a TES system one sees that several supporting technology applications are implied that enable such services to deliver the benefits above. In seeking to understand the scope of TES it is helpful to have a taxonomy

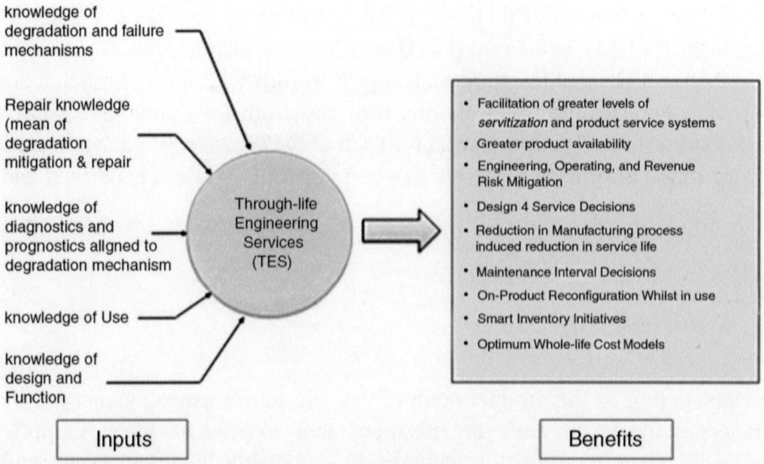

Fig. 2.3 Inputs to, and benefits of, through-life engineering services

for the concept. The literature offers no such single holistic insight however to the breadth or depth of TES and its component applications by way of service delivery systems. In order to illustrate the scope and content of TES the following illustration is offered (Fig. 2.4).

The diagram illustrates that TES enabled product support systems are the sum of many technical applications. Whilst all these elements have significant bodies of

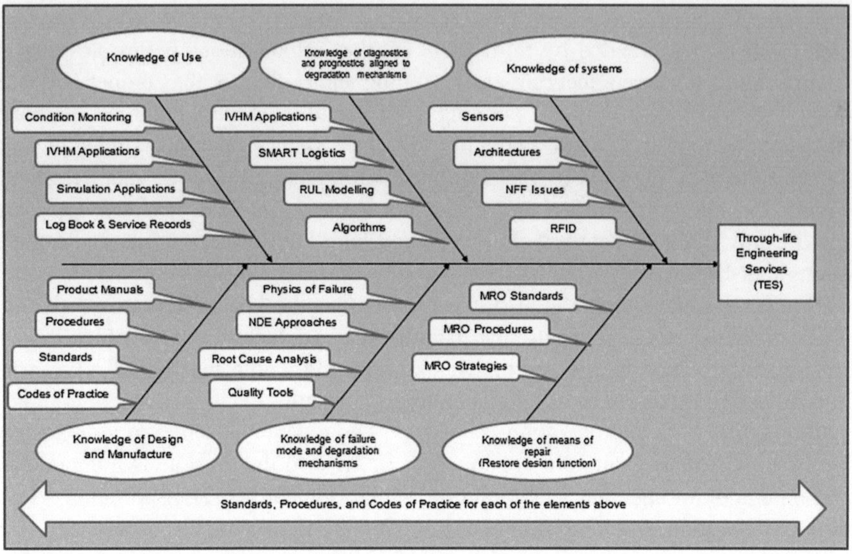

Fig. 2.4 Initial taxonomy for through-life engineering services

literature relating to each of them, with the exception of Integrated Vehicle Health Management (IVHM), and Condition Based Management approaches which could be described as philosophies, they each appear in silos with few approaches seeking to look at holistic whole life solutions resulting from the cumulative sum of the bodies of knowledge. The foundation for each of these concepts is the acquisition of data from which information, knowledge and hopefully wisdom is derived and then applied.

2.3.1 Knowledge of Use

An understanding of the product code of use and performance against pre-defined parameters is important when organisations seek to offer extended support. This understanding was traditionally achieved by reviewing historical usage and performance records supported by maintenance, repair, and overhaul data. This is increasingly being delivered today by Condition Based Monitoring (CBM_1) and Condition Based Management (CBM_2) techniques which see the use of sensors being fitted to the product, with data being harvested intermittently by either download from on-product data storage systems, or via transmission using telecommunication and satellite technologies. IVHM generic solutions are the pinnacle of this type of application. The concept is far more than just condition monitoring but has the potential to offer a paradigm shift in business operations and support strategies. An often cited example of such a solution is the aerospace sector and particularly the OEM's supplying gas turbines. Here the engines are fitted with on-board sensors which transmit engine data at pre-determined times and positions within the flight curve (take-off, mid-cruise, and landing). These packets of data are then cleansed and trended by third party support organisations before arriving in control rooms where engineering and operating teams monitor the engines and take mitigating actions to engine operating configurations during flight as required. Decisions relating to existing operating life, maintenance intervals, strategies, and logistics are also informed by the data, information and knowledge acquired. With aircraft the operating profiles are also recorded and the loads on an engine are significantly different between short haul and long haul operators (i.e. cycle times throughout life).

Similar product support is also observed within the UK Rail Sector with suppliers of rolling stock. Typically the manufacturer produces the train which is then subsequently owned by the Rolling Stock Operating Companies (ROSCO'S) who in turn lease the train to the Train Operating Companies (TOC's) through complex leasing agreements and operating franchises. The revenue is achieved increasingly by the provision of availability contracting and risk to the revenue is mitigated by on-board monitoring, one such solution being Bombardier Transportation PLc's *Orbita*™ solution.

This approach also facilitates the use of simulation techniques using real time acquired data. This can be used in scenario assessments thereby facilitating better understanding of the products operating performance, degradation mechanisms and reaction to defined operating parameters. Knowledge of remaining useful life, better strategic decisions relating to product support solutions and business models are also facilitated.

2.3.2 Knowledge of Diagnostics and Prognostics Aligned to Degradation Mechanisms

For any product support system to operate effectively, knowledge of how to detect, diagnose and predict when a product should be maintained or overhauled is essential. Whilst IVHM generic systems monitor the *dynamic* as used situation and *real or near real time* condition, under pinning this is the ability to understand the data obtained and through the use of modelling and bespoke algorithms, decisions relative to diagnostics and prognostics are made. Implicit in this is the requirement to have in-depth understanding of the degradation mechanisms that can occur given a set of operating parameters and the rate at which such degradation would develop, evolve and propagate. This understanding and knowledge is the sum of the inputs acquired whilst the product is in use, and also from the MRO function who see the product in the maintenance facility when a requirement for repair or replacement after failure is needed. Such knowledge and the ability to effectively predict critical thresholds of degradation which trigger mitigation events are is also analogous to a KANBAN signal which in turn can trigger smart logistics and component replacement throughout the supply chain. However knowledge of the means of conducting diagnostics and prognostics relative to understanding degradation is far more than just evaluating condition based data from IVHM generic systems. The ability to undertake effective diagnosis and offer a prognosis relative to RUL requires in-depth understanding of the degradation mechanism observed.

2.3.3 Knowledge of Failure Modes and Degradation Mechanisms

The knowledge of failure modes and degradation mechanisms comes from the application of many tools and techniques which relate to the body of knowledge found within the literature and the acquisition, analysis, and synthesis of MRO data acquired from practitioners in the field which is both explicit and tacit in nature. There are many techniques that can be applied to aid diagnostic and prognostic decisions. These include such ‘quality tools’ [42, 43] as *ISHIKAWA* approaches, *root cause analysis* [44], understanding the ‘*physics of failure*’ [45], and the application of product testing and non-destructive testing approaches (X-ray, CT Scan, Thermography etc).

When considering the degradation of mechanical components knowledge is required not only of the in-service operating parameters, but also the inherent degradation and failure mechanisms at component feature level if design is to be able to mitigate product failure. Typically an understanding of the high levels of degradation taxonomy is a pre-requisite to delivering an effective service delivery system in support of the physical product which includes fatigue, creep, fracture mechanics, corrosion, stress, strain etc., which can be caused by such parameters as mechanical and/or thermal stress, environment, temperature, pressure, impact, wear/abrasion, and other operating parameters.

2.3.4 Knowledge of Means of Repair (Restore Design Function)

Whilst the understanding of diagnostics, prognostics, and the degradation mechanisms which may occur are essential foundation elements to the service delivery system facilitated by MRO and CBM initiatives, they add little value unless they are aligned to the means of repair in order to restore the 'as designed' function of the product. The explicit and tacit knowledge of the MRO processes used is also required if rectification of degradation or failure is to be done in a timely and cost effective way.

2.3.5 Knowledge of Design and Manufacturing

The role of MRO is directly linked to both the design and manufacturing functions. Design for function is the foundation of product development. Traditionally requirements are identified from which a specification is defined and agreed. The resultant design is deemed to be successful if it meets the attributes defined in the specification which are primarily function and performance based. The outputs of the design process normally include approved General Arrangements, Detail Designs, Service Layouts (Pneumatic, Hydraulic, and Electrical), Bills of Materials, Bills of Process, Gauge and Inspection Plans, Reporting Schedules and Instructions, and Service Instructions. These outputs are generally in both hard and soft copy formats supported by Product Manuals and Procedures which in turn are guided by standards and guidance notes etc. These outputs inform both the manufacturing and service functions.

A TES approach has the potential to close the loop and use service knowledge to drive design. This is seen as being important when considering whole life approaches where the design life of the product could be 20–30 years. When this lens is applied the cost of service throughout the product's life becomes significant for complex products and can be many times the cost of manufacture. With greater levels of servitization and the emergence of availability contracting the risk to the revenue stream for the manufacturer makes the efficient acquisition and assimilation of service knowledge ever more important. The effective use of TES and its elements becomes a game changer for organisations competing using these business models with a major aerospace manufacturing OEM stating that over 50 % of its revenue is now achieved from its service activities. The ability to fully understand the design, manufacturing, and through-life service interaction and alignments are seen therefore as essential in order to increase product availability and reduce whole life costs.

2.3.6 Knowledge of Systems

Supporting these dimensions of TES is the role of systems and system engineering. The dynamic monitoring and transmission of performance data is based upon sensor applications, data handling systems, data transmission systems, and decision algorithms. Within the TES related references to date the majority of *technical* contributions to the literature focus on these aspects of the application with system architectures featuring prominently although the majority base these on moderated systems based upon the open-system architecture for condition based maintenance (OSA-CBM). The literature also discusses at length the issues relating to the interface of new systems to federated and legacy systems to be found in products which are significantly through their design life. All too often system failure occurs at the interfaces in the system rather than within the subsystem itself.

Significantly, issues relating to false error codes with a given system are also worthy of significant study and this research and practitioner supplied solutions are essential facilitators of efficient and effective service delivery systems. The ability to understand and identify the presence of ‘false positive’ error codes in such solutions is very important when seeking to effectively mitigate or reconfigure remote and/or autonomous operating systems in support of product performance, thus reducing risk.

2.3.7 The Role of Standards, Procedures, and Codes of Practice in TES

Underpinning all of the above is the role of standards, procedures, guidance notes, and codes of practice. Whilst there are a plethora of such guidance for the supporting aforementioned technologies which form the constituent elements of TES, there are no such contributions that deal with TES directly. Shaw and Tasker [46] seek to address this issue with their ongoing work with the UK-BSI. Whilst research into, and the development of, such standards is very much a current work in progress no explicit accreditations are obtainable for TES with the nearest related standard being PASS55 [47] which deals with and offers 28 aspects of good Asset Management supported by the ISO 55,000 series of standards. This is seen as a major opportunity for the community of practitioners and future research as '*first past the post*' has the potential to define the concept and the terms of accreditation.

2.4 Conclusion

This chapter has informed the reader of the evolutionary process of service support for manufactured complex engineering products from basic MRO activities triggered by component failure to real time condition based maintenance and management facilitated by IVHM generic applications. Through-life Engineering Services (TES) was introduced as a holistic concept around 2010 and sought to "*guarantee the required and predictable performance of an engineering system...(or product)... throughout its expected operational life with the optimum cost*" [27]. Whilst there have been numerous contributions to the literature relative to TES no definitive identity has emerged. This chapter has, in consideration to other product support concepts and initiatives, sought to offer a definition of TES which addresses its *content*, *dimension*, and *purpose*. Whilst further work is to be done by the community of researchers and practitioners in the field to develop further the definition offered and generate a taxonomy for the concept, the author offers this contribution as an identity from which the concept can develop.

The following chapters written by experts in the field, and the contents therein, serve to illustrate the scope and depth of TES and the ability of the concept to facilitate effective Product Service Systems through a process of *servitization* whilst mitigating the risk to the revenue obtained by the use of the product (or its availability) through the application of technology. The author suggests that TES' are the foundation of a Technology Enabled Service Delivery System.

References

1. Mont O (2000) Product service systems; final report for IIIEE, Lund University, Lund
2. Tukker A, Tischner U (2006) Product-services as a research field: past, present and future. Reflections from a decade of research. J Clean Prod 14(17):1552–1556
3. Baines TS, Lightfoot HW, Evans S, Neely A, Greenough R, Peppard J et al (2007) State-of-the-art in product-service systems. Proc Inst Mech Eng Part B J Eng Manuf 221:1543–1552
4. Vandermerwe S, Rada J (1988) Servitization of business: adding value by adding services. Eur Manag J 6(4):314–324
5. Neely A (2008) Exploring the financial consequences of the servitization of manufacturing. Oper Manage Resour 1:103–118
6. Treacy M, Wiersema F (1997) The discipline of market leaders: choose your customers, narrow your focus, dominate your market. Expanded addition edn: Perseus Book Group, New York

7. Cohen MA (2007) Power by the hour: can paying only for performance redefine how products are sold and serviced? Available at http://knowledge.wharton.upenn.edu/article.cfm?articleid=1665. Accessed Nov 2007
8. Pinjala S, Pintelon L, Vereeche A (2006) An empirical investigation on the relationship between business and maintenance strategies. Int J Economics 104(1):214–229
9. Waeyenburgh G, Pintelon L (2002) A framework for maintenance concept development. Int J Economics 77(3):299–313
10. Aurich JC, Fuchs C, Wagenknecht C (2006) Life cycle oriented design of technical product-service systems. J Clean Prod 14:1480–1494
11. Aurich J, Fuchs C, Barbian P (2004) An approach to the design of technical product service systems. Industrie Manage 20(5):13–16
12. Johnstone S, Dainty A (2008) Integrating products and services through life: an aerospace experience. Int J Oper Prod Manage 29(5):520–538
13. Savage G (2010) Dealing with crisis–solving engineering failures in Formula 1 motor racing. Eng Fail Anal 17:760–770
14. Baxter D, Roy R, Doultsinou A, Gao J, Kalta M (2009) A knowledge management framework to support product-service systems design. Int J Comput Integr Manuf 22(12):1073–1088
15. Holguin L (2005) IEEE conditioned based maintenance (CBM). Autotestcon 2005:188–193
16. Chatterjee SK (2008) Condition-based maintenance management: enhances reliability. Chem Eng 115(12):46
17. Gulledge T, Hiroshige S, Iyer R (2010) Condition-based maintenance and the product improvement process. Comput Ind 61(9):813–832
18. Jennions IK (ed) (2011) Integrated vehicle health management: perspectives on an emerging field. SAE International, Warrendale
19. GE's IVHM technology for Business Aviation. Available at: http://ge-aviation.jesiii.net/2013/08/26/watch-aircraft-health-management-system-ivhm-business-aviation-ge-aviation. Accessed 06 Jan 2012
20. National Aeronautics and Space Administration (NASA) (1992) Research and technology goals and objectives for integrated vehicle health management (IVHM). NASA-CR-192656
21. Aurich JC, Fuchs C, DeVries M (2004) An approach to life cycle oriented technical service design. CIRP Annals-Manuf Technol 53(1):151–154
22. Harrison A (2006) Design for service: harmonising product design with a services strategy. In: Proceedings of GT2006 ASME turbo expo 2006: power for land, sea, and air
23. Jagtap S, Johnson A (2010) Requirements and use of in-service information in an engineering redesign task: case studies from the aerospace industry. J Am Soc Inf Sci Technol 61 (12):2442–2460
24. Jagtap S, Johnson A (2011) In-service information required by engineering designers. Res Eng Design 22:207–221
25. Roy R, Shehab E, Hockley C, Khan S (eds) (2012) In: Proceedings of the 1st international conference on through-life engineering services: enduring and cost-effective engineering support solutions. Cranfield University, Cranfield
26. Roy R, Tiwari A, Shaw A, Bell C, Phillips P (2013) In: Proceedings of the 2nd International through-life engineering services conference, 5 and 6 Nov 2013. Elsevier, London
27. Roy R, Shaw A, Erkoyuncu JA, Redding L (2013) Through-life engineering services. Meas Control 46(6):172–175
28. Tukker A, Tischner U (eds) (2006) New business for old europe: product-service development, competitiveness and sustainability. Greenleaf Publishing Ltd, Sheffield
29. Davenport TH, Prusak L (1998) Working knowledge: how organisations manage what they know. Harvard Business School Press, Cambridge
30. Jagtap S, Johnson A (2011) In-service information required by engineering designers. Res Eng Design 22(4):207–221
31. Easterby-Smith M, Lyles MA (2011) Handbook or organizational learning and knowledge management, 2nd edn. Wiley, New york

32. Nonaka I (1994) The dynamic theory of organizational knowledge creation. Organ Sci 5 (1):14–37
33. Zack MH (1999) Managing codified knowledge. Sloan Manag Rev 40(4):45–58
34. Nonaka I, Toyama R, Konno N (2000) SECI, Ba, and leadership: a unified model of dynamic knowledge creation. Long Range Plan 33:5–34
35. Stokes M (2001) Managing engineering knowledge: MOKA methodology for knowledge based engineering applications. MOKA Consortium, London
36. Gupta S, Garg D (2011) Comparison of semantic and syntactic information retrieval system on the basis of precision and recall. Int J Data Eng 2(3):93–101
37. Mountney S (2009) Acquisition and shaping of innovative manufacturing knowledge for preliminary designcranfield university. Cranfield, Bedfordshire
38. Szykman S, Sriram RD, Regli WC (2001) The role of knowledge in next generation product development systems. ASME J Comput Inf Sci Eng 1(1):3–11
39. Doultsinou A (2010) Service knowledge capture and re-use in support of design. PhD Thesis, Cranfield University, Cranfield, UK
40. Hauschild M, Jeswiet J, Alting L (2005) From life cycle assessment to sustainable production: status and perspectives. CIRP Annals-Manuf Technol 54(2):1–21
41. Service Provision for Industrial Product Service Systems (2012). In: Proceedings of the 1st international conference on through-life engineering services: enduring and cost effective engineering support solutions
42. Adrian N, Snow D (2007) Quality tools, teamwork lead to a boeing system redesign. Qual Prog 40(11):43
43. Xu J, Xun Xu, Xie SQ (2010) A comprehensive review on recent developments in quality function deployment. Int J Prod Quality Manage 6(4):457–494
44. Tan KH, Platts K (2003) Linking objectives to actions: a decision support approach based on cause-effect linkages. Decision Sci 34(3):569–594
45. Pecht M, Dasgupta A (1995) Physics-of-failure: an approach to reliable product development. J Inst Environmental Sci 38(5):30–34
46. Through-life Engineering Services Standards Development (2012). In: Proceedings of the 1st international conference on through-life engineering services-enduring and cost-effective engineering support solutions
47. Institution of Asset Management (2014) What is PASS%. Available at https://theiam.org/products-and-services/pas-55/what-pas55. Accessed June 2014

Chapter 3
Through-Life Engineering Services: The NedTrain Case

Leo A.M. van Dongen

Abstract Investing in all operational management aspects is of great importance to the proper management and maintenance of capital assets. The primary maintenance processes are to be up to the task and to have the organization of the production plants, statements of work, planning and work preparation well in order. Continuous improvement presupposes an atmosphere where presenting improvement proposals, reporting, and removing unsafe working conditions is encouraged as a matter of course. By conducting the technical management of trains in maintenance close to the workshop floor, staff members are able to improve on the performance of the trains. Correct and proper information and the related necessary information systems allow for the analysis of performance and the implementation of improvement measures wherever required. Over the past few years, NedTrain has implemented major improvements to the performance of rolling stock in the Netherlands. Plans to perform maintenance activities closer to the transportation process and more closely attuned to dynamic maintenance demand await implementation in the near future.

3.1 Introduction

In the late 1990s, the Netherlands Railways (NS) was the organization responsible for infrastructure construction and maintenance as well as passenger & freight transport services. In response to the passing into effect of new European regulations, the construction and maintenance of the railways infrastructure were separated from transport services. NS was split into four organizations, namely, the for-profit NS Group and three infrastructure management task organizations, Railinfrabeheer, Railned and Verkeersleiding. The latter three were later brought together under the

L.A.M. van Dongen (✉)
NedTrain Director of Fleet Services and Chair of Maintenance Engineering, University of Twente, Enschede, The Netherlands
e-mail: leo.vandongen@nedtrain.nl; l.a.m.vandongen@utwente.nl

L. Redding and R. Roy (eds.), *Through-life Engineering Services*, Decision Engineering, DOI 10.1007/978-3-319-12111-6_3

ProRail umbrella. The national government appointed ProRail to construct, manage and maintain the railway infrastructure, and ProRail is responsible for allocating rail capacity and traffic control.

The wide range of commercial activities of the old NS were all brought under NS Group, which hived off all interests—such as railway construction, freight transport and telecommunications—not forming part of the core business. The core competences for passenger transport services reinforce each other. These competences include rolling stock maintenance, train station operations and hub development.

A new focus appeared aimed towards return on investment (ROI), with a context of high uncertainty about the future position of the company. The reorganization enforced by general management provided insufficient attention to operational processes. The result was a decrease in quality of service and a drop in punctuality to below the 80 % standard.

A new NS Board was installed, which formulated five quality improvement targets: (i) punctuality, (ii) providing service and information, (iii) improving social security, (iv) providing sufficient transport capacity and (v) offering clean trains and stations.

The national government invested heavily in railway maintenance, construction and renovation. NS was awarded the main rail network concession and over two billion euros were invested in new rolling stock. Transport services on the regional rail networks were put out to public tender by the regional governments and the contracts were awarded to market parties like Connexxion, Syntus, Arriva and Veolia.

Over the past few years, operational improvements have been realized to such an extent that the frequency of train movements on the main corridors has doubled and that punctuality has increased to 94 percent. The customer is king, multimodal transport from door to door, predictable daily performance and sustainability: these are currently the main national working themes for NS.

This chapter outlines the broad approach to rolling stock asset management, technical management and maintenance at the intersection of product, process and technology. The remainder of this chapter is as follows. Section 3.2 discusses the NedTrain rolling stock maintenance concepts and existing challenges. Next, Sect. 3.3 begins the discussion of maintenance improvement by presenting the role of the maintainer in the design of new trains. The following five sections focus on performance improvements achieved by NedTrain. Section 3.4 discusses placing operational processes in order during the period of from 2005 to 2010. Section 3.5 presents the period from 2010 to 2015, when placing technical management close to the workshop floor is high in the agenda. Next, Sect. 3.6 shows the focus within the period from 2013 to 2015, when the fundamental prerequisite to improving performance is the clear provision of information. Section 3.7 shows the parallel developments in rolling stock management of the period 2013–2015. Future improvement programs are discussed in Sect. 3.8. Next, Sect. 3.9 discusses the NedTrain life cycle logistics research program. Section 3.10 discusses how NedTrain invests in people. Finally, Sect. 3.11 summarizes this chapter's main points.

3.2 The Company

NedTrain is the part of NS Group responsible for the cleaning, maintenance & service, and overhaul of rolling stock. Acting in its capacity of Maintenance Integrator, NedTrain works 24/7 on ensuring that all trains are safe and reliable to operate, at the lowest possible cost. NedTrains approximately 3,000 employees manage to maintain an approximately equal number of cars [1]. Figure 3.1 shows how their tasks are divided, and we explain this as follows.

- **Daily maintenance** takes place at 35 service locations. Thorough cleaning of inside and outside, safety inspections (clearly specified for each type of stock), and minor repairs where necessary.
- **Short-term maintenance** is carried out at three workshops for national traffic, and one workshop for international and high speed traffic. All 750 electric multiple units (EMUs) operated by NS are withdrawn for compulsory maintenance and inspection after 50,000 to 90,000 km of running service. This short cycle maintenance includes check-ups to and replacement of brake linings, wheel axles, pneumatic components, filters, oil inspections, and exchange of parts that reached replacement age. All workshops allow for easy access to the roof and under-floor equipment. It is of course important that all cars are returned to service as soon as possible.
- **Parts overhaul**. NedTrain avails of two workshops for the overhaul of repairable parts, like bogies, wheelsets, traction motors, compressors and pantographs, for example. Smaller parts like brake valves, door control systems and electronics are overhauled to the extent they can functionally return to operations in an 'as new' condition. High-quality performance and short turnaround times are key in this process.

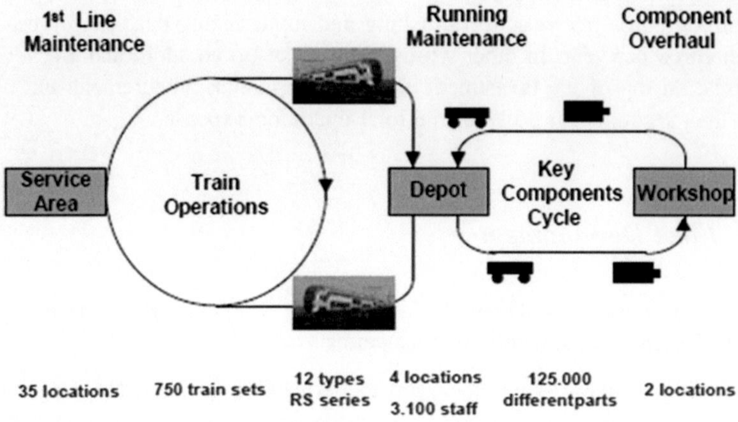

Fig. 3.1 NS/NedTrain maintenance logistics

- **Refurbishment and modernization**. Rolling stock requires long term maintenance when the train has reached half of the operational life. This includes maintenance work on the electrical, mechanical and hydraulic technology. Additionally, work on the train body is performed, both on the inside and on the outside of the train. Depending on the relevant functional and/or statutory requirements, maintenance work may involve anything from making small adjustments to full modernization.Work may include the retrofitting of air conditioning (HVAC), toilet systems with bioreactors, passenger information systems and new signalling systems.

3.2.1 High Costs of Rolling Stock

As is common in most heavy industry branches, maintenance requires investment. Over the entire life cycle of rolling stock, a multiple of the initial purchase or replacement cost is spent on cyclical maintenance, refurbishment, modernization and life extension. Finding the right balance between investment and other operational costs is crucial for NS.

A passenger car costs about two million euros, meaning that replacing its entire current fleet of some 3,000 cars would cost NS some six billion euros. The total value chain of national passenger transport service amounts to 1.75 Billion euros per year. Using round numbers, over the past few years, 100 coaches—worth 200 million euros in total—were supplied each year. As this concerns a recurring spending pattern, the expenses may be (conveniently) classified as depreciation charges.

As investment costs make up a sizeable portion of total expenses (12 %), it would at first glance make sense to keep them to a minimum. However, the reality is that the costs of maintaining the fleet more than double the purchasing expenditures. Cyclical maintenance and overhaul of repairable parts cost an approximate 300 million euros per year. Refurbishing and modernizing coaches represent 100 million euros per year. In other words, they make up an additional 24 % of total expenses, on top of the investment costs. Rolling stock procurement and maintenance thus accounts for 36 % of the total operating expenses.

3.2.2 Fleet Development

In managing the fleet, NedTrain distinguishes the following stages in the life cycle of trains for each type of rolling stock series:

- **Investing** in new rolling stock and modernization projects. Drawing up functional or technical specifications, assessing tenders, supplier selection, contract award, project management, participation in design freezes and supply quality control.

- **Phasing-in** new or modernized rolling stock. Training staff, ordering initial stockof spare parts, drawing up the required maintenance documentation, introducing the new trains and solving any teething problems.
- **Guaranteeing** and optimizing core fleet performance (with a minimum of constructive modifications). Drawing up maintenance frameworks, monitoring expenses, performance and incidents, drawing up life cycle plans, and laying down maintenance and overhaul volume annual plans.
- **Extending** operational life. Detailing (when desired from a fleet development perspective) a set of technical and maintenance measures to extend safe operational life.
- **Phasing-out**. Drawing up and managing a phasing-out schedule for specific units of the stock type concerned, including plans for the optimum rotation and use of capacity remaining in capital intensive spare parts.

NedTrain draws up a life cycle plan for each rolling stock series. This enables the company to create maximum value at controlled risks and costs. Also, the life cycle plan allows making the right operational decisions over the various stages of the operational life of the train types. This concerns the whole life cycle, including the period after refurbishment and the planned end-of-life withdrawal of the train series. Attention is provided to compliance and obsolescence as well as common technical and economic ageing.

Each year, current performance, bottlenecks and points for improvement are assessed for every stock type and all life cycle plans are updated accordingly. Based on these assessments, short-term objectives and specific required actions seasonal adjustments, interim upgrading of systems or parts, or even preparation for full renovation are determined for each stock type.

The information contained in the life cycle plans for the existing fleet also includes relevant suggestions for the long-term development of the entire fleet of rolling stock, new procurement initiatives and any modernization projects to be carried out being derived from the expected development in the transport sector and the desired future functionality of the fleet.

3.3 Design of New Trains

Before the changes in the railways took place, the technical sector was the dominant partner in the design process for rolling stock. National railway operators specified the requirements for the design of their rolling stock in detail. The requirements were part of their technical contract documents. The rolling stock manufacturing industry was fragmented, and the design and construction work was commonly contracted out to various contractors and subcontractors. Separate companies manufactured and supplied the trains subsystems, for example, the car body, traction equipment, braking systems, door control systems, wheel sets and bogies, gearboxes and signalling systems.

Providing attention to RAMS and LCC was perhaps implicitly of great importance, but it was important nonetheless. Co-operation followed the so-called Rhineland Model, dominated by the technical sector and sometimes pursuing technical innovation for its own sake. The client contributed its experiences when assessing the construction drawings in the design stage. This happened despite the fact that the equipment was technically specified to great detail in advance. Design freezes were made during design review and final construction was the result of agreement. The subcontractors transmitted their substantive technical knowledge of the equipment to the owner/user via the principal contractor. Therefore, the client possessed all the knowledge required to draw up the operating and maintenance instructions based on the documentation provided by suppliers. Teething troubles were solved jointly by all parties during the introduction stage. The other side of the coin was that the client, being substantively involved and responsible, was also required to make some warrantee commitments.

The privatization of public rail sector has focused newly privatized companies ever more on their core activities. The end customer's interests are paramount and rightly so! and full attention is provided to passenger service development. The design and maintenance of rolling stock plays as smaller part. Trains are designed, built and supplied by major companies and consortia, each of which developing and offering new systems and equipment based on their own (or third-party) market awareness and technical experience.

The result is that the order for and construction of new trains is being thought of as a one-off and isolated investment project. Meanwhile, there is a very real chance that the final design will not properly fit the requirements of the production organization (transport and maintenance services). The (future) stakeholders should participate during all project stages, namely, specification, tender assessment, provisional design, engineering, construction and introduction. This should involve participation by way of various project and design freezes, training and introduction, familiarization, and solving teething troubles, to name a few.

The quality of the design should be brought in balance with its repercussions for the maintainer of the product [2]. On the one hand, design savings should not be later compensated for by more maintenance being required. On the other hand, the quality of a well thought-out and possibly bit more expensive design can be supported with lower maintenance costs, leading to optimum Life Cycle Costs (LCC).

Figure 3.2 shows that intrinsic characteristics of capital assets are laid down in the initial design, and in the event of modernization or upgrading. Maintenance remains crucial for the entire life cycle of equipment. From initial idea, to the business case, the investment decision, the procurement, the design, the construction, the delivery, the use, the cleaning and repair, up to the final withdrawal and scrapping of the train [3].

The life cycle of rolling stock generally exceeds the average employment term of a NedTrain employee. This means that any single member of staff (manager, designer, operator or mechanic) faces the benefits and burdens arising from the work of their predecessors. Therefore, every member of staff has to work, and act, from an inner sense of responsibility towards their successors. This is the key to success.

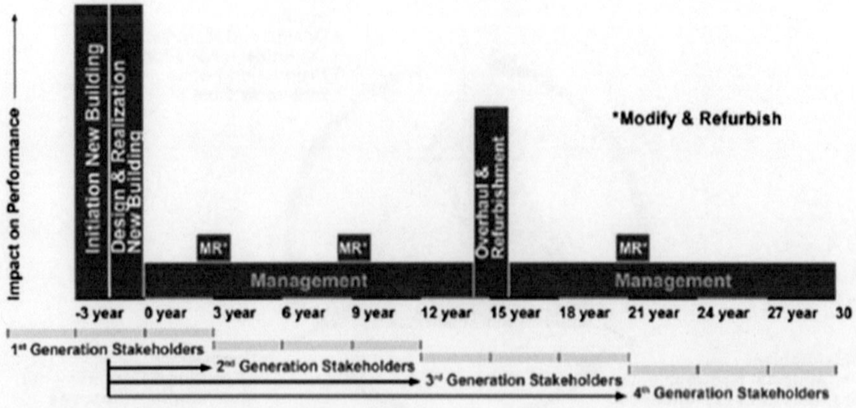

Fig. 3.2 Design and construction determine LCC

3.3.1 System Integrator and Maintenance Integrator

In general, technical innovation on the suppliers side, takes place in collaboration and cooperation with various parties. These parties may be knowledge institutions, engineering firms, architects, specialists, subcontractors and system integrators. The operator and the maintenance organization perform their work where both sides of the project meet: they join forces to construct the desired product with the technology provided to them. Their experience with the production process enables them to introduce new technology into the product design. Naturally, the basis of professional asset management is connecting all segments in the innovation circle comprising technology, process and product. The maintenance organization has a large role to play in this respect.

Maintenance is sometimes considered a necessary evil. An understandable viewpoint wherever the necessity for basic maintenance cannot be removed. However, the previous paragraphs will have made clear that suppliers, system integrators, operators and owners have a lot to gain from the knowledge and expertise available within the maintenance organization.

If the maintenance engineer were to play more prominent a role during the entire life cycle of capital assets, the interplay between supplier, owner and maintenance provider would be more balanced. The existence of such a new balance, shown in Fig. 3.3, would make responsibilities more clear. The result would be a more balanced. The existence of such a new balance, shown in Fig. 3.3, would make responsibilities more clear. The result would be a target-oriented approach to integral capital assets management, as considered from a maintenance perspective being included in the decision process [3, 4].

The equipment owner/user benefits from proper functional performance at low cost, both concerning capital and operating expenditures consider energy consumption, availability, reliability and maintenance. The maintenance organization is

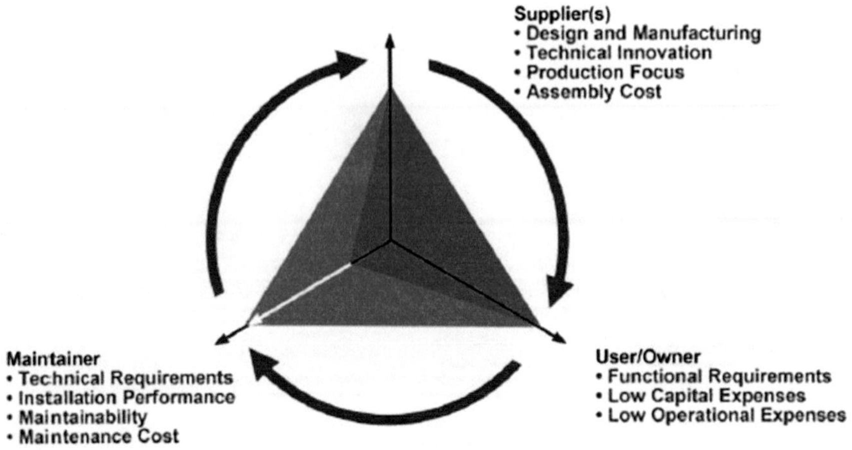

Fig. 3.3 Integral approach of life cycle performance

able to 'translate' these functional requirements into technical requirements for the system integrator. Both the performance and maintainability of the equipment are then taken into account. Using the technology available to it, the contractor designs and manufactures the equipment at the lowest possible cost. It goes without saying that an integral improvement can only be attained by way of interactive consultation and joint decision-making.

The modern maintenance organization therefore acts as 'maintenance integrator' in the chain, linking the technical disciplines, bundling information flows, connecting links and acting as director in capital asset management.

3.3.2 New Sprinter Light Train

Between 2009 and 2012, 131 new EMUs (in configurations of four and six coaches) were introduced for intensive regional rail services on the Dutch rail network. This was an investment worth 880 million euros. In this project, functional specifications formed the cornerstone of the procurement process. However, it needed maintenance improvements.

At one point during the engineering stage, the NS and the Bombardier/Siemens consortium decided to halt the project for a few weeks in order to implement a number of modifications to the design. This allowed improving the maintainability of the trains. Although the original design was well developed in terms of assembly, it had not automatically resulted in a train that could be efficiently and effectively maintained. For instance, the orientation of the compressor underneath the floor needed to be rotated for it to be replaceable with a lift table in the pit track. Special tools were developed to disassemble and replace the traction motor from underneath the train. When the trains were introduced in active operation, their maintenance

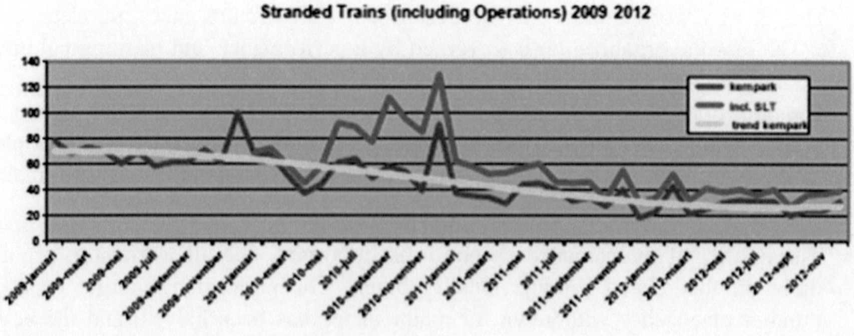

Fig. 3.4 Reliability improvement (2010–2015)

schedule was optimized in mutual consultation. Ultrasonic testing of the axles would not be performed at regular time intervals, but based on the number of kilometres driven.

As shown in Fig. 3.4, the teething troubles of the new stock were solved during the introduction stage. This was in cooperation with the consortium. Today, the new Sprinter Light Train performs as well as the rest of the fleet's rolling stock and has entered the 'guaranteeing' life cycle stage. The exact extent of the first overhaul, to be held after 6 years of operational service, is presently being determined in mutual consultation. To a large extent, it will be based on the experience obtained by NedTrain in the maintenance of rolling stock over the years.

Companies like NedTrain are therefore of great importance for the sector in realizing optimum performance and acquiring all necessary information. Maintenance depots for their part require the technical documentation possessed by the contractors and subcontractors. In short, all parties to the project absolutely need to continue their cooperation for the full life cycle of the train.

3.4 Operational Processes in Order (2005–2010)

Rolling stock performance cannot be guaranteed just by technically perfect designs and maintenance concepts. The result is largely dependent on managers, operators and mechanics. There must be emphasis in the importance of clear and specific procedures, working regulations, working conditions, good housekeeping, job descriptions and delineation of tasks and responsibilities. Achieving good results also requires a culture that encourages members of staff to identify and report unsafe working conditions, technical problems or a lack of clarity in the organization. Also this requires openness to take appropriate measures, not just within their own working environment, but within the hierarchy of the entire organization. Often, dangers lie in routine behaviour, where an accumulation of inaccuracies due to pressure from external factors can have major consequences.

This is why work should only be performed if all circumstances are clear and safe. The less the organization is governed by a performance and blaming culture, the sooner risks will be identified and tackled.

The poor operational performance during the early 2000s led NedTrain to adopt a primary maintenance process improvement programme. Guided by the principle of 'doing more with fewer but better means', the programme contained three main targets:

- The quality of maintenance execution is important, and all workshops report daily on the related Product Quality Index. The performance target for the number of coaches withdrawn for maintenance has been halved, and the new target is set at 200 coaches. To actually achieve these targets, production managers are provided with all space and assistance to set up the processes as clear-cut (Kaizen, Lean Working & First Time Right) as possible.
- Safety is a must. NedTrain has invested a great deal in its stated goal of having 'everyone return home in good health, everyday'. This program involves risk identification and assessment, safety training and reporting. Weekly performance indicator reports and regular inspections are being conducted by the NedTrain Board. This underlines its importance.
- A lot of attention is being paid to clearing the backlog in training sessions, to enable staff to properly do their work. This includes developing training manuals, fault trees and fault correction procedures. Social innovation (ideas thought up by the workforce) is also encouraged by having staff members present their improvement plans directly during a 'road show'. The ideas are directly assessed by company managers and, where applicable, allocated an implementation budget.

This programme for the improvement of the primary process is accompanied by investments to the amount of 200 million euros in new facilities and equipment: A new building for rolling stock refurbishment in Haarlem, a new maintenance depot for international trains in Amsterdam, train washing installations and upgrading of 5 major service locations, and a new production plant in Tilburg for overhauling repairable parts. In this period, two old workshops in Zwolle and Amsterdam have been closed.

All these efforts have led to major performance improvements within a span of 5 years. The number of cars taken out of service for short-term routine maintenance decreased from 450 to 200, resulting in savings amounting to 30 million euros per annum. The NedTrain headcount in this period dropped from 3,800 to 3,100, while the kilometre performance of the NS vehicle fleet was retained at 650 million kilometres driven per annum. This corresponds to a decrease in maintenance costs from 52 to 45 eurocents per car kilometre. The number of Lost Time Incidents (LTIs) decreased from 15 to 0.5 for every million hours worked. This means a reduction from 60 to 2 incidents per year for NedTrain as a whole. The overall performance, the attention provided to the members of staff and their ability to contribute to process improvement has resulted in the employee satisfaction rating increasing from a 6.3 to a 7.2 and, in 2014, up to a 7.5 on a ten grade scale. The improved NedTrain performance resulted in a decrease of rolling stock dis-punctuality in daily services from 3.2 to 0.7 %.

3.5 Technical Management Close to the Workshop Floor (2010–2015)

Once the investment decisions have been made and the train has been designed, the Failure Mode Effect and Criticality Analysis (FMECA) can be used to determine the intrinsic quality of the design. This original quality can be maintained within acceptable margins. Performance of the new stock is measured, registered and analysed to guarantee Reliability, Availability, Maintainability and Safety (RAMS) through custom maintenance work, employing techniques from Reliability Centred Maintenance and Risk Based Inspection.

Managing equipment as described above for the short and medium term, demands close cooperation between production and the technical staff. The basis is to be found in soundly organized use and maintenance according to the approaches of Kaizen, Six Sigma, Lean Working and First Time Right. Technical deficiencies can only be properly identified when the operational processes are in order. Technical support services of the production organization should be available on the workshop floor, allowing for maintenance/reliability engineers, operators and mechanics to obtain a better understanding of each other's work and increase their cooperation in analysing the relevant equipment failure behaviour, solving short-term issues and putting long-term measures in place, maintaining a high equipment performance standard. Improvement proposals can be formed based on approaches like Root Cause Analysis (RCA) and Failure Modes, Effects and Criticality Analysis (FMECA). Adjustments to the design of the equipment or the maintenance concept and improvement of the maintenance process execution quality are brought to the attention of the managers responsible.

Only when the operational processes are in order can it be clearly determined what the 'technical' causes of failure of the rolling stock. This is why NedTrain some years ago established rolling stock teams for each stock type at each maintenance depot. Teams were setup with a rolling stock manager (acting as the 'owner'), a reliability engineer, a configuration engineer, a business analyst, and, as a 'permanent guest', a maintenance engineer, responsible for safeguarding the integrity maintenance standards. They are responsible for developing operational reliability and the technical part of the asset management conducted by the owner/transporter. The team determines the deployment, performance, maintenance, modification and cost requirements and needs on the basis of the life cycle of the train series.

The reliability engineers are responsible for setting the operational reliability targets of the stock type assigned to them by conducting analyses of the failure behaviour in the systems and the parts within those systems and implementing improvements and solutions. They start up operational reliability sessions with a focus on train components. Any suspect parts identified during those sessions are taken to discussion with the suppliers. Based on the findings of the teams, feedback is provided to suggest modifications to the design. Furthermore, they provide adjustments to the maintenance concept and improvements to quality of maintenance execution. Adjustments are brought to the attention of the responsible managers, the maintenance engineer and the production manager of the maintenance depot.

A rolling stock operational reliability key performance indicator was defined. This produced a clear image of the contribution of the rolling stock teams to the operational reliability. The KPI closely relates to a direct source of nuisance for the customer: a stranded train due to technical failure. This does not so much concern trains breaking down on the main line which only happens very occasionally but trains unable to depart within ten minutes following a failure notification. NedTrain provides particular attention to preventing recurrence of the problem and to First Time Right repairs on the basis of sound fault tree analyses. Figure 3.4 shows that NedTrain has managed to reduce the number of stranded trains from 80 to 30 a month. The figure also displays the fail rate of the Sprinter Light Train as compared to the core vehicle fleet. Following the 2010–2011 introduction period, the failure rate of this train series stabilized at the level of the rest of the fleet.

Establishing the rolling stock teams in the maintenance depots has taken technical management out of the company headquarters and placed them right on the workshop floor. This places production and technical management at the same level.

3.5.1 Continuous Improvement of Fleet Performance

Rolling stock performance management combines technical, administrative and management tasks. These tasks are required to sustain equipment, or return it to such a state that it can perform its intended task. Maintenance is required because the equipment gets into an ever worse state due to degeneration of parts and systems. In an ideal world, the state of any piece of equipment is always known or predictable, allowing for maintenance work to be performed at exactly the right time. But a lack of information, insufficient awareness and logistical and operational limitations ensure that reality does not live up to the ideal.

For modern rolling stock, an initial maintenance concept, providing an overview of clustered tasks, is drawn up on the basis of Reliability Centered Maintenance (RCM) and Failure Mode Effect and Criticality Analysis (FMECA) at the same time the technical design is drafted. This concept includes daily and short-term maintenance and the schedule of light and heavy maintenance for the longer cycle. This is complemented by a description of all the aspects required to perform the work in terms of staff (knowledge, experience or applicable degrees), information and documentation, company facilities and equipment, and proper parts.

The maintenance tasks are divided into preventive and corrective maintenance. Preventive maintenance is subdivided into two categories: maintenance dependent on period of active use and maintenance dependent on actual condition. If maintenance is dependent on period of active use, it is the use irrespective of the actual condition of the equipment that determines which inspections and work need to be performed, and at which intervals. Otherwise, the actual performance determines replacement, repair and adjustment activities as well as the amount of kilometres driven, hours in active operational use or start cycles. In most cases, such maintenance activities are scheduled on the basis of experience and investigations into degradation behaviour.

In the case of condition based maintenance, the actual condition is determined by way of equipment inspection and compared against established standards. This determines which parts or system needs replacement, repair or adjustment activities. Inspections are scheduled periodically, but repair is only scheduled if required.

Corrective maintenance is required in the case of equipment failure or defect. If a defect does not have any direct consequences to the primary operations of the equipment, it is repaired during the next scheduled maintenance. In the worst case, a failure needs to be rectified immediately, which negatively impacts the availability of rolling stock.

Rolling stock behaviour is registered: for example, performance, user data and failure behaviour are all noted down and laid down in logs by hand or directly into digital information systems. Figure 3.5 provides an overview of the failure behaviour of the double deck intercity trains, a stock type composed of 176 four-car and six car units (868 cars) covering a distance of some 300,000 km per annum. Timely information on their maintenance requirements allows for the scheduling of activities to be performed. In addition, analysis of the data allows for establishing links between the maintenance performed and the deployment of the equipment. Based on these progressively more detailed insights, the maintenance plan can be continuously upgraded. A maintenance concept might seem to be a static piece of work, but it can be adjusted on the basis of experience obtained. For instance, more

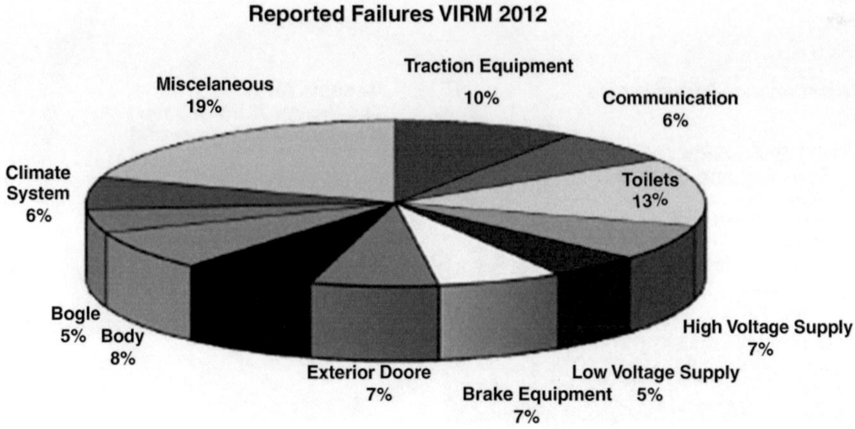

Fig. 3.5 Monitoring and reporting is key

preventive maintenance work might be scheduled, so as to improve reliability. Or a well-considered decision might be reduced to cut back on maintenance within the systems technical limits to save costs. Maintenance research (e.g., degeneration as related to use) allows for a sound scheduling of preventive maintenance work on mechanical parts. The state of such parts can also be directly determined by physical measurements performed on the equipment. For instance, the thickness of brake linings is relevant for maintenance work dependent on present state.

Relevant experience can be drawn on during refurbishment or modernization to make technical improvements to the equipment. If necessary, the construction may be subjected to modifications.

Figure 3.6 provides an overview of these control loops:

- Design for Maintenance. On the strategic level (long-term decisions and investments), the main objective is to have the optimum resources in place: suitable equipment, correct configuration information, adequate infrastructure, machinery & tools, an initial maintenance concept, and a solid operator and mechanic training programme. In short: does the organization have all the right resources available?
- Maintenance Engineering. The medium-term targets are all about performing proper maintenance. Registered failures and malfunctions are analysed, trends are monitored and points for improvement are determined. The maintenance concept is adjusted where necessary, performance quality is discussed with the production manager and the quality of parts with the suppliers. This concerns tactical improvements in the long-term equipment management.

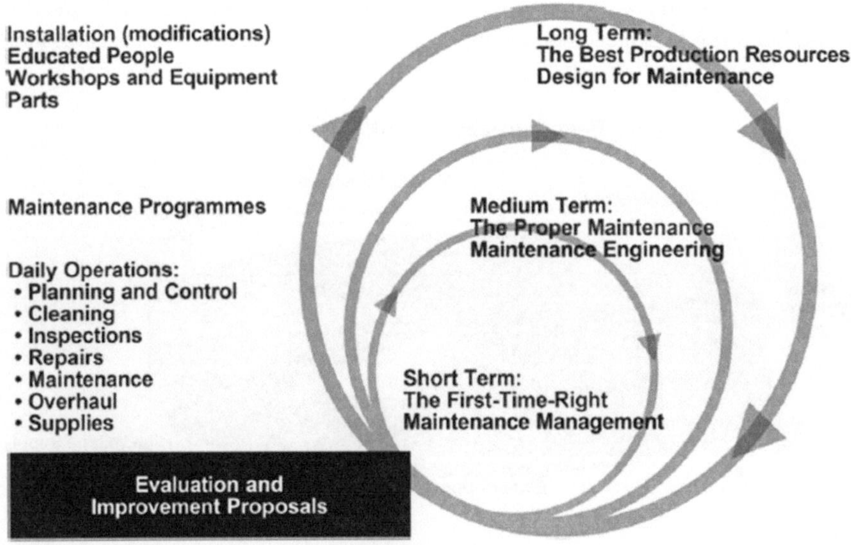

Fig. 3.6 Optimization of maintenance approach

- Maintenance Management. The scheduling of all the required work activities in the short term: inspection, cleaning operations, maintenance, repair work, parts procurement and timetabling personnel.

In other words, rolling stock performance management is more than just the actual maintenance work. By performing the strategic and tactical improvement activities detailed in the above, 'maintenance' shifts from being an unavoidable cost item to a creator of added value to the company.

3.6 Clear Information Provision: A Fundamental Prerequisite (2013–2015)

Efficient, effective and predictable business operations require the availability of a sound company management and information system. In addition, this is the paramount precondition for the proper management of the capital assets.

Tightened legislation also requires a higher level of maintenance process demonstrability and predictability. In order to have such in place, not only do the basic processes need to be in order, but a number of administrative affairs also need to be properly organized. This includes a certified maintenance concept, maintenance process standardization, scheduling and planning, training and certification of staff, maintenance performance quality check-ups, controlled release of rolling stock reported finished, development of rolling stock failure elimination strategies, and registration of all work performed.

All this requires a dedicated ERP system for all maintenance activities. In 2013–2014 NedTrain replaced its legacy system that consisted of dozens of isolated IT applications. A new core system uses Maximo for maintenance management, production planning in maintenance depots and service locations and SAP for parts logistics, finance management and HR management.

All mechanics have been equipped with iPads, allowing for easier communication of scheduled work, registration of failures, of activities by subject and time, and for the consulting of relevant documentation.

3.7 Rolling Stock Management (2013–2015)

The implementation of many improvements these past few years (operational processes, performance of the rolling stock and implementation of Maximo and SAP) has allowed NedTrain to achieve the following:

- Stronger customer-orientation by implementing the right focus on stock types and guaranteeing such in internal business process management;

- Tightening of financial guidance and control over the performance of rolling stock for all the various departments within the company;
- Control over the company chain in addition to the technical focus of the asset management, asset based budgeting are being introduced in order to optimize performance and costs.

The scope of rolling stock management covers maintenance, cleaning, repair work and modifications. On the strategic level, the market, the company business strategy, the fleet development plan and the use of rolling stock set the tone for the (annual) development and budgeting activities. The related Business Plan process translates the product, numerical, temporal and performance level requirements by way of the performance management concept into work packages, capacity allocations and budget allocations. The production departments are responsible for determining staffing levels, such on the basis of the expected workload.

For each step in the process of the repeating planning, performance of maintenance, performance measurement and after-care cycle, its contribution to the final performance of the rolling stock is measured by relevant performance indicators. In the end, NedTrain is able to determine per stock type how these processes contribute to safety, operational reliability, availability, comfort, customer satisfaction and stock type costs.

Various key functions are responsible for the various fields of work performed by NedTrain on each train series:

- **Maintenance Engineer**: optimization of the maintenance concept and specifications in accordance with standards and the law (License to Operate);
- **Rolling stock manager**: contract management and plan-do-check-act (PDCA) as concerns performance and cost;
- **Production manager**: planning, execution, final check-ups before releasing back into active use, after-care and staffing level variance (production efficiency).

Its approach of disciplined registration and analysis of relevant information and costs allows NedTrain to further improve its performance and bring its asset management to a higher level in the future.

3.8 Maintenance Closer to Operations (2015–2020)

NedTrain has three maintenance depots for national passenger transport trains, spread over the country (Fig. 3.7). MUs are allocated to one of these three depots and are brought in for short-term routine maintenance after set intervals of some 75 to 120 days, depending on stock and usage. These maintenance jobs are divided into preventive predictable and scheduled maintenance work and work to be planned for each case depending on actual state. The maintenance work scheduled

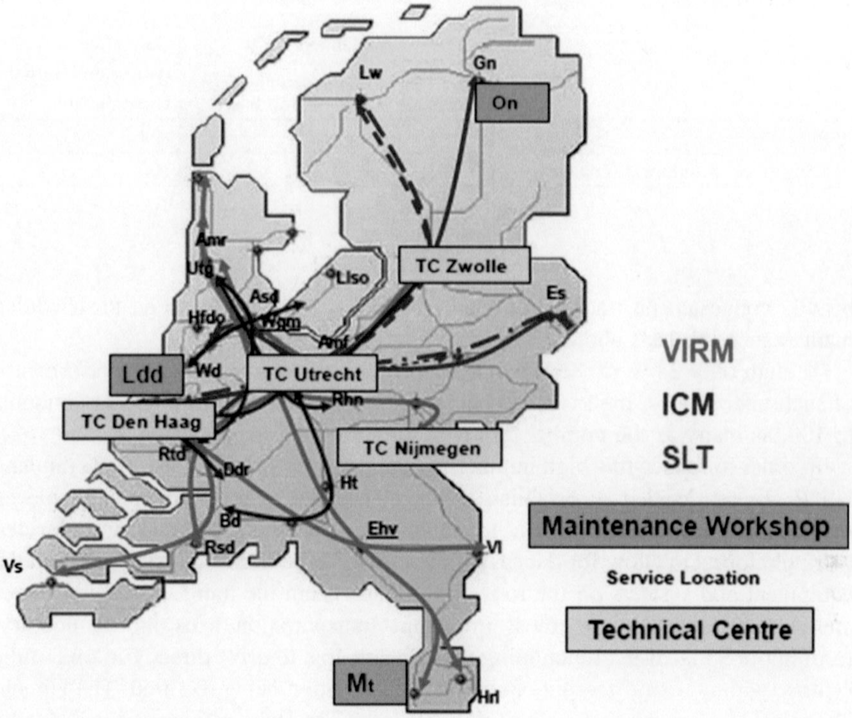

Fig. 3.7 Overview of locations in network

in advance is based on either experience or targeted maintenance research and is often required because of physical degeneration. This type of maintenance work accounts for some 45 % of all short-term routine maintenance activities and the associated amount of cars removed from active use is 50. The actual state-dependent maintenance (including postponed corrective maintenance) is planned on the basis of baseline inspections when the rolling stock is withdrawn: this type of work accounts for some 31 % of all maintenance activities and the associated amount of cars removed from active use is also 50. Refer to Table 3.1.

Over the period between these relatively major maintenance jobs, inspections of and small repair work on the rolling stock are performed at the service locations. Major defects and repairs that require specific equipment also require the rolling stock to be removed from passenger service, and sent to one of the three workshops for an unscheduled maintenance job. Given the locations of those workshops and the high utilization of the rail network, such routing of empty trains is costly and has long lead time. Due to the high train density on the Dutch railway network, malfunctioning rolling stock can usually only be brought to the maintenance depots by night. Next, the repair job needs to be slotted in for performance on one of the

Table 3.1 Maintenance categories

	Preventive & planned (workshops)		Corrective (workshops/yards)
	Predictable	Condition based	Unpredictable
Efforts (%)	45	31	24
Rolling stock withdrawal (coaches)	50	50	100

heavily congested pit tracks. Removal from active service due to an unscheduled maintenance job lasts about 3 days.

Though only 24 % of NedTrain's staff capacity is required for the performance of such interim jobs, the associated number of cars removed from service amounts to 100, as many as the number removed for scheduled maintenance.

In order to reduce this high number of cars removed from service, NedTrain and NS Reizigers decided to establish technical Centres at four strategically placed service locations. Each Centre is made up of a hall with a pit track and elevated work platforms to allow for the easy access to and ergonomic and safe work on the equipment and systems on the roof of and underneath the trains. The four centres are located alongside the most important transportation axes in the country, resulting in 95 % of malfunctioning trains being able to drive directly to one of the centres, with no extra steering or special transportation being required. The Utrecht Technical Centre was opened in Spring 2014; the The Hague, Nijmegen and Zwolle centres will start operations shortly. This investment (of 25 million euros) will lead to a reduction in the removal of rolling stock from active service by over thirty cars (worth 60 million euros), savings that can be cashed in in the next round of ordering rolling stock.

By means of early inspections of the trains at these locations, a higher percentage of maintenance to be performed at the three depots can be scheduled in advance, also leading to a reduction of removal from service. In connection therewith, NedTrain has launched a 'No Withdrawals During Rush Hour' pilot project with a subset of the double deck intercity train series. The maintenance package of this stock is split up into smaller packages, allowing for better balancing between transportation use and maintenance work and the uninterrupted deployment of the stock during peak time. It goes without saying that the availability of properly trained staff, specific equipment, a sound registration system and the careful use thereof, smart maintenance scheduling and rolling stock are logistic preconditions for the success of the pilot. The implemented improvements detailed in the above allow for experimentation in this pilot. The goal is to improve rolling stock performance through sound cooperation and deployment of the specific expertise of all transportation and maintenance chain partners.

3.8.1 Real Time Monitoring Enabling Dynamic Maintenance Scheduling

Modern trains are increasingly equipped with mechatronics and digital train management systems. As a result, they have so much information 'internally' stored that by reading out the data a wealth of information on the technical state of the equipment can be retrieved. In addition, sensors allow for targeted monitoring where necessary: an example would be the use of vibration meters in monitoring crucial bearings. Monitoring is about more than identifying a failure or malfunction. Provided the development of relevant parameters is registered and analysed, it also provides new insight into the equipment degeneration behaviour before it actually malfunctions. Monitoring can be performed on the spot, but also from a distance, using (wireless) data communications. These techniques allow for determining the actual maintenance required and scheduling the work: specific repair and maintenance work in the short term and clustered routine jobs at any time suitable to production over the long term. Reliability engineers (responsible for equipment operational reliability) are able to provide repair and maintenance advice from the control room even before a malfunction actually occurs.

Monitoring and automatic data processing result in the gathering of more accurate data than was possible in the past. Existing double deck intercity trains are equipped from build with an on-board Train Management System for 15 systems with 3,000–4,000 sensors depending on train length and build series.

A pilot project for 54 trains involved the installation of a train-to-shore connection to download existing on-board failure reports, operational events (doors opening/closing, coupling), counters (compressor operating hours etc.) and a number of sensor measurements from the TMS. Information on GPS location and time is also stored. The information is stored in a generic database for use by different user groups, for various applications [5].

The first application is an operational dashboard for the NedTrain control room that displays the current status of the trains. Business rules are used to filter important events. Setting the right business rules was essential to prevent information overload of alerts here no immediate action is required. Priority 1 events are highlighted and require immediate attention by control room staff. For all priority 1 events the dashboard provides advice on the failure elimination strategy. The strategy combines urgency (immediate and of route, etc.), locations where the failure can be eliminated (taking into account facilities, equipment, staff competence and spare parts available) and repair advice (short-term repairs or reference to documentation like fault trees, spare parts and equipment most likely to be required).

A second application of the information is for the maintenance depot to use it to schedule both preventive and corrective maintenance jobs. The MU is brought in for maintenance between peak hours once per month. Prior to the train arriving at the depot, its failure history is analysed and compared to the fleet average. Based on this analysis, work orders are prepared, allowing for ordering the spare parts most likely to be required and timetabling the required staffing capacity. Should the time necessary to complete the repair works not fit in the maintenance slot, the repair job is rescheduled so as to guarantee the timely delivery of the rolling stock for peak service use.

The data are also presented in an analytical dashboard for use by maintenance and reliability engineers. This allows for trend analysis over the entire fleet and allows for data-driven performance improvements and reviewing maintenance schedules based on performance data. Currently, the maintenance schedule is still largely a preventive maintenance schedule based on fleet performance. The ambition for the future is to make the maintenance tasks dependant on actual performance and state of individual MUs.

The first results indicate that removal from service due to additional maintenance being required can be reduced by 12 % on average in response to the advance scheduling of the repair job. The throughput time of door control system and HVAC repairs can be reduced by 35 %. Learning points:

- Business rules and action plans are essential in order to prevent a 'data tsunami';
- Implementation of the business process is key, the IT application simply serves as a support tool;

The implementation of large-scale Real Time Monitoring requires establishing the layout of the control room, as well as identifying what smart analysis and scheduling software is required for the specific processing of large amounts of information on the state of trains and translating it into scheduling information, without this negatively impacting practicability. In other words, the underlying logistics might prove to be the greatest challenge!

Maintenance is no longer static, mechanic, reactive to events already occurred, a matter of logs, oil and greasy hands, and a cost item. It is becoming electronic, flexible, dynamic, proactively aimed at the future, a matter of digital management technology, and a creator of added value. Whether an organization is ready for a dynamic, flexible maintenance programme, actually requires a fixed schedule for the entire equipment, or would benefit most from a mixed model with a fixed maintenance schedule for the heavier work combined with actual state-dependent maintenance, depends on the capabilities and aims of the organization.

3.9 Life Cycle Logistics R&D Program

'Progress' entails more than simply implementing some obvious process improvements. NedTrain is committed to continuous product improvement and has therefore established a research programme in joint cooperation with the three Dutch technical universities (Delft, Eindhoven and Twente). An interdisciplinary team of doctoral candidates, students and staff is working on the following subjects:

- Strategic decisions: maintenance as an aspect of rolling stock selection and design choices, including topics like accessibility of technical systems, diagnostics systems, repair strategies, etc.

- Tactical maintenance organizations: structure of the maintenance concept, dimensioning production capacity and inventories, allocating work packages to specific locations, cooperation with suppliers.
- Operational management: dynamic allocation of task schedules based on actual rolling stock maintenance requirements on the one hand and operational feasibility on the other.

3.10 Investing in People

Organizational success requires more than investing in resources, methodologies and materials. A well-trained workforce is instrumental as well. This is why NedTrain pays a great deal of attention in providing training programmes and refresher courses, covering a range of topics, including process awareness, working with electrical equipment, safety on and around the tracks and specific rolling stock technology. Collaboration with various educational institutions is sought out on all sides. Young academics are recruited within the framework of the research programme referred to above, and in addition, NS offers trainee programmes, aimed both at general management and at specific areas of work like Finance, IT, Logistics and Technology. In a 2-year period, these recently graduated trainees are set to work on various projects and hold various positions so as to get a solid grip on the company. Afterwards, when proven suitable, they come to hold a regular position and continue their career development with the company.

NedTrain cooperates with a number of universities of applied sciences in offering additional vocational refresher training. In cooperation with the Twente Regional Training Centre, NedTrain has set up the 'TechniekFabriek' vocational school, with branches in Amsterdam and Zwolle, recruiting 50 mechanics annually. Students at this school follow the Mechatronics secondary vocational education programme, combined with a number of programmes related to specific rolling stock. The first year of the 'TechniekFabriek' programme mainly covers theory, with some ventures into practice, while the second year mainly entails putting everything into practice at a NedTrain facility, coupled with a few weeks of classes in school. Programme graduates are offered a position at one of the NedTrain locations. In this manner, NedTrain is constantly re-training its staff. This increases people's professional skills towards the high level of technological mastery required to work with the fleet of rolling stock.

3.11 Summary

In the last decade, NedTrain has worked hard in order to bring the performance of the rolling stock fleet on a higher level. Figure 3.8 shows the evolution of these performance improvements. The use of rolling stock entails paying the initial

Fig. 3.8 Summary of performance development at NedTrain

investment cost many times over in maintenance. Asset management is an ongoing process and requires open and continuous cooperation between the parties involved, both in the chain with suppliers and builders and when outsourcing management, maintenance and modernization work.

Well-organized operational maintenance processes are an absolute necessity for keeping rolling stock performance at the desired levels. This is the prerequisite for high quality targets in the maintenance execution and a low maintenance withdrawal of the fleet. The maintenance depot's technical staff need to be as close as possible to the primary process in order to register and analyse stock behaviour and consequently make adjustments to its use, technical build, maintenance concept or the quality of the performance of maintenance work. In this way, a high rolling stock reliability is being guaranteed.

The introduction of Maximo and SAP as information, production and planning system facilitates the improvement of NedTrains performance. In addition to the technical asset management, asset based budgeting on the maintenance execution chain is being worked on.

Modern management technology increasingly allows maintenance depots to measure the technical state of all relevant systems in real time, to determine the necessity for maintenance and to predict any malfunctions. The challenge of getting the underlying maintenance scheduling and logistics processes in order is the greater one.

Successful asset management becomes more likely through innovation at the intersection of product, process and technology. Yet more important even than technological innovation is improving the human aspects in the chain.

Acknowledgments The author gratefully acknowledges the support of the Lloyd's Register Foundation (LRF). LRF helps to protect life and property by supporting engineering-related education, public engagement and the application of research. The author wishes to thank J.E. Parada Puig for his operational support to prepare this chapter.

References

1. van Dongen LAM, Lutters D, van Houten FJAM (2011) Maintenance as an integrated optimization criterion in development life cycles. In: 44th CIRP conference on manufacturing systems, CIRP, Madison, Wisconsin
2. Mulder W, Basten R, Jauregui Becker J, van Dongen L (2013) Work in progress: developing tools that support the design of easily maintainable rolling stock. Procedia CIRP 11:204–206. doi:10.1016/j.procir.2013.07.034
3. van Dongen LAM (2011) Maintenance engineering: maintaining the links. Innaugural Lecture on appointment as Chair of Maintenance Engineering
4. Parada Puig JE, Basten RJI, van Dongen LAM (2013) Investigating maintenance decisions during initial fielding of rolling stock. Procedia CIRP 11:199–203. doi:10.1016/j.procir.2013.07.033
5. Mooren FP, van Dongen L (2013) Application of remote condition monitoring in different rolling stock life cycle phases. Procedia CIRP 11:135–138. doi:10.1016/j.procir.2013.07.050

Part II
Data, Diagnostics and Prognostics

Chapter 4
The Impact of No Fault Found (NFF) on Through Life Engineering Services

Christopher J. Hockley

Abstract Faults cause system downtime and require resolution before the system can be put back into service. When the fault cannot be replicated or diagnosed successfully, the effort causes wasted man-hours and reduced availability of the system. The fault is then designated by any number of descriptors such as No Fault Found, Re-Test OK or Cannot Duplicate. The diversity of the taxonomy is a problem in itself which often masks the true costs of fault resolution and maintenance. The situation, whether real or perceived therefore has a cost for organisations, but often these costs are not evident either at the operational level and certainly not usually throughout the supply chain or the support organisation. This chapter will highlight the problems caused by faults that cannot be resolved and the many causes which are varied and diverse. It will cover the impact and the through-life costs of the No Fault Found problem and show the hidden costs and the impact they cause to operations. The chapter will not seek to be a comprehensive treatise on the problem as it is a vast and complex one; it will be an introduction to the problem and seeks to lead the reader to the more detailed publications on the subject that are now in preparation at the EPSRC Centre. Whilst examples of best practice and solutions will be covered, the chapter will describe the complexity of solutions and mitigations, and highlight the research that is necessary to produce design solutions in the future.

4.1 Introduction

Faults that cannot be replicated or confirmed on diagnosis, generally known as NFF, are an aspect that is well known to those providing maintenance and asset support in all industries. Its impact on maintenance effectiveness can be highly damaging,

C.J. Hockley (✉)
Centre for Defence Acquisition, Defence Academy of the UK, Cranfield University, Shrivenham SN 6 8LA, UK
e-mail: c.hockley@cranfield.ac.uk

L. Redding and R. Roy (eds.), *Through-life Engineering Services*,
Decision Engineering, DOI 10.1007/978-3-319-12111-6_4

causing loss of availability of the assets and causing high costs to those who provide through-life support. In some areas the actual level of NFF is unknown, or at least not visible, and yet it is essential that there is an understanding of the causes and the solutions if maintenance and support is to be cost effective through life.

4.2 No-Fault-Found (NFF) Taxonomy

One of the issues that has surrounded the achievement of a better understanding of NFF is that there is a plethora of descriptions and phrases that mean the same; Cannot Duplicate (CND), Re-Test OK, Tested Satisfactorily (TSat) and Fault not Found (FNF). Indeed the last one, FNF has a slightly more positive sound. NFF suggests a resignation that perhaps there was not a fault in the first place, whereas FNF perhaps has a more positive interpretation that the fault was there but unfortunately cannot yet be found? The surfeit of terms has probably confused the subject and reduced the opportunity for resolutions although there have been occasions over the past few decades where the subject has had periods under the microscope.

4.3 Understanding NFF

The most understanding of NFF comes from the airline industry. A common interpretation or "definition" might be:

> A reported fault for which the root cause cannot be found.

This suggests that the maintenance engineer has failed in achieving a satisfactory and successful fault diagnosis so it implies that there is perhaps a problem with the diagnostic technique or process. Yet this ignores the occasions where perhaps there was never a fault in the first place or an intermittent fault where there may be difficulty in reproducing the conditions when the fault originally occurred. Perhaps there might have been misreporting or confusion in reporting the fault symptoms. Some of the only published information is available in a US airline industry generated standard ARINC 672 [1] which defines NFF as:

> Removals of equipment from service for reasons that cannot be verified by the maintenance process (shop or elsewhere).

However, what these definitions or explanations miss are the aspects of maintenance and logistic effort that are spent on chasing and diagnosing faults without achieving a satisfactory and successful solution. So perhaps we should also highlight diagnostic success which will involve finding the root cause of a fault or establishing that there really isn't a root cause. So perhaps a more comprehensive interpretation of NFF is:

> Removals of equipment from service for reasons that cannot be verified by the diagnostic process and which thus causes unnecessary maintenance and logistic cost.

Such a definition brings into focus the additional aspects of a fault that may or may not remain on the equipment or might generate logistic costs by the removal of items or components which are consigned to deeper bench testing further down the support chain. Such effort will have transport and packaging costs and further maintenance costs at the next levels of maintenance.

4.4 The Consequence of NFF

The impact or consequence of NFF occurrences are seen both in operational effects and of course general costs which can be many and varied but start with the initial maintenance and diagnostic effort. How long this takes will depend on operational pressure to return the asset to service. In commercial aircraft operations, faults that cause delays or cancellations are usually unacceptable as it affects reputation and may also involve customer compensation claims. In other industries pressures come about for similar reasons; for trains there are similar timetables that must be maintained. In the military when on operations, similar pressures apply but in peacetime more time is generally available for diagnosis. However, even in peacetime, there are often real or perceived pressures to keep as many assets serviceable as possible. These pressures will often cause several components to be changed in an approach that is seen as ensuring that the component at fault is removed. Whilst one of the components might be the cause, now the others will also enter the test and repair chain and will result in a NFF assessment. There is clearly a cost for testing these components as well and faults may even be found, but which are actually not anything to do with the original fault symptoms. There are also the associated logistic costs of having these assets in the repair chain.

The amount of data available in some organisations is often far too much and it then masks the true level of NFF and the true costs that these are causing to the organisation. This may be a result of the maintenance recording system that is in place for the organisation as this will determine how easy it is to be aware of the man-hour cost of NFF. Unless it is possible to assess the right data, which would be the fault, the work carried to resolve the fault, the man-hours spent in its resolution and any other pertinent information, it will be very difficult to assess the true costs of NFF.

One of the interesting aspects of the cost of NFF also results from human nature which creates an attitude where there is an inability or perhaps an unwillingness to recognise the costs. Human nature is such that managers will often not want to highlight the true costs in the belief that it will reflect badly on their management effort. If they highlight the cost of NFF, they risk the view that they should have done something about this cost before.

4.5 Common Causes and Impact of NFF

The causes of NFF are many and varied ranging from the technical to the procedural. Before tackling the many and varied causes of NFF, it is perhaps necessary to develop the earlier definition of NFF as:

> a reported fault for which the root cause cannot be found, and is therefore a diagnostic failure.

However, this implies that we merely have to achieve diagnostic success? In fact the issue is surely:

> to identify the root cause of the fault so that the right maintenance relates the symptom to the fault and so delivers the right maintenance solution.

Or in other words it implies being able to relate the symptoms to the root cause so the right maintenance action provides the right solution. Nevertheless, NFF causes are much wider and diverse than this interpretation suggests. So to establish what the causes are we must first consider the variety of causes. In the past it has been usual to classify NFF into 3 categories:

- ***Intermittent***—This category is perhaps self-explanatory and usually manifests itself as imperceptible faults in connections such as solder joints that flex and crack perhaps caused by vibration.
- ***Integration***—This category includes a component or sub-system that is working successfully when tested, but indicates the presence of faults when incorporated with other systems.
- ***Testing***—The third common category concerns Built-in-Test (BIT) and Built-in-Test Equipment (BITE) and testing in a repair facility.

These BIT/BITE tests are incorporated on-board as part of the routines where a fault is indicated but the BITE fails to diagnose the root cause and isolate the right component to replace. The test equipment used in the repair facility may also fail to isolate the fault.

The existence of three classes or categories has now been questioned by the research at the EPSRC Centre for Innovative Manufacturing in Through-life Engineering Services. The research has evaluated the variety of causes in different industries, many of which can be industry specific. So we can add the following:

- Poor design of both systems and the testing.
- Lack of communication or mis-communication.
- Poor diagnostic methods or wrong processes.
- Poor or lack of training.
- Operational pressure.

These all can produce NFF; nevertheless, where one of the three original categories perhaps describes an identifiable root cause, these additional classes or categories might all be classified under a fourth category of Human Factors.

However, that is perhaps too general as there are still various root causes that contribute to these NFF occurrences. Poor design of the diagnostic routine can indeed be classified under the testing category, but the root cause was poor or inadequate design. Similarly operational pressure is created by humans and the organisation's need to provide serviceable assets so the root cause is human factors in its broadest sense.

As it can be seen it is not easy to just group all the causes into just three of four categories. There is, however, another way of classifying therefore which is to identify causes as merely General Causes and Technical Causes.

4.6 General Causes

General causes might be grouped as follows:

- Organizational,
- Culture,
- Procedures and rules,
- Technical inefficiencies,
- Workforce behaviour.

These causes or groupings have many inter linking aspects and it is worth exploring each of them but also adding some additional topics that also need to be classified under general causes.

4.6.1 Organisational and Culture

Organisational issues in general are found to be inextricably linked with culture, at least the culture of the organisation rather than individual cultural aspects. The organisation is often seen as too bureaucratic, big and cumbersome to be able to change; it is often the case that the organisation may not even recognise that there is a problem. There is also the organisational culture issue which exists in some industries because there is no cross-functionality, employee empowerment and encouragement to identify the root causes of reported faults. The organisation has a culture problem in effect. In these cases the organisation has not recognised the importance of having enough information available within the organisation to demonstrate the costs involved in NFF and the need for any change. Costs, however, are often very difficult to establish and there may not even be any enthusiasm to find out the costs, again signifying a culture problem. There are even considerations in some organisations that more faults, whether solved or not, are a good thing, as it gives their own maintenance and supply organisations the extra work which then justifies their existence. Nevertheless, there are now many contracts which provide more pressure on the managers of individual fleets to improve efficiency especially with contracts for through-life support and availability that are now prevalent in many industries; the belief is that such contracts will incentivise the provider to drive out as much waste as possible which should therefore encourage greater effort on reducing the occurrence of NFF.

4.6.2 Procedures and Rules

With procedures and rules there are differences depending on the industry. In the aircraft industry there are different design criteria for military and for civil aircraft. Military aircraft are built to safe-life criteria which means there is no redundancy, whereas civil aircraft are built to fail-safe standards where there is redundancy in order to increase safety. Thus rules and procedures have to be appropriately crafted in order to reflect these different design criteria. In other industries such as rail and nuclear, similar safety criteria will also apply and generate rules and procedures that may be overly conservative and mask the effective analysis of faults. Another aspect of rules and procedures occurs with big organisations in that they are so big that effective resolution of the NFF costs becomes hidden by the bureaucracy and too many other rules and procedures. It can be seen that procedures and rules are also inexorably linked therefore to organisational and culture.

Perhaps simpler to recognize though are certain aspects of rules and procedures that must be dealt with by the maintenance personnel and will contribute to rates of NFF. These are:

- **Incomplete documentation**. When documentation such as maintenance schedules or specific diagnostic routines have not been created successfully, the process might be described as incomplete and generate an unsuccessful diagnosis or NFF. There may be good reason for this such as all the operating conditions were not clear to the original writer of the procedure.
- **Wrong processes applied**. In similar vein the diagnostic schedules and guidance might direct maintenance personnel to the wrong process merely because the actual operating conditions are now different to those expected or designed for. Similarly there may be deliberate misapplication of processes through misinterpretation and assessment of symptoms that are presumed to exist rather than actually exist. A lack of proper training could also contribute here of course.
- **Poor or incorrect instructions and procedures**. In this case the schedule and the guidance available is just not good enough for the particular fault situation. It may even be the case that the guidance is actually incorrect but it is unfortunately not yet obvious that the guidance is incorrect. Until experience is gained such errors will keep perpetuating an NFF.

4.6.3 Operational Pressure

Time pressures on maintenance operations are another source of organisational problem. In some organisations there is a crucial need to keep availability as high as possible and return equipment to service as quickly as possible. Demanding levels of availability provides overwhelming pressure on maintenance staff to provide successful and quick diagnosis and maintenance recovery. This means that some maintenance "on spec" is often the quickest solution even though it involves removing several Line Replaceable Units (LRUs). Such action then causes NFF at the next level of repair but has solved the original fault. Consequently the operational pressure has caused 3 LRUs to be changed as this was the quickest solution rather than providing a test set and performing the diagnosis on each of the 3 LRUs in turn. The result, however, is that there is now one faulty LRU, presumably in the repair chain and two that will, or should, be NFF. Of course the equipment is now returned to serviceability but it involved changing 3 boxes where only one was actually at fault. The costs are now in the repair chain and often not as visible and certainly not attributed to the original fault.

4.6.4 Technical Inefficiencies

Under the heading general causes, we consider technical inefficiencies to be another group of causes which relate to both the organisation and the procedures used rather than technical issues in an engineering sense. They could range from the inability to get test equipment and diagnostic processes altered when they are wrong, to the inability to track units that might continually cause problems. Again the existence of data and information within the organisation is key to identifying components or assets that are continually entering the repair chain and still exhibiting faults. Such items are known as rogue units and should be isolated and removed from the supply chain as they consume resources and cause needless expenditure. Tracking and identifying these rogue units requires a technically efficient system based on gathering the right and pertinent information. Many organisations have the information but either do not realise they do, or have just not set up a system to collate the information—i.e. technically inefficient. Test procedures and diagnostic routines often need improvement as the work-force gains experience and knowledge. Such routines were designed on the drawing board by the designer and may well need modification in the light of in-service usage and experience. Yet such changes may not be actioned for all sorts of reasons, but not having an easy and technically efficient way of proposing and approving such changes, is another case of technical inefficiency; it also encourages the final cause described as workforce behaviours.

4.6.5 *Workforce Behaviours*

Workforce behaviours bring us back to culture and to a certain extent organisations as they will influence the individual culture of the workforce. However, we must now consider the culture of the individual. Workforce behaviours is a clear human factors issue and one example is where, for example, the workforce have developed easier and quicker diagnostic routines, shot cuts perhaps. These attitudes are referred to as "Norms". This means that personnel have subtlety changed the way they do things and it has become the "normal" way of doing the work. There is always a reluctance then to alter what has become a well-accepted procedure. However, what must be questioned is whether the "norm" produces successful fault resolution or does it generate a "tested satis" result—a NFF? Another example of behaviour is that of course people don't like change or being told they have been doing things incorrectly. There may also be implied criticism that there have been high costs of NFF if these costs are suddenly exposed. Reluctance to change in general also links back to culture, organisation and procedures within the organisation. Solutions that are seen as possibly disruptive to people's usual working practices are seen as counter-productive and unnecessary; they are also sometimes seen as a challenge to people's technical skill and ability. Resistance to change is a well-known behavioural trait that often precludes the necessary recognition and acceptance that it is the organisation that needs to change. At the individual level though, people are often reluctant to admit that their behaviour and culture might be part of the problem and there is then an inevitable reluctance to change. Maintenance personnel might take short cuts because they "know best" or might make incorrect assumptions. Admission by an individual that they have made an error and used the wrong diagnosis, is a cause of NFF that is difficult to correct though.

4.7 Additional Subject Areas that Contribute to NFF

Having looked at what is classed as general causes and their impact, it is clear that there are other causes which need to be considered and whilst they are not causes in their own right, how well or badly they are done, will have an impact and generate or contribute to the NFF situation. It is perhaps easier to describe these subject areas.

4.7.1 *Training*

We have identified inadequate training or lack of diagnostic tools as a cause that can contribute to levels of NFF. Training has to be effective particularly if complicated diagnostic routines are involved. NFF, its causes and implications are not covered in most maintenance training programmes and thus maintenance personnel are not appraised of the costs and implications. A reduction in skill levels also clearly contributes to NFF as the personnel do not have the necessary skills and understanding to find the root cause of the fault. The lack of proper diagnostic tools also affects the level of maintenance effectiveness and can be linked to training in that the personnel do not have the skills and resources to reduce the levels of NFF.

4.7.2 Communication and Miscommunication

This is often seen as lack of communication perhaps within the organisation and thus it links back to organisational aspects that are far from ideal. It may also manifest itself at higher levels such as the fleet manager or Original Equipment Manufacturer (OEM) not sharing information; perhaps a particular fault is occurring in the same equipment being used elsewhere in the world and a solution or better diagnostic process is available but needs to be communicated better. Such a lack of communication between experts in the organisation might mean vital information is not passed on to those who might then solve a regularly reported fault but on different equipments. Sharing information is vitally important at all levels if NFF is to be reduced. It also links back to culture as there needs to be a culture and the commitment to share knowledge and information at all levels and in particular between designers, manufacturers, service providers and of course the operators. Whilst this may sound easy to say, it needs systems within the organisation that enable the right information to be shared quickly and effectively between all stakeholders. There may well be a small cost to enable such a system or to resource it effectively, but once the costs of NFF are able to be determined, the cost of the system or resource can be evaluated to show its cost benefit. At the individual and personal level, communication is also important in order that a common understanding is achieved of the symptoms and description of the fault. It is often the case that people's perceptions and understanding differ which means that the fault will not be properly diagnosed. A lack of good communication between maintenance personnel, perhaps on shift changeover might cause the new shift to misdiagnose the problem.

4.7.3 Lack of Historical Data

A large amount of costs of NFF are caused within the repair and support chain. Many components that perhaps have an intermittent fault continually populate the repair chain because the test equipment or repair organisation cannot reproduce the fault, often because the tests are not reproducing the environment or usage conditions that occurred when the fault was first recorded. By having the historical data available on both the item or component and its history, these items with repeat arising history can be extracted and excluded from the supply and repair chain. They can then be subject to more detailed repair and overhaul and not returned to stock unless a real and positive fault that can be isolated to the original fault symptoms can be found.

4.7.4 Supply Chain Effect

The supply chain effect occurs when a particular fault (fault X) usually leads to the replacement of a particular component (component Y); the root cause however in most cases, is in component Z. The supply chain, however, sees an increased usage of component Y and so forecasts that more stock of component Y is required. Because there is a good stock level of component Y, maintenance staff believe that it is perhaps because of a high fault rate. This is sometimes known as the "Phantom Supply Chain" and is certainly an impact on operations.

4.8 Technical Causes

Technical causes are similarly diverse and varied but can generally be seen to originate at the design stage.

4.8.1 Undefined or Inappropriate Performance Measures

If the performance level is not defined correctly or adequately, then a fault might be recorded or defined by one person whereas another might not classify the same symptoms as a fault and declare there is no fault for that level of performance.

4.8.2 BIT/BITE and Testing

BIT levels may be set too low. In this case false alarms are created when the BIT indicates a fault, but actually no fault exists, or certainly not one that would affect mission success. The ability of the BIT/BITE to detect a fault in the first place must be designed in correctly. Similarly the ability of BITE to apply a reliable fault diagnostic routine has to have been designed into the BITEs. Discrepancies and faults in test procedure are also prevalent. Such faults in the procedures might exist but they are not obvious and lie unidentified but nevertheless generate NFF.

4.8.3 Erroneous Repairs

This situation usually exists when there are intermittent faults which are by their very nature difficult to detect during test as the environment and usage cannot be replicated. When an item is subject to test and repair at deeper levels of repair, more sophisticated test equipment is usually being used with more sensitive tolerances; in this case faults may appear that were nothing to do with the original intermittent fault, but these dormant faults will now be repaired and the original intermittent fault will still exist and still lie undetected. The component then returns to service with every possibility of the original fault quickly happening once again in service but now fitted to a different system or platform. In this case the vital history for the component is unlikely to be visible.

4.8.4 Information on Usage and Operating Environment

A lack of information on the usage and operating environment when the fault appeared is a common occurrence in many industries. It provides vital clues though perhaps as to vibration or temperature conditions when the fault first appeared. An inability then to reproduce the operating environment often occurs when the system is put on test. Clearly in a lot of situations it is impossible to reproduce the usage or environment, even with an environmental chamber, but unless the actual environment and usage conditions can be provided, many faults cannot be reproduced. Intermittent failures are often caused by stresses that are not able to be reproduced on test. These failures can only be shown if the same, or similar stresses and environment, can be provided as when the original fault occurred.

4.8.5 Inadequate Design

The actual design of the component or system and its integration into the platform needs to be such that easy on-board test and diagnosis is facilitated, or it is easy to facilitate an off-board test and diagnosis. This clearly requires designers to be thinking in-service support when they design the systems and platforms in order to make such on-board and off-board testing as easy as possible. Part of the design process then requires fault models and fault trees to be properly created such that root causes are determined. Poor and inadequate design or the inability to analyse and identify the true cause of potential in-service problems will generate higher NFF levels.

4.8.6 Interactions Between Software in Integrated Systems

Software intensive systems can also be most difficult when it comes to understanding faults. The understanding of the interactions between different integrated systems and software can be particularly challenging for maintenance personnel. Reliance on test equipment and diagnostic routines produced by the OEM is almost certainly the only solution and if the routines declare NFF there is little alternative but to accept the technical solution provided by the OEM's routine.

4.8.7 New Technology Adoption

The adoption of new technologies carries risk and so there is often a reluctance to adopt them and uncertainty over their performance. The reluctance might be associated with the need to re-design a system or due to the problem of providing the necessary data handling and/decision making infrastructure. Health and Usage Monitoring Systems (HUMS) and Condition Monitoring (CM) technologies are available and would assist in providing usage and environmental information for when the fault occurred. However such technologies require analysis for decision making both on-board and off-board. They are nevertheless very beneficial if designed in at the start rather than added later when equipment is already in service.

4.9 Cost

What is not in doubt is that any and all of these general or technical causes will result in increases in cost and a significant wastage of effort and resources. NFF problems manifest themselves in many and varied ways and will continually affect the efficient and effective application of Through Life Engineering Services. It is thus vital to identify how these problems affect stakeholder requirements. What are the problems that most need solving? Are there impacts on System Safety? How do these things affect system safety and above all what are the effects on Life Cycle Costs in order that we can highlight the true cost that needs to be reduced.

Cost is clearly a major aspect that must be identified and initial studies show a strange reluctance to collect the necessary data that will enable costs to be identified. Perhaps this is understandable though as it may provide an unwelcome shock to an organisation when there are no immediate solutions. It is generally accepted that the operating and support phase typically accounts for 60–80 % of the whole life cost (WLC) of the equipment. The importance of understanding an equipment's WLC therefore is vital. The costs that contribute to NFF will also occur at the different levels of support and may be easier to collect and identify at some levels and more difficult at others. It is instructive therefore to look at the problem of identifying costs at the various levels.

4.9.1 First Level of Support

At the first level of support, the equipment is being operated and maintained and there is a cost for investigation and diagnosis whether something is removed or not from the main operating platform. Downtime and manhours expended are invariably available and collected. Depending on how the maintenance recording is organised, identifying which maintenance effort is associated with faults that are not found may not be as easy as first thought. Nevertheless if the recording is effective and well detailed, the costs then manifest themselves as follows:

- No fault is found despite a great deal of effort but nothing is actually replaced or changed and no positive cause of the reported fault is identified. The main equipment is declared serviceable and all tests passed. A NFF should then be recorded in the paperwork and be obvious for costing purposes. However this type of fault is often recorded merely as "test carried out satis" or similar and the actual fault then re-appears subsequently in next mission or soon after and further work is necessary. Only by detailed historical analysis will the costs be obvious as a NFF cost and must of course include the original work when the fault was first reported.
- The wrong solution or diagnosis may be applied which then shows that the fault is apparently fixed but it then re-appears subsequently on the next mission. The cost should then also include the cost of the first occurrence.
- A speculative replacement may be made usually due to operational pressure and perhaps involving several equipments being changed; it shows that some positive corrective maintenance action has been taken and the fault is solved at the main equipment. However, whilst the fault maybe fixed at the main equipment, the fault is now in one of the equipments removed which are sent to a subsequent maintenance level where a NFF will occur in all but one of the speculatively removed equipments.

4.9.2 At the Second and Subsequent Levels of Support

At subsequent levels of support there will be wasted man-hours in unsuccessful or ineffective maintenance effort. This might be:

- Testing for faults that do not in fact exist following speculative replacement at the first level.
- Testing for faults where the cause is an intermittent fault that cannot be replicated on the test bench because it is caused by vibration or excessive loads or temperatures in the operating environment.
- The application of inappropriate or inadequate tests such as the wrong procedures, or the test parameters that are set at a too high a level to find fault.

4.9.3 *In the Supply Chain*

Whatever work has been done at first and second levels of repair, any associated with NFF will have effort expended in the supply chain, where there will be wasted man-hours and costs for transportation and supply activity. Costs can be divided as follows:

- When faults are being continually attributed incorrectly to one component, stock replenishment happens more frequently because demands are higher. More components are in the repair loop, albeit many with no genuine faults. However more stock is purchased to replace items in the supply chain. There is thus a capital cost of purchasing extra components.
- The increased numbers in the repair loop all need transport, handling and storage when actually these items should not be in the repair and maintenance loop.

4.10 Diagnostic Success

Clearly one of the central aspects of solving the NFF issue is to ensure diagnostic success. As we saw in the definition at the start of this chapter, diagnostic success is a major factor in solving this issue of NFF. It is therefore instructive to look at some evidence from research done by one of the leaders in the field, Copernicus Ltd who are specialists in the field of intermittent fault detection. The evidence is from a study of diagnostic success rate for avionics and shows that diagnostic success can be relatively low at perhaps barely more than 40 %. The study [2] shown in Fig. 4.1, includes all types of faults within the system, i.e. hard faults and intermittent faults. Diagnostic failures account for over 50 % of all occurrences. The 'Functional Test Only' portion covers the common situation where maintenance

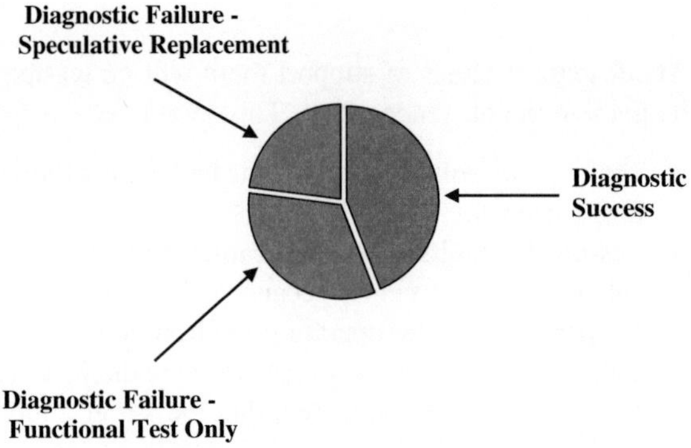

Fig. 4.1 Diagnostic success rates for avionics system repairs

personnel cannot confirm a fault, but by confirming there is no fault during a functional test, are able to declare the equipment serviceable. The 'Speculative Replacement' portion covers the many reasons already described in this chapter.

4.11 Mitigation of NFF

Consequently, it is clear that identification of the root causes and the origin of the fault will mean diagnostic success will have been achieved. Without this diagnostic success the NFF issue will not be resolved. There is, however, a surprising lack of mitigation to address the problem despite the NFF situation being known for many decades. There is some best practice and principles that have been adopted in the aviation industry in particular, but even here it is patchy and different cultures and operating models determine the attitudes to NFF. The following are some of the more popular mitigation processes and solutions:

- Identifying BIT deficiencies and reviewing BIT trigger levels.
- Establishing better transfer processes for NFF data and delivering it as information.
- Reviewing test and diagnostic processes for their effectiveness.
- Applying and proving the effectiveness of diagnostic tests before commissioning.
- Providing linkage between an in-built health monitoring capability and diagnostic ability.
- Provide more fault tolerant systems during design.
- Identification and recognition of false alarms.
- Management policies and procedures to identify, cost and reduce NFF.

4.12 Conclusion

A literature review and initial research at the EPSRC Centre for Innovative Manufacturing in Through-life Engineering Services has identified that there is certainly an on-going problem with NFF. It has shown that organisations need to change their culture and behaviours in order to avoid the high costs that are generated by NFF. Taxonomy is also critical in achieving the culture and behavioural changes required. By describing the problem as NFF a certain attitude of resignation is engendered that suggests the effort to resolve the fault was a waste of time and resources as opposed to the more positive attitude of acceptance of the fault that must be resolved as described by Fault not Found (FNF). Describing the event as FNF provides a positive outcome where diagnostic success is still required and further action is still required. Organisations first need to recognise and accept the need to address the problem by identifying the true costs of NFF and then employing a range of proactive measures to address the problem which might include the following:

- Identify procedural, process and behavioural issues that need to be changed.
- Develop process and procedural arrangements to address the management of NFF rates, starting with the most prolific components and systems.
- Develop processes to track NFF costs and impact throughout the repair chain.
- Develop processes to identify and track rogue units.
- Develop health monitoring approaches where possible to detect, characterize and locate NFF intermittent failures and deliver a fault localisation mechanism and demonstrator at the board and sub-system level.
- Devise strategies and methodologies appropriate for the organization to mitigate the number of NFF occurrences.

References

1. ARINC Report 672 (2008) Guidelines for the reduction of No Fault Found (NFF). Avionics Maintenance Conference, Aeronautical Radio Inc
2. Huby G, Cockram J (2010) System integrity approach, RAeS A&W, 27 Aug 2010

Chapter 5
Holistic Prognostics

Charlie Dibsdale

Abstract Prognostics is the determination of condition and remaining useful life (RUL) at any time in a machine's operations. Prognostics are based in probability, and best practice would be to provide the forecast of UL with indications of certainty where the population of similar prognostic events allows. It is also best practice to describe the impact and effects of loss of function, along with the estimated times for recovery, along with the resources which are required. Prognostics are often misunderstood with several approaches being possible to calculate RUL for machinery. This chapter describes and extends a known model (Hines 2008) for prognostics, covering all of the approaches in the model and describes how prognostics should be measured (Saxena et al. in Int J Prognostics Health Manage 2153–2648, 2008, Int J Prognostics Health Management 1:20, 2010). A brief treatment for the difficulties of validation is illustrated due to lack of failure data, along with an outline of how these deficiencies might be addressed with the advent of the information revolution and big data.

5.1 Definitions

Prognostics is closely bound with, and used with maintenance. The purpose of maintenance is to preserve functions required by stakeholders of physical assets in defined operating contexts and environments. This chapter uses and extends the language to describe a taxonomy of maintenance that is derived from reliability centered maintenance (RCM) [2], which breaks down maintenance in the following manner (Fig. 5.1).

The description of each type of maintenance is introduced in the table below along with how prognostics are used or has relevance. Some of the descriptions use terminology introduced in Fig. 5.2 (Table 5.1).

C. Dibsdale (✉)
Controls and data services Ltd., Moor Lane, Derby DE24 8BJ, UK
e-mail: charlie.e.dibsdale@controlsdata.com

L. Redding and R. Roy (eds.), *Through-life Engineering Services*,
Decision Engineering, DOI 10.1007/978-3-319-12111-6_5

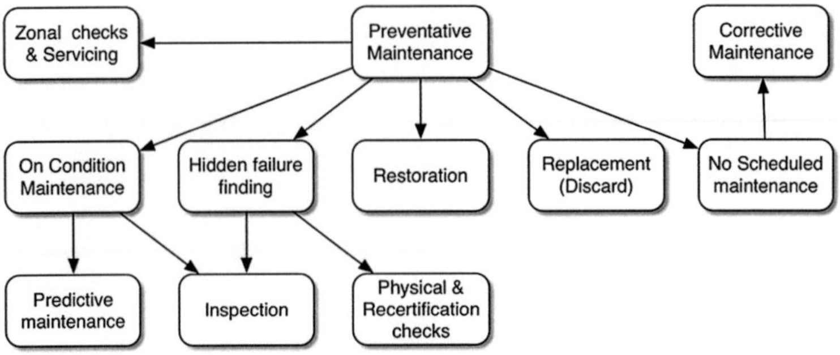

Fig. 5.1 A taxonomy of maintenance types

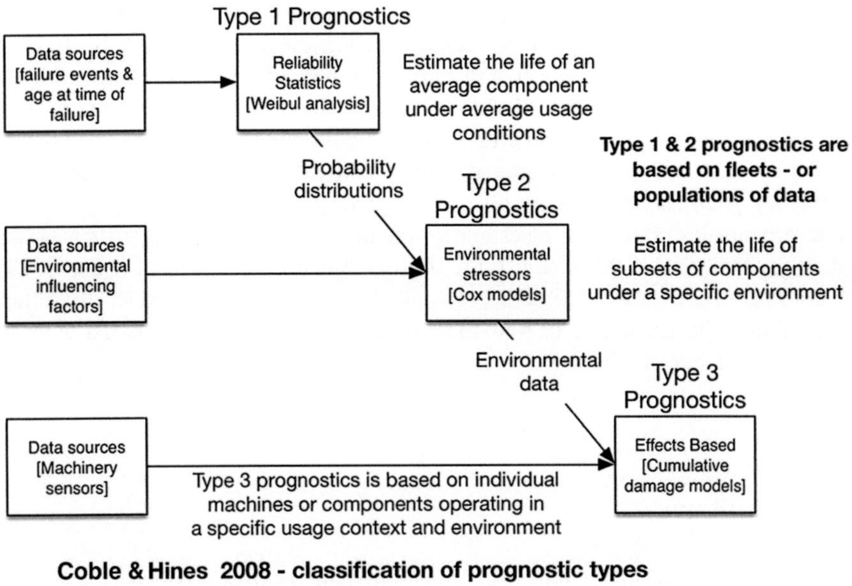

Fig. 5.2 A classification of prognostic types

5.1.1 The Full Range of Prognostics

This section describes the different approaches possible to use with prognostics, in determining Remaining Useful Life (RUL). The basic categorization of prognostics is extended and elaborated from the definitions presented in the PHM society [1]. The following diagram illustrates the Hines prognostic method types:

It is the Hines [1] model that is described and extended in this chapter.

Table 5.1 A taxonomy of maintenance types and their relationship with prognostics

Maintenance type	Brief description	How this relates to prognostics
Preventative maintenance **S**	Seeks to prevent loss of function or avoid consequences of the loss of function	Prognostics is a vital component of preventative maintenance
Corrective maintenance	Conducted to restore functionality, after its loss	Prognostics may be used to forecast when corrective maintenance may be required
Zonal checks and servicing	Where operators or maintainers carry out walk downs in machinery rooms, and conduct simple servicing checks (cleanliness—filters—lube levels etc.)	Experienced maintainers and operators are excellent human sensors for on-condition maintenance. Experts can prognose machinery by judging their state.
On condition maintenance	Observation of a physical asset to determine health or condition	Prognostics is applied after diagnosis of an asset being in a partially failed state to determine remaining useful life
Inspections and scheduled sampling	A subset of on condition maintenance using manual inspection or periodic manual sampling to determine health or condition	Once health or condition is established prognostics determines remaining useful life, where wear out is the predominant failure pattern
Predictive maintenance	A subset of on condition maintenance which uses continuous sampling from fixed sensors with automation of determining health or condition	Prognostics is used to determine remaining useful life after diagnosis of incipient failure or after determination of condition
Hidden failure finding	Used to discover functional loss which may not be apparent to operations or maintenance staff in their normal duties	The timing of hidden failure finding tasks are related to the probability of failure (type 1 and 2 prognostics)
Physical or recertification checks	Is conducted to find hidden failures that will not otherwise be observed. In some circumstances these checks may restore risk probabilities to a level which certifies continued operation (e.g. nuclear plant)	The timing of physical checks are related to the probability of failure over time
Restoration	Restores condition in parts that have reached an age where the probability of loss of function gets unacceptably high regardless of condition (on condition maintenance is not feasible or cost effective)	Only applicable where loss of function is age related. Acceptable probability (useful life) is determined by type 1 or type 2 prognostics.

(continued)

Table 5.1 (continued)

Maintenance type	Brief description	How this relates to prognostics
Replacement or discard	Replaces parts that have reached an age where the probability of loss of function gets unacceptably high regardless of condition (on condition maintenance is not feasible or cost effective)	As per restoration
No scheduled maintenance	A deliberate policy of allowing assets to run to loss of function, and then apply corrective maintenance. Preventative maintenance is not practical, effective or cost effective	Data may be collected to determine the pattern of failure and measure achieved reliability to apply preventative maintenance if circumstances change, or reliability us not as required

5.1.2 Prognostics Type 1: Reliability Statistics

Many organizations have extensive inventories of machinery where the cost of gathering data about failure events and operating time is not practical, and simple determinism or reliance of manufacturers data for Mean-time-between-failure (MTBF) is used. MTBF implies a constant failure rate, which is highly unlikely to occur in the real world, but even with this some organizations base preventative maintenance periodicities on MTBF calculations. If a machine with a random exponentially decaying conditional probability of failure, if Preventative maintenance (PM), is applied at a calculated MTBF interval then 63 % of the fleet population would have already failed. In RCM the use of MTBF as an input to calculate the periodicity of restoration and discard type PM is rejected. The concept of Useful Life Limits, which is used to determine PM periodicity, will be described below.

The most commonly applied method of prognostics applies to failure modes that may be categorized as age related, where machinery wears out, and the probability of failure increases over usage or time. The statistics apply to a fleet of similar machines or components.

A powerful means of enhancing reliability statistics is to apply a Bayesian technique that enables a Weibul characteristic to be updated as new failure event data is supplied.

5.1.3 Prognostics Type 2—Environmental Factors

Type 1 prognostic models may be enhanced by the addition of environmental influencing factors that may change the rates of degradation or probability of inception of failure. These factors may include the effects of environmental vibration, exposure to corrosive or abrasive materials, extremes of temperature or exposure to moisture and salt. Other factors include those associated with operating contexts, where peak lopping (only using an asset to satisfy cyclic peak demands) may be more prone to fatigue (due to temperature cycles) or base loaded rotating machine may be more prone to wear out.

Type 2 prognostics may also be enhanced by Bayesian networks, which use the current priori knowledge as a baseline and updates this as new data is received as the posterior knowledge. A typical type 2 prognostics model is a proportional hazards models which are a class of statistical survival models. The Weibul distribution can be parameterized as either a proportional hazards model or an Accelerated failure time (AFT) model.

Survival models consist of a hazard function, which is the probability of failure over time, along with the effect parameters, which influence how the hazard varies.

5.1.4 Prognostics Type 3—Individual Machine Effects

Type 3 prognostics are focused on estimating the RUL for individual machines, or any of their components, although statistics from a fleet of similar assets may be used, taking advantage of separating the data by environmental factors in order to determine certainty for a new observed event. A common form of prognostics is to infer accumulation of damage by monitoring transients. A common form would be a temperature transient correlated with exchange rates to infer low cycle fatigue life up to a life limit.

Prognostics associated with on condition maintenance may use a state approach which may be derived from expert knowledge, or fused methods, which diagnose the potential failure state at different times. An example may be a rolling element bearing failure with three on condition techniques applied; first where laboratory oil debris analysis detects bearing race debris at about 10 μm size, and some time after vibration analysis observes raised energy peaks resonating with the ball passing rate, and finally as the bearing is close to functional failure bearing or lube oil temperature rises. These three events or state changes may be used as a state based prognostics approach with expected times learned between the events.

Other prognosis could be inferred by using the same symptoms of failure as used in the diagnosis of the partially failed state, but symptoms may not behave intuitively, and such things as vibration energy may improve as the condition worsens, because of factors like structural cracking may change the resonant behaviour.

5.1.5 Measuring Failure Mechanisms or Direct Effects

The introduction of 'failure mechanisms' prompts the need to define terms. A failure mode is the manifestation of a state or condition that may lead to a loss of function required by equipment's stakeholders. An example of a failure mode may be a quantified leak (a loss of containment function).

A failure mechanism is the underlying physical or chemical process that drives a failure mode. Examples of failure mechanisms include erosion, corrosion etc. In the lifecycle of a failure mode, it might be that the underlying failure mechanisms change or combine (such as erosion and corrosion acting simultaneously). The failure mechanisms that are currently the predominant drivers of failure modes are known as 'forcing functions'.

An environmental stressor may indicate the presence of a hazard such as the presence of acid rain, which may increase the probability that a failure mechanism may be more likely to initiate and or increase rates of degradation.

A 'root cause' is synonymous to a failure mode, but because of the laws of cause and effects, the relationship is recursive and linked. A well-known method of 'root cause' identification is asking "five whys", which perfectly describes the iterative relationships. There may be a large number of iterations before a root cause is identified. A root cause is therefore defined as a causal factor that is in the power of the asset owner or maintainer to prevent; and may be prevented by redesign, change in operation or maintenance within the power or ability of the stakeholders. In other words a 'root cause' is something machinery stakeholders have the ability to prevent. The following influence diagram illustrates the relationships (Fig. 5.3).

The ability to directly measure failure mechanisms or their direct effects compared with observing symptoms of a failure mode improves the ability to prognose a failure and allows decoupling from operating up to a probability of failure for age related failures to shifting toward the natural MTBF for that set of components. This means that for components that exhibit wear out as their predominant failure modes,

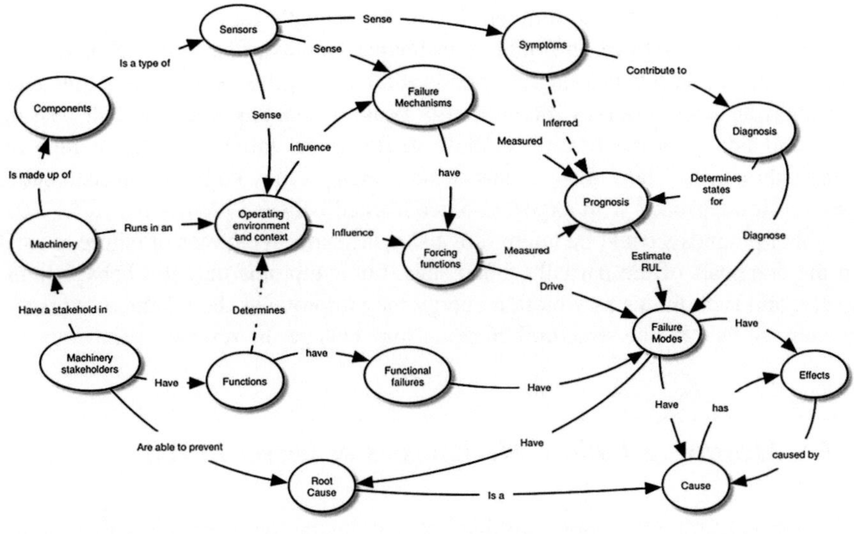

Fig. 5.3 Concepts and relationships diagram for a prognosis system

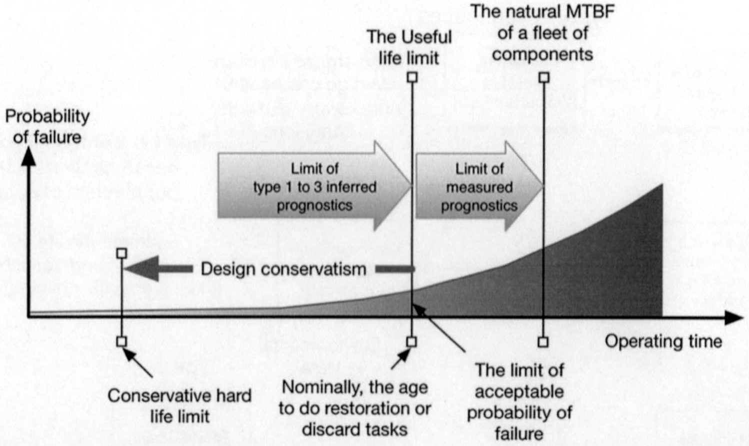

Fig. 5.4 For age related failure patterns—shows the possible benefit of applying Type 4 measured prognostics

it is possible to leave components operating for longer than inferred prognostic methods. This is illustrated in Fig. 5.4.

Another significant advantage of measured prognostics for assets exhibiting wear out, is it is possible to maintain the statistics of failure, because the items are not withdrawn from service for restoration or replacement before their failure events.

5.1.6 A Proposed Extension to the Prognostics Model

The [1] Hines model of three types of prognostics may be extended by considering the difference between methods that 'infer' and 'measure' RUL. Inference of life used (Fig. 5.5).

5.1.7 Metrics for Prognostics

In order to specify prognostics needs, an operator and/or maintainer (O&M) of a machine will need to determine the priority of the failure modes.

If prognostics are based on 'type 4 measured', the fundamental requirement is what RUL is necessary for asset O&M to minimize the impact and consequences of failure. There will be a natural time frame for the physical degradation process, and the ability to diagnose the system is in a partially failed state. The basic needs are based on

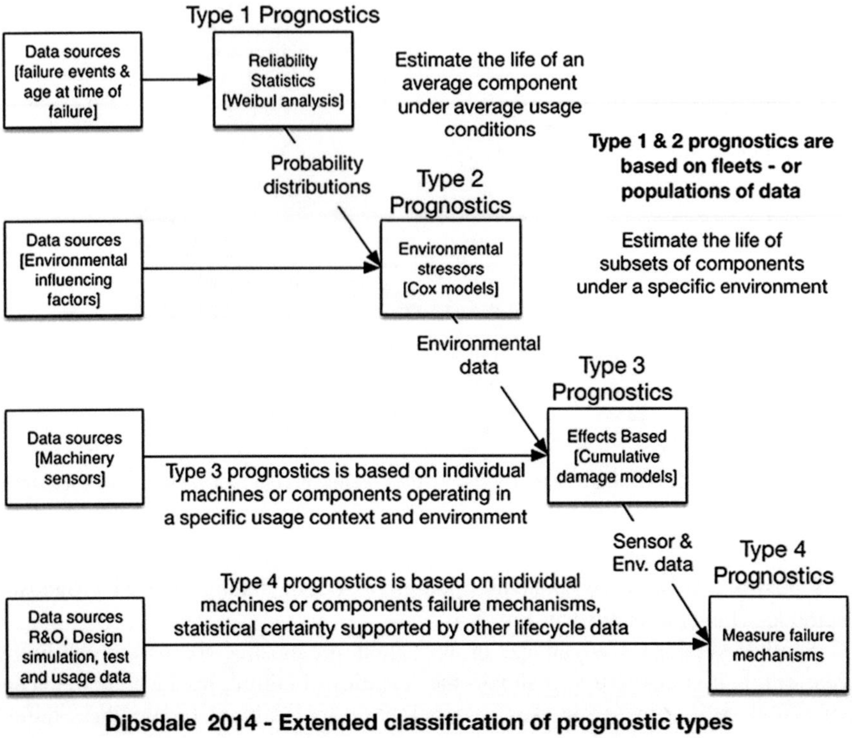

Fig. 5.5 Proposed extension of the Hines prognostic classification model

- Are the machine and the failure mode able to be prognosed? Are there suitable sensors and access to their data, sampled at the appropriate rate? Have the appropriate prognostic methods and algorithms been selected and deployed? The knowledge of likely RUL for a given machine in an operating context, is most practically determined from expert knowledge form an experienced operator or maintainer.
- How much lead-time does O&M need to plan recovery? The wise maintainer will also identify opportunistic Planned Maintenance that may increase the probability of survival for the following operational period (or used to extend the following planned outage time).
- How much lead time does O&M need to predispose resources necessary to execute the most rapid and efficient recovery? Special skills or special to type tools (which may also require their own calibration to be in date) may need to be considered.
- How much time is necessary for operations to select the best time for withdrawal of the machine, to minimize schedule disruption, and/or minimize production loss, traded off against the accumulation of damage and the initiation of further incipient failure.

- If the RUL is long enough, O&M will consider what risks exist to keep the machine in operation until the next planned outage. Considerations should also be given to run machinery at reduced stresses or loads, which might extend the RUL, or reduce risks of catastrophic failure.
- Where a machine is installed in a facility (such as a power station or oil and gas platform), then it is likely that a proportion of other machinery may be out or service for defect repair or Planned maintenance. The O&M function will also need to consider overall risks at the facility level, as the combination of machines out of service might increase risks of such things as environmental pollution or other process quality risks that may be unacceptable. This kind of facility risk management may be calculated using fault-tree-analysis (FTA) (Fig. 5.6).
- What risks do we incur of catastrophic failure the closer we run machinery to its loss of function limits? If we are able to sample data at a high enough continuous frequency we may be able to determine 'shock events' which may cause catastrophic failure if we run a machine close to its functional failure limit. The following graph illustrates the concepts. The analysis of abnormal shock events may be possible utilizing Extreme Value Theory (EVT) (Fig. 5.7).

A more formalized set of Prognostics metrics has been defined [3, 4], which whilst definitive and detailed does require knowledge of the "Ground Truth" of prognostics, which provides a baseline against which to measure prognostic performance. As discussed in the section below.

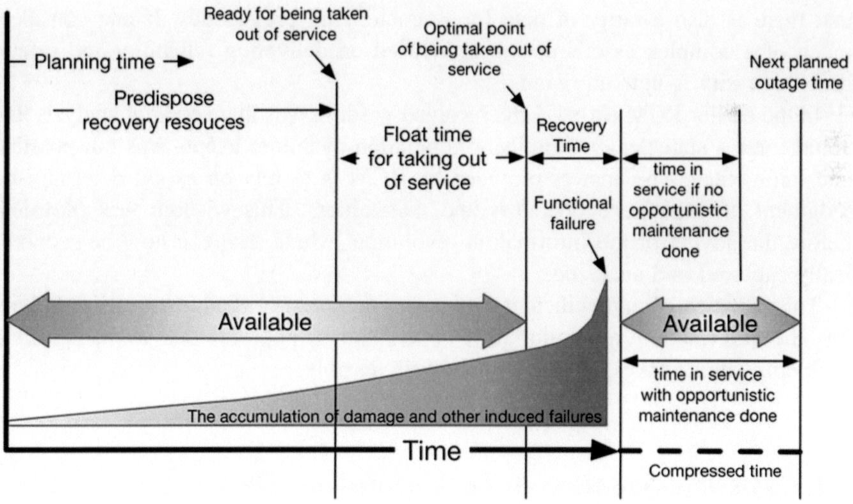

Fig. 5.6 The effective utilization of prognostics to maximize availability, and minimize maintenance costs

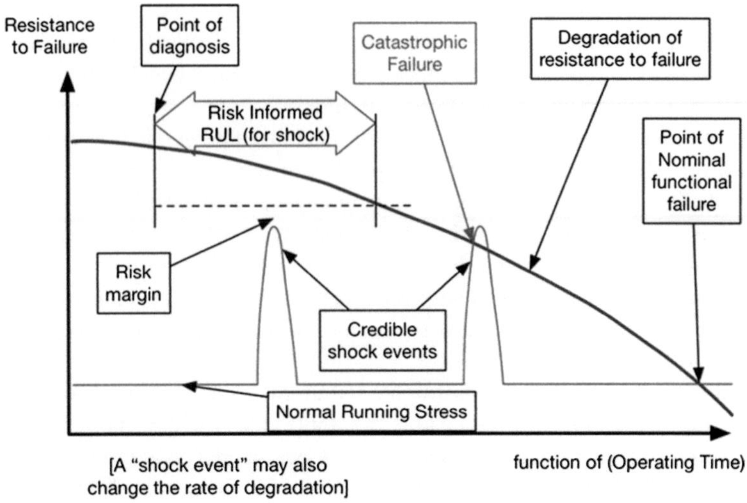

Fig. 5.7 The impact of shock events on selecting a reduced risk RUL

5.1.8 The Problems with Validation and Prognostic 'Ground Truth'

There is a generic problem recognized In Reliability Centered Maintenance (RCM), where the higher the impact of failure the less failure events are suffered. It follows that there is also a rarity of data for characterizing prognostics. If one considers much of a complex asset's design is focused on delivering reliability and safety, then this rarity is not surprising.

In the earlier RCM texts [2] the received wisdom was that actuarial analysis that could form a statistically significant population of failure events was not possible and impractical. The answer provided by RCM is to rely on expert deterministic judgment of experience operators and maintainers. This wisdom was prevalent before the advent of the Information revolution, where data can now be economically captured and analyzed.

This situation of more efficient data gathering does not change the rate of failure, but the information revolution now opens other opportunities to address the problem, which makes an actuarial analysis feasible.

5.1.9 Possible Solutions to Lack of Failure Data for Validation

Many organizations do not conduct Weibul analysis associated with reliability engineering because the cost and practicality of collecting failure data, conducting root cause analysis and capturing age data is prohibitive. If advanced condition monitoring systems are fitted and run, it is relatively cheap and easy to collect cumulative running hour or cyclic data as part of the data acquisition.

The use of physical models has already been mentioned as a method of estimating remaining useful life. Many organizations use high fidelity simulations such as finite element analysis for design purposes. These simulations should be enhanced to include aging models, and be able to accept failure conditions, and simulate signals. These models may also be used to optimize the specification and siting of sensors for Condition monitoring purposes.

Much data about complex assets may be possible to collect through manufacturing, where correlations (using 'associative data mining' techniques) are possible to determine likely life variation in-service.

There is an opportunity to accurately measure 'condition' of assets as they come through major overhaul and repair. This has the potential to establish an accurate baseline of condition at a known age (assuming advanced condition monitoring systems are used). A small selection of rejected components (which are intended to be scrapped) should either be deeply inspected, or run to destruction in order to collect further data.

Other prognostic data may be collated when conducting manufacturer's component, assembly or equipment tests. This may be part of a reliability growth program or a form of accelerated life testing conducted during design and development. The data gathered from these tests are useful, but may be deficient because tests may not include the effects of operational environmental factors.

As the concept of the 'internet of things' and 'big data' emerges, it will be possible for a rich environmental data set to be gathered, and correlated with machinery operations and the influences on the rate of degradation.

In order to access this wider range of data enterprises in a market sector will need to be willing to share data, where IP and sensitivity may be accommodated. This implies emergent behaviour in changing value flows and relationships in market verticals. We have observed changing business models as manufacturers offer services associated with their products. Many of these manufacturers are also making investments to own or control the maintenance and repair functions. This allows them to gain easier access to the data.

5.2 Conclusion

This chapter has taken an established model for prognostics and argues for and justifies its extension. A distinct forth class of prognostics is proposed, that has other areas where there are inherent problems in validating prognostics due to the lack of corroborative failure data, have been explored and practical means of providing alternative evidence are suggested. These alternative means of providing evidence are now possible as emergent ICT technology makes it practical and cost effective to simulate, collect, manage and analyze huge quantities of data. This is one way in which maintenance may evolve and take advantage of the advent of 'big data' and the 'internet of things'.

References

1. Coble JB, Hines JW (2008) Prognostic algorithm categorization with PHM challenge application 1st international conference on prognostics and health management (PHM08), Denver, CO
2. Moubray J (1997) Reliability-centred maintenance. Industrial Press Inc, New York
3. Saxena A, Balaban J, Saha B, Saha S, Goebel K (2008) Metrics for evaluating performance of prognostic techniques. 1st international conference on prognostics and health management (PHM 08), Denver CO
4. Saxena A, Celaya J, Saha B, Saha S, Goebel K (2010) Metrics for offline evaluation of prognostic performance. Int J Prognostics Health Manage, 1(1), p 20, ISSN 2153–2648

Chapter 6
Ultra Low Carbon Vehicle Management Based on Telematic Monitoring

M.J. Knowles and D. Baglee

Abstract Ultra Low Carbon Vehicles (ULCVs) include fully electric and hybrid vehicles utilising various combinations of battery systems and hydrogen based technologies. ULCVs offer many organisations with large vehicle fleets an opportunity to reduce the cost of their operations. One of the major concerns regarding the operation of ULCVs is uncertainties relating to the ongoing cost of vehicles due to issues such as replacement of major parts e.g. traction batteries, and ensuring the vehicles' lifespan is maximised. Telematic monitoring equipment is widely used in fleet vehicles but is rarely used to its full potential and does not match the benefits exhibited by remote condition monitoring systems used in other sectors. Such technology has the potential to address many of the concerns relating to ULCV operation. A number of barriers exist, however, to the wider use of telematic systems including integration with other vehicles systems, management of large volumes of data and issues relating to the effect on the finite power supply of an idle vehicle. In this chapter we will examine the underlying technology of vehicle tracking and monitoring systems and investigate the technical, organisational and operational enablers which are required to exploit fully the benefits of telematic monitoring.

M.J. Knowles (✉)
Department of Computing, Engineering and Technology, University of Sunderland, DGIC, St Peters Way, Sunderland SR6 0DD, UK
e-mail: michael.knowles@sunderland.ac.uk

D. Baglee
Institute for Automotive and Manufacturing Advanced Practice (AMAP), Univesity of Sunderland, The Industry Centre, Colima Avenue, Sunderland SR5 3XB, UK
e-mail: david.baglee@sunderland.ac.uk

L. Redding and R. Roy (eds.), *Through-life Engineering Services*, Decision Engineering, DOI 10.1007/978-3-319-12111-6_6

6.1 Introduction

Long term opportunities for the automotive industry often centre on light weighting, new materials, low emission engines and data and communications. This final area addresses the increased availability of in-vehicle information and the growing use of on-board electronic systems such as drive-by-wire systems. A significant opportunity exists to improve human-machine interfaces and driver information systems using a range of telematic systems to improve safety, while reducing congestion which leads to reduced fuel consumption and pollution. Improved information should reduce the cognitive burden on drivers, who now have access to complex vehicle systems providing more information, therefore the interface and the delivery of information must suit different driving situations. The development of high resolution satellite navigation systems with accurate positioning systems will provide drivers with information on their location and by utilising this technology and integrating it with other data sources, such as traffic congestion or accident information, safety could be improved.

Ultra Low Carbon Vehicles (ULCVs) are becoming increasingly attractive to a variety of private and commercial operators due to their potential to reduce whole life cost as well as improving environmental impact. The majority of leading automotive Plug-in electric vehicle (EV) manufacturers including Nissan, Renault and Peugeot are developing a range of telematic systems with, initially, the aim of alleviating range anxiety and to provide drivers with information on the location of the nearest charging stations. Recently, however, telematic systems are providing the driver with a range of information previously not available to electric vehicle drivers including local road conditions and local weather. In addition Nissan have developed a telematic system which interacts with smart phones or smart tablets to allow the in-car heating and air-conditioning systems to be activated remotely while the vehicle is plugged in and charging.

Fleet Operators are now using telematic systems to their advantage, especially where the electric vehicle fleet is large and often spread over a number of towns and cities. Telematic systems are used to track vehicles to provide information on location, miles driven, state of battery charge, and in certain situations the system can be used to lock and unlock car doors utilising text and email technologies. Systems developed by BMW Germany are now capable of contacting the emergency services after an incident.

In this chapter we will examine a range of aspects relating to the application of telematics to ULCV operation. Section 6.3 provides an overview of the technical functionality used telematic applications and the developments in technology which have led to an increase in use. Section 6.4 describes the benefits available to organisations choosing to use telemetry systems while Sect. 6.5 addresses the barriers to such use. Section 6.6 describes the ULCV specific applications of telemetry which are likely to become popular in coming years while Sect. 6.7 outlines some potential benefits on a wider scale which might come about following an increase in the use of telematics.

6.2 Technical Aspects of Telemetry Systems

The development and application of telematic monitoring systems is underpinned by the current state of the art of the relevant electronic hardware. In this section we will consider this from two perspectives. Firstly the functionalities which may be required to create a telematics solution are described. Subsequently the advances in hardware which have enabled the wider application of telemetry systems are described.

6.2.1 Telemetry System Functionality

Telematic systems offer a range of potential functionality. The configuration used in particular circumstances is governed by the needs of particular applications. Areas of variability include:

Position Tracking. The cost of Global Positioning Satellite (GPS) receivers has reduced substantially in recent years. The ability to monitor the position of a vehicle and derive factors such as speed is a central requirement for many telemetry applications. This is beneficial in terms of providing knowledge on the location of a vehicle, the style in which it is being driven [1, 2], and for providing context for other types of data e.g. determining whether harsh deceleration occurred due to the presence of traffic signals.

Data communications. While the term 'telematic monitoring', in the strictest sense, refers to the remote monitoring of data captured within a vehicle, systems which purely record data can often fulfil similar functions and so are worthy of consideration here. The low cost of data transmission, which has come down over recent years, means that systems that only log data are becoming less common in many applications. The implementation of data transmission varies between applications with some systems transmitting data in real time with a minimum of onboard storage, while other systems are able to store a predefined quantity of data for transmission at regular intervals (e.g. every 15 min). The ability to buffer data does improve the reliability of transmission since data is not lost if the vehicle is unable to establish a connection, subject to the duration of the transmission blackout and the size of the available storage. Data can also be transmitted to vehicles for either interaction with vehicle systems or presentation to the driver.

Vehicle connectivity. Many telematic solutions link to systems on board the vehicle. This has the benefit of allowing information on the condition and performance of vehicle systems to be reported. Furthermore such systems can, in some instances, allow data to be passed to the vehicle allowing some degree of remote control of onboard systems.

6.2.2 Technological Enablers

The rise in the use of telematic systems has been underpinned by a series of technological developments which have reduced the cost of such systems [3] including:

The emergence of low cost microcontroller platforms which provide much of the necessary hardware to create a data capture and management system including a microprocessor and associated hardware such as voltage regulation and clock generation, as well as a range of I/O connectivity. Such platforms are supported by software tools which enable the rapid prototyping and design of the necessary firmware required for such applications.
The increased availability and reduced cost of GPS technology, underpinned by the availability of modules which can easily and simply be integrated with microcontroller platforms supported by software libraries.
The increased ease with which data can be transmitted from remote locations. The increasing popularity of GSM based mobile phones, often with data connectivity, means that coverage is high. Furthermore a range of hardware modules are available which support rapid integration with microprocessor platforms.

These advances mean that development of a system which can capture position from GPS data and transmit this back to some central location is now within the capabilities of enthusiastic amateurs. Prior to this the creation of such systems needed large development teams of electronics and software specialists. Furthermore the cost of such units means that they are viable for production systems as well as prototypes and design works. The use of 'off the shelf' also makes the production of systems viable at low volumes and avoids the need for investments in manufacturing infrastructure.

The compatibility of such systems with rapid product development and low volume manufacturing has enabled Small to Medium Sized Enterprises (SMEs) to develop their own bespoke remote tracking and monitoring systems. Prior to this, major investment would be required to establish production runs of the necessary electronic systems. Large volumes would be required to recoup this investment, making any such enterprise too high risk for most SMEs.

6.3 Operational and Organisational Drivers for the Use of Telemetry Systems

The increase in popularity of telemetry systems has come about as a consequence of both technological 'push' and a market 'pull'. A number of motivations exist for organisations to use telematic systems. These include:

Fuel savings. The style in which a vehicle is driven has a substantial effect on its efficiency. This is particularly true for ULCVs [1, 2].

Improving driver safety. Speed has been shown to be the biggest single factor when assessing risk [4].
Providing data in the event of an incident. Should an accident occur, data gathered from telematic systems can be of significant value in determining the cause and liability.
Managing the availability of vehicles. GPS tracking data can be used to measure the amount of time vehicles spend undergoing maintenance etc. based on the time spent in depot locations.
Managing insurance premiums. Insurance companies are increasingly using telemetry systems to determine premiums based on a more realistic assessment of risk. This means that savings are possible, particularly in low risk situations, since empirical factors such as gender/age of driver, vehicle type etc. do not correlate perfectly with risk [3]. Since ULCVs typically have a higher upfront cost it is reasonable to conclude that insurance costs will be higher making this a significant factor.

6.4 Barriers to Increased Use of Telematics

A number of barriers exist to the successful realisation of the benefits outlined above. As can be seen from the discussion above, the removal of technological and organisational barriers to progress can lead to a step change in the nature and success of particular technological applications.

The current barriers which need to be addressed include:

Access to on board data systems within vehicles to enable meaningful and useful information to be accessed. Vehicle Original Equipment Manufacturers (OEMs) are almost always reluctant to share this type of data.
An understanding and framework to support the processing of the vast volumes of data which could be produced and an understanding of where this processing should be performed.
The development of applications, frameworks and protocols to allow vehicle telemetry data to integrate with other systems allowing vehicle telematics to take its place in the 'Big Data' ecosystem
The need for adequate protection of individual privacy in these applications

These issues will be considered in the following subsections.

6.4.1 On-Board Systems

A substantial barrier which exists to the use of telemetry systems to collect data from vehicle systems is the availability of documentation on the data which is available on vehicle CAN bus. The CAN bus carries information between various systems using a standard protocol laid out in standards such as SAE J1939 [5] and ISO 11898 [6]. While the communication protocol is standardised, the content of the messages follows proprietary protocol which is unique to each vehicle manufacturer and regarding which they are reluctant to share information, even with major fleet buyers [7]. The result of this is that many organisations wishing to install this type of telematic systems need to determine the codes manually by operating the various vehicle functions and observing the messages on the bus in a time consuming and costly fashion. A standardised interface does exist in the form of the diagnostic port offered on almost all widely available vehicles using the ODBII (USA) or EOBD (Europe) protocol which were standardised in 1996 and 2001 and which are compulsory for vehicles sold in their respective territories. Some telematic systems are capable of relaying this data, however the range of information available is limited to certain fault conditions and does not encompass all of the information which may be required by vehicle users and operators.

Some telematic system operators are marketing solutions which claim to offer information on the degradation of vehicle components e.g. EV batteries. In many cases, however, the degradation is estimated using empirically based models linking degradation to use and does not involve measurement of equipment condition.

6.4.2 Data Processing Considerations

Many telemetry and tracking systems generate large volumes of data. Transmitting this data to a central location and storing it remains an issue, although recent reductions in the cost of data storage and transmission partially offset this cost. Processing large volumes of data remains a challenge however. One potential means of offsetting this is to perform some of the necessary processing and filtering on board the vehicle.

6.4.3 Privacy Concerns

Privacy remains a major concern for many potential telemetry system users. While many automated tracking technologies are already in existence [8, 9], many individuals remain concerned regarding the potential use of such systems to invade the privacy of citizens [10].

One potential solution to this issue is to use data in a fashion which does not involve raw position data being transmitted. Options here include:

Onboard use of geofencing. Vehicle position is only reported if it breeches certain geographic conditions, an approach known as 'geofencing'. These could be permanently encoded within a tracking system or downloaded periodically from a central repository.

Derivation/calculation of key parameters. This could include information used for a wide variety of purposes:

- Aggregation of braking data to inform replacement rates
- Online monitoring and estimation of the condition of vehicle traction batteries
- Monitoring driving style to indicate training needs
- High level data on the utilisation rate and patterns of the vehicle
- Data regarding the charging patterns and energy consumption of the vehicle.

While such approaches protect the driver they substantially reduce the potential value of vehicle tracking systems. Potential applications where the aggregation of data from multiple vehicles such as those described in Sect. 6.7 below are completely reliant on the combination of a significant volume of data.

6.5 ULCV Specific Applications of Telematics

The ever increasing provision of telematic data equipment within vehicles creates a range of possible new applications. These are outlined in the following subsections.

6.5.1 Management of Charge and Range

One of the key enablers for electric vehicle uptake is the provision of adequate charging infrastructure and the ability to provide adequate access. The specific factors affecting commercial use of EVs and similar types of ULCV include:

Consistent and regular operational patterns. The regularity or otherwise of operational patterns allowing vehicles to be charged at a sufficient frequency to ensure continued operation and availability. Fixed route vehicles offer the greatest scope to ensure this.

Availability of appropriate infrastructure in the correct locations. Commercial operators are unlikely to be willing to use freely shared or public infrastructure since this cannot be guaranteed to be available when required which could lead to scenarios where a vehicle requires charging and is forced to join a queue of some sort. For this reason private charging systems are likely to be required and in order to manage and optimize their utilization it is likely to be necessary to investigate and assess the calling patterns of the vehicles.

The time taken to charge vehicles. Commercial operators are likely to favour a minimization in the number of vehicles required for a particular set of operational requirements. This requires the maximization of the availability of each vehicle. For this reason the time taken to charge a vehicle is likely to be critical. Rapid Charge systems can address this problem but they are, however, extremely costly to install due to the cost of the equipment, the requirement for a high power electricity supply and, in many cases, the need for planning consent. Furthermore concerns exist regarding the effect of rapid charging on battery life.

The use of telematic data to mitigate against these issues is seen as a key means of enabling wider commercial use of EV technology. The ability to dynamically monitor the state of charge (SoC) and remaining range of vehicles allows the possibility of dynamically managing planned routes and charging opportunities to ensure that operations can be maintained.

In order for such systems to be realised it will be necessary to ensure that the interconnectivity of telematic data sources can be expanded through the provision of streamed data and simplified Application Programmers Interfaces (APIs) allowing organistions to interlink their tracking and planning systems into new bespoke configurations which meet their operational needs.

6.5.2 Telematic Condition Monitoring

Vehicle condition is a critical issue for fleet operators in a number of respects:

- Ensuring the ongoing reliability of vehicles.
- Prolonging the life of the vehicle.
- Protecting the value of vehicles for future resale.
- Reducing operational costs such as maintenance.

Telematic systems can support the optimization of vehicle condition in a number of ways:

- Driving practices which may affect the condition of the vehicle e.g. harsh braking which will lead to accelerated brake wear.
- Monitoring of the condition of key components via either dedicated sensors or interface with onboard systems.
- Monitoring use patterns and levels to ensure that maintenance schedules are optimized to provide maximum reliability for a minimum cost level.

Historically, vehicle maintenance has occurred in two ways: planned preventive maintenance and reactive maintenance which occurs in the event of failures. In general, the construction of EVs and similar vehicle types make them generally reliable meaning little reactive maintenance is required. Preventive maintenance takes the form a scheduled works based on either time or mileage intervals. The efficiency of such an approach depends on the optimal intervals being selected. If the intervals are too long then failures are likely to occur whereas if the intervals are too short then additional cost and scheduled downtime is accrued. Condition based maintenance (CBM) uses estimates of the condition of particular components and systems to determine the optimal time to perform maintenance. Selecting the optimal condition measurement at which to perform maintenance requires an estimate of the expected life of components i.e. a prognostic estimation.

ULCV systems which could benefit from the use of such an approach include brake systems—these can be monitored directly using sensors to ascertain condition measurements. Alternatively the utilization rate of braking systems can be measured to provide a more accurate estimate of wear rate than that which might be afforded based on duration or mileage data.

Traction batteries are significant component of EVs and PHEVs. The dominant battery technology used in such vehicles is based on a lithium ion chemistry due to its high energy density. A major disadvantage of this type of battery is the capacity degradation which occurs due to both ageing and use. The cost of batteries and their criticality on the operational range of the vehicle makes their capacity a significant parameter to be managed. This makes monitoring the condition of electric vehicle batteries a significant potential application for ULCVs.

6.6 Wider Benefits of Telematic Monitoring

The widespread gathering of telematic and tracking data offers the potential for a range of benefits besides those directly applicable to the vehicle operator. These are summarised in the following subsections.

6.6.1 Smart Traffic Management

The use of rich telematic data captured from vehicles offers the potential to employ so-called 'smart-traffic management' schemes. In this way traffic signals can be dynamically controlled based on a more complete range of data than is currently possible. The types of information which could be utilised in this way include:

Vehicle position and speed. This data could be used to indicate the level of traffic flow in particular situations.
Intended destination and routing. The use of such data can be used to dynamically adjust routes and traffic signals to adapt to particular variations in traffic flows and particular popular destinations.

A potential application of the latter type of data listed above could be found in situations where the traffic flow varies at particular times. An example of this would be large employment centres which operate a shift based system, e.g. Hospitals, manufacturing plants and other major employment centres such as city centres. Major sports and conference venues would offer potential for such systems. Adapting traffic flow to increase capacity has been used for a number of years [11] but is currently based on fixed patterns and lacks the ability to adapt dynamically to current and developing trends and patterns.

6.6.2 Removing Road Signs

Maintaining roadside signage comes at a significant expense. It is estimated that in the UK alone there are over 200,000 signs indicating speed restrictions alone [12]. Furthermore recent years have seen an increase in the use of intelligent signage to include systems such as matrix signs which can display a range of information to drivers. An extension of this type of technology is the use of active traffic management systems which can open and close lanes ad change speed limits.

In-cab interfaces offer the potential for such information to be relayed directly to the driver without the need for roadside infrastructure. This offers substantial benefits in terms of cost and reduction in environmental 'clutter' produced by such signage.

In-cab signalling is already being implemented in the rail industry and its roll-out is underpinned by the development of international standards such as the European Rail Traffic Management System (ERTMS) [13]. Such standards offer considerable benefits in terms of allowing interoperation of vehicles and reducing the cost of hardware.

6.6.3 The Development of Autonomous Vehicle Technology

The use of in cab technology and dynamic routing is seen by many in the automotive and road transport industry as an important intermediate stage in the development of fully autonomous vehicles. A range of systems already exist in isolation which could contribute towards autonomous operation including:

- Collision avoidance systems [14]
- Automated parking aids [15]
- Automated steering systems such as those used in guided busways [16]

Combining these technologies which suitable telematics systems enables autonomous control as has already been demonstrated in a range of prototypes [16]. As these vehicle control technologies mature, achieving truly autonomous operation of vehicles becomes an issue of infrastructure and integration.

6.7 Conclusions

ULCVs are rapidly becoming more and more attractive to vehicle fleet operators due to their potential to reduce whole life costs as well as supporting the efforts of many organisations to improve their environmental performance. Significant barriers still exist however. Uncertainty regarding cost and reliability/availability remain significant barriers to fleet operators. Telematic monitoring systems offer considerable opportunity to address many of these barriers and mitigate against them.

The wide range of current and potential applications for telematic monitoring systems means that the recent trend of increasing levels of use is likely to continue. It is reasonable to assume that vehicle manufacturers are increasingly likely to incorporate such systems at the point of manufacture—particularly for commercial vehicles targeted at fleet operators. Such a trend would go some way to addressing some of the barriers relating to interconnectivity between telematics and existing vehicle systems. If, however, the full potential for telematic monitoring systems is to be realised then it is crucial that standardisation of interfaces occurs at a vehicular level to allow certain data to be read from vehicle systems. This is particularly true for ULCVs where considerable variation is perceived to exist in the lifespan of certain key components such as traction batteries.

In addition, if the full potential of telematic vehicle monitoring is to be realised, both in terms of enabling the use of ULCVs and optimising their lifespan, as well as for the range of additional applications listed in this chapter, then it is critical that accessibility and usability of the collected data is addressed. Greater use of standardised interfaces to data sources which offer a suitable degree of flexibility to suit individual requirements of particular applications is thus strongly recommended.

References

1. Bingham C, Walsh C, Carroll S (2012) Impact of driving characteristics on electric vehicle energy consumption and range. IET Intell Transp Syst 6:29–35
2. Knowles M, Scott H, Baglee D (2012) The effect of driving style on electric vehicle performance, economy and perception. Int J Electr Hybrid Veh 4(3):228–247
3. Desyllas P, Salko M (2013) Profiting from business model innovation: evidence from Pay-as-You-Drive auto insurance. Res Policy 42:101–116
4. Bolderijk J, Steg L (2011) Pay-as-You-Drive vehicle insurance as a tool to reduce crash risk: results so far and further potential. International Transport Forum Discussion paper No. 2011–23
5. SAE, Recommended Practice for a Serial Control and Communications Vehicle Network, vol. J1939, ed. 2005th, Warrendale Pa
6. ISO (2003) Road vehicles—controller area network (CAN), vol 11898
7. Thayer P (2009) Vehicle dynamics: driving consumer value. Available: http://www.uktelematicsonline.co.uk/html/telematics_articles.html. Cited 14 Apr 2014
8. Clarke R, Wigan M (2011) You are where you've been: the privacy implications of location and tracking technologies. J Location Based Serv 5(3–4):138–155
9. Gurney R, Rhead M, Ramalingam S, Cohen N (2012) Working towards an international ANPR standard—an initial investigation into the UK standard. In: 2012 IEEE international Carnahan conference on security technology (ICCST), pp 331–337
10. Warren I, Lippert R, Walby K, Palmer D (2013) When the profile becomes the population: examining privacy governance and road traffic surveillance in Canada and Australia. Curr Issues Crim Justice 25(2):565–584
11. Papageorgiou M, Diakaki C, Dinopoulou V, Kotsialos A, Wang Y (2003) Review of road traffic control strategies. Proc IEEE 91(12):2043–2067. doi:10.1109/JPROC.2003.819610

12. Paice R (2009) Metric signs ahead: the case for converting road signs to metric units. UK metric association report, http://www.metric.org.uk/sites/default/files/MSA.pdf. Cited 30 Mar 2014
13. UNIFE 2013 http://www.ertms.net/. Cited 30 Mar 2014
14. Llorca DF, Milanes V, Alonso IP, Gavilan M, Daza IG, Perez J, Sotelo MA (2011) Autonomous pedestrian collision avoidance using a fuzzy steering controller. IEEE Trans Intell Transp Syst 12(2):390–401
15. Revathi G, Dhulipala V, Sarma R (2012) Smart parking systems and sensors: a survey. In: 2012 international conference on computing, communications and applications (ICCCA), pp 1–5
16. Cervero R (2013) Bus rapid transit (BRT): an efficient and competitive mode of public transport. Institute of Urban and Regional Development Working Paper 2013–01. http://escholarship.org/uc/item/4sn2f5wc. Cited 30 Mar 2013

Chapter 7
A Weak Signal Detection Method Based on Stochastic Resonances and Its Application to the Fault Diagnosis of Critical Mechanical Components

Niaoqing Hu, Bin Fan, Lei Hu, Zhe Cheng, Lun Zhang, Guoying Zhao and Fengshou Gu

Abstract This chapter presents studies on the enhanced detection of characteristic signals from critical mechanical components such as bearings by the nonlinear effect of stochastic resonance (SR). In the past decades, classical stochastic resonance (CSR) method has been extensively studied to enhance the fault detection of these critical mechanical components such as bearings and gears. Based on CSR theories, the main content of this chapter includes two parts. The first is aiming at identifying the component characteristic frequencies in the spectra, SR normalized scale transform is proposed based on parameter-tuning bistable SR model, which leading to a new method via averaged stochastic resonance (ASR) to enhance the result of incipient fault detection. Then, rather than achieving the improvement of the signal-to-noise ratio (SNR) by increasing the noise intensity, a new approach is developed based on adding a harmonic excitation with a frequency based on the system's Melnikov scale factor to the system while the noise is left unchanged. The effectiveness of this method is confirmed and replicated by numerical simulations. Combined with the strategy of the scale transform, the method can be used to detect weak periodic signal with arbitrary frequency buried in the heavy noise. In addition, the chapter also presents the case study of applying these methods for the enhancement of fault characteristic signals in detecting incipient faults of roller bearings.

N. Hu (✉) · B. Fan · L. Hu · Z. Cheng · L. Zhang · G. Zhao
Key Laboratory of Science and Technology on Integrated Logistics Support, School of Mechatronics Engineering and Automation, National University of Defense Technology, Changsha 410073, Hunan, People's Republic of China
e-mail: hnq@nudt.edu.cn

F. Gu
Centre for Diagnostic Engineering, University of Huddersfield, Huddersfield, Yorkshire, UK

L. Redding and R. Roy (eds.), *Through-life Engineering Services*,
Decision Engineering, DOI 10.1007/978-3-319-12111-6_7

7.1 Introduction

Fault detection of critical mechanical components, such as bearings and gears of a power train in complex machines like helicopters and wind turbines, is one of the important tasks in the field of running reliability. It is well known that the evaluation of the dynamic behavior of the mechanical components solely depends on the quality of the measured signals. The factors such as the influence of their transmission path, the transmission medium, the ambient environment, change of internal dynamics etc., degrade the measured signals, leading to measurements with low signal-to-noise ratio (SNR). In many cases, the useful information is buried in the noise seriously so that it can be hardly recovered by conventional method. It means that the early signatures of possible potential faults may be missed, which would lose the chance to prevent the catastrophic failures induced by the mechanical components like helicopters. Thus, the detection issue of early symptoms of a dynamic mechanical component fault is essentially a topic of weak signal detection.

Therefore, weak signal detection under heavy background noise becomes more and more important for early detection of fault. Over the last decades, it is commonly concerned by scientists and engineers to detect and enhance weak target signal more expeditiously and precisely in noise environment with special restrictions. Therefore, some notions concerning weak signal detection were recommended. Several approaches have been applied to weak signal detection, such as chaotic resonator [1, 2], difference resonator [3], wavelet analysis [4], holospectral analysis, high order statistics, Hilbert-Huang transform and so on [5]. Furthermore, an enhanced detection solution for weak signals based on stochastic resonance theory has been presented, which can detect a weak signal in the presence of heavy noise from a very short data record [6].

In this chapter, two new enhanced detection approaches are presented based on extended stochastic resonance theory to detect weak signal in the presence of heavy noise. One is 'averaged' enhanced detection strategy based on normalized scale stochastic resonance model [5] to detect weak signal. The other effective SNR enhancement is achieved by adding a harmonic excitation with frequency based on the system's Melnikov scale factor to the system while the noise is left unchanged.

Although the principle and property have been illustrated previously [7–9], key issues deeply related to SR and its application in weak mechanical signal processing will be discussed. Through a kind of normalized scale transform, the frequency range of weak signal SR model can detect is extended from low frequency to relative higher frequency. Based on normalized scale transform, a new method via averaged stochastic resonance (ASR) is presented to enhance the result of rolling element bearing fault detection furtherly.

Furthermore, in classical SR, the SNR can be improved by increasing the noise. But the approach by increasing the noise is counterintuitive and unwieldy. According to Melnikov theory, for a wide class of systems, deterministic and stochastic excitations play qualitatively equivalent roles in inducing chaotic motions with escapes over a potential barrier [10]. Such motions therefore possess common qualitative features that suggest the extension of SR approaches beyond classical SR, so that the SNR can alternatively be improved by keeping the noise unchanged and adding a deterministic excitation which is close to the detected signal and selected in accordance with Melnikov theory, rather than by increasing the noise.

7.2 Fundamental of SR and Normalized Scale Transform

7.2.1 Fundamental of SR

The study of stochastic resonance in signal processing has received considerable attention over the last decades. In the context, stochastic resonance is commonly used as an approach to increase the SNR at the output through the increase of the special noise level at input signal. For a class of multistable system with noise and a periodic signal, the improvement of SNR achieved by increasing the noise intensity is known as stochastic resonance (SR) which will be referred to classical SR. The essence of the physical mechanism underlying classical SR has been described in [5–9]. Considering the motion in a bistable double-well potential of a lightly damped particle subjected to stochastic excitation and a harmonic excitation (i.e., a signal) with low frequency ω_0. The signal is assumed to have small amplitude that, by itself (i.e., in the absence of the stochastic excitation), it is unable to move the particle from one well to another. It is denoted that the characteristic rate, that is, the escape rate from a well under the combined effects of the periodic excitation and the noise, by $\alpha = 2\pi n_{tot}/T_{tot}$, where n_{tot} is the total number of exits from one well during time T_{tot}. The behavior of the system will change when increasing the noise while the signal amplitude and frequency are unchanged. For zero noise, $\alpha = 0$, as noted earlier. For very small noise, $\alpha < \omega_0$. However, as the noise increases gradually, the ordinate of the spectral density of the output noise at the frequency ω_0, denoted by $\Phi_n(\omega_0)$, and the characteristic rate α increases accordingly. Experimental and analytical studies show that, until $\alpha \approx \omega_0$, a cooperative effect (i.e., a synchronization-like phenomenon) occurs wherein the signal output power $\Phi_s(\omega_0)$ increases as the noise intensity increases. Remarkably, the increase of $\Phi_s(\omega_0)$ with noise is faster than that of $\Phi_n(\omega_0)$. This results in an enhancement of the SNR. The synchronization-like phenomenon plays a key role in the mechanism as described in [10].

At present, the most common studied SR system is a bistable system, which can be described by the following Langevin equation

$$\dot{x} = ax - bx^3 + A\sin(\omega_0 t + \phi_0) + \Gamma(t) \tag{7.1}$$

where $\Gamma(t)$ is noise term and $\langle \Gamma(t), \Gamma(0) \rangle = 2D\delta(t)$, $A\sin(\omega_0 t + \varphi_0)$ is a periodic driving signal. Generally, it is also written as the form of Duffing equation

$$\ddot{x} = -\beta\dot{x} + ax - bx^3 + A\sin(\omega_0 t + \phi_0) + \Gamma(t) \tag{7.2}$$

where β is the damping coefficient.

7.2.2 Normalized Scale Transform of SR Model

Equation (7.1) has two stable solutions $x_s = \pm\sqrt{a/b} = \pm c$ (stable points) and a unstable solution $x_u = 0$ (unstable point) when $A = D = 0$, here the potential of Eq. (7.1) is given by

$$V(x) = -\frac{1}{2}ax^2 + \frac{1}{4}bx^4 \tag{7.3}$$

The height of potential is

$$\varDelta V = V(0) - V(c) = \frac{a^2}{4b} \tag{7.4}$$

When adding the modulation signal, potential function is

$$V(x,t) = -\frac{1}{2}ax^2 + \frac{1}{4}bx^4 - Ax\cos\omega_0 t \tag{7.5}$$

For a stationary potential, and for $D \ll \varDelta V$, the probability that a switching event will occur in unit time, i.e. the switching rate, is given by the Kramers formula [11]

$$r_0 = (2\pi)^{-1}[V''(0)|V''(c)]^{1/2}\exp(-\varDelta V/D) \tag{7.6}$$

where $V''(x) \equiv \mathrm{d}^2V/\mathrm{d}x^2$. When a periodic modulation term $A\sin\omega_0 t$ is included on the right-hand-side of Eq. (7.1), it leads to a modulation of the potential Eq. (7.5) with time and an additional term $-Ax\cos\omega_0 t$ is now present on the right-hand-side of Eq. (7.5). In this case, the Kramers rate Eq. (7.6) becomes time-dependent:

$$r(t) \approx r(0)\exp(-Ax\sin\omega_0 t/D) \tag{7.7}$$

Which is accurate only for $A \ll \varDelta V$ and $\omega_0 \ll \{V''(\pm c)\}^{1/2}$. The latter condition is referred to as the adiabatic approximation. It ensures that the probability density corresponding to the time-modulated potential is approximately stationary (the modulation is slow enough that the instantaneous probability density can 'adiabatically' relax to a succession of quasi-stationary states). The slow modulation means that the signal to detect is confined to a rather low frequency range and small amplitude. It is well known that the characteristic frequency reflecting mechanical system state exceeds the range of limit, so how to detect the high frequency signal is of great importance in weak characteristic signal detection of mechanical system.

To overcome the low frequency limitation of SR, a scale transform needs to be introduced to shift the frequency of interest into the range in which SR operates. Considering the bistable system modeled by Eq. (7.1), where A is the amplitude of the input signal, $\omega \gg 1$ is its frequency, $\Gamma(t)$ is Gaussian white noise with the correlation $\langle \Gamma(t) \rangle = 0$; $\langle \Gamma(t), \Gamma(0) \rangle = 2D\delta(t)$, and D is the noise intensity, when a and b are positive real numbers, a variable substitution can be carried out by

$$z = x\sqrt{b/a}, \quad \tau = at \tag{7.8}$$

Substituting Eq. (7.8) into Eq. (7.1) yields:

$$a\sqrt{\frac{a}{b}}\frac{\mathrm{d}z}{\mathrm{d}t} = a\sqrt{\frac{a}{b}}z - a\sqrt{\frac{a}{b}}z^3 + A\cos\left(\frac{\omega_0}{a}\tau + \phi_0\right) + \Gamma\left(\frac{\tau}{a}\right) \tag{7.9}$$

where the noise $\Gamma(\tau/a)$ satisfies $\langle \Gamma(\tau/a)\Gamma(0) \rangle = 2Da\delta(\tau)$. Therefore

$$\Gamma\left(\frac{\tau}{a}\right) = \sqrt{2Da}\xi(\tau) \tag{7.10}$$

where $\langle \xi(\tau) \rangle = 0$, $\langle \xi(\tau), \xi(0) \rangle = \delta(\tau)$.

Substituting Eq. (7.10) into Eq. (7.9) results in:

$$a\sqrt{\frac{a}{b}}\frac{\mathrm{d}z}{\mathrm{d}t} = a\sqrt{\frac{a}{b}}z - a\sqrt{\frac{a}{b}}z^3 + A\cos\left(\frac{\omega_0}{a}\tau + \phi_0\right) + \sqrt{2Da}\xi(\tau) \tag{7.11}$$

Equation (7.11) can be simplified into

$$\frac{\mathrm{d}z}{\mathrm{d}t} = z - z^3 + \sqrt{\frac{b}{a^3}}A\cos\left(\frac{\omega_0}{a}\tau + \phi_0\right) + \sqrt{\frac{2Db}{a^2}}\xi(\tau) \tag{7.12}$$

Equation (7.12) is a normalized form and equals to Eq. (7.1). The frequency of the signal after the scale transform is $1/a$ times of which before transform. Hence, through the chosen of larger parameter a, a high frequency signal can be normalized to lower one to satisfy the request of the theory of SR.

During the numerical simulation, the variance σ^2 is used to describe the statistical property of the white noise. As the noise intensity D is influenced by sample step h, the actual value $D = \sigma^2 h/2$. If RMS value of the noise is $\sigma_0 = \sqrt{2D/h}$ before transform, the intensity of the noise will change to $2Db/a^2$ after the transform. In addition, because the sample frequency descends, the sample step becomes a times of the original sample step. Therefore, the RMS of the noise after transform is $\sigma = \sqrt{2Db/(a^2 \cdot ah)}$. The ratio of the noise RMS after the transform to which before the transform is

$$\sigma/\sigma_0 = \sqrt{b/a^3} \tag{7.13}$$

It is easy to be seen that, after the transform, the signal and noise are amplified $\sqrt{b/a^3}$ times.

7.2.3 Averaged SR Model with Normalized Scale Transform

Taking $y(t)$ as the sampled signal or the envelop signal of raw signal with Hilbert transform (as described in Sect. 4.1), Eq. (7.12) can be written as

$$\frac{dz}{d\tau} = z - z^3 + \sqrt{\frac{b}{a^3}} y(\tau) + \sqrt{\frac{2Db}{a^2}} \xi(\tau) \tag{7.14}$$

Then Eq. (7.14) can be solved numerically J times with different $\xi_i(\tau)$ $(i = 1, 2, \ldots, J)$ to obtain $z_i(\tau)$ $(i = 1, 2, \ldots, J)$, then an average is carried out by

$$\bar{z}(\tau) = \frac{1}{J} \sum_{i=1}^{J} z_i(\tau) \tag{7.15}$$

Through this average procedure a more reliable detection result can be obtained, which is a general operation based on the Monta Carlo principle.

7.2.4 Model Validation Using Simulated Data

To evaluate the performance of the scale transform proposed a numerical simulation study is carried out based on a mixed signal to be enhanced through the model of bistable system with parameters $a = b = 1$, $A = 0.5$, $f = 0.1\,\text{Hz}$, $\sigma = 5$, $f_s = 20\,\text{Hz}$, $N = 2{,}000$.

Figure 7.1a, b shows the mixed signal and its spectrum, while Fig. 7.1c, d gives the output of the bistable system and the spectrum of the output signal. From Fig. 7.1d, it can be seen that although the input $\text{SNR} = 20\log(A/\sigma) = -20\,\text{dB}$, there is a clear spectrum line at $f = 0.1\,\text{Hz}$, and the noise attenuation is obvious.

If the signal frequency is changed into $f = 1\,\text{kHz}$, according to the transform principle, the parameters $a = b = 10^4, f_s = 200\,\text{kHz}$ and $N = 2{,}000$ can be used for frequency shift. The mixed signal will be amplified by $\sqrt{a^3/b} = 10{,}000$ times. Applying SR enhancement to this mixed signal after the transform produces results as shown in Fig. 7.2.

As it can be seen in Fig. 7.2, the signal and spectrum is consistent with that of Fig. 7.1 although they show differences in the scales the time domains and frequency coordinates. The noise components are greatly suppressed, and the

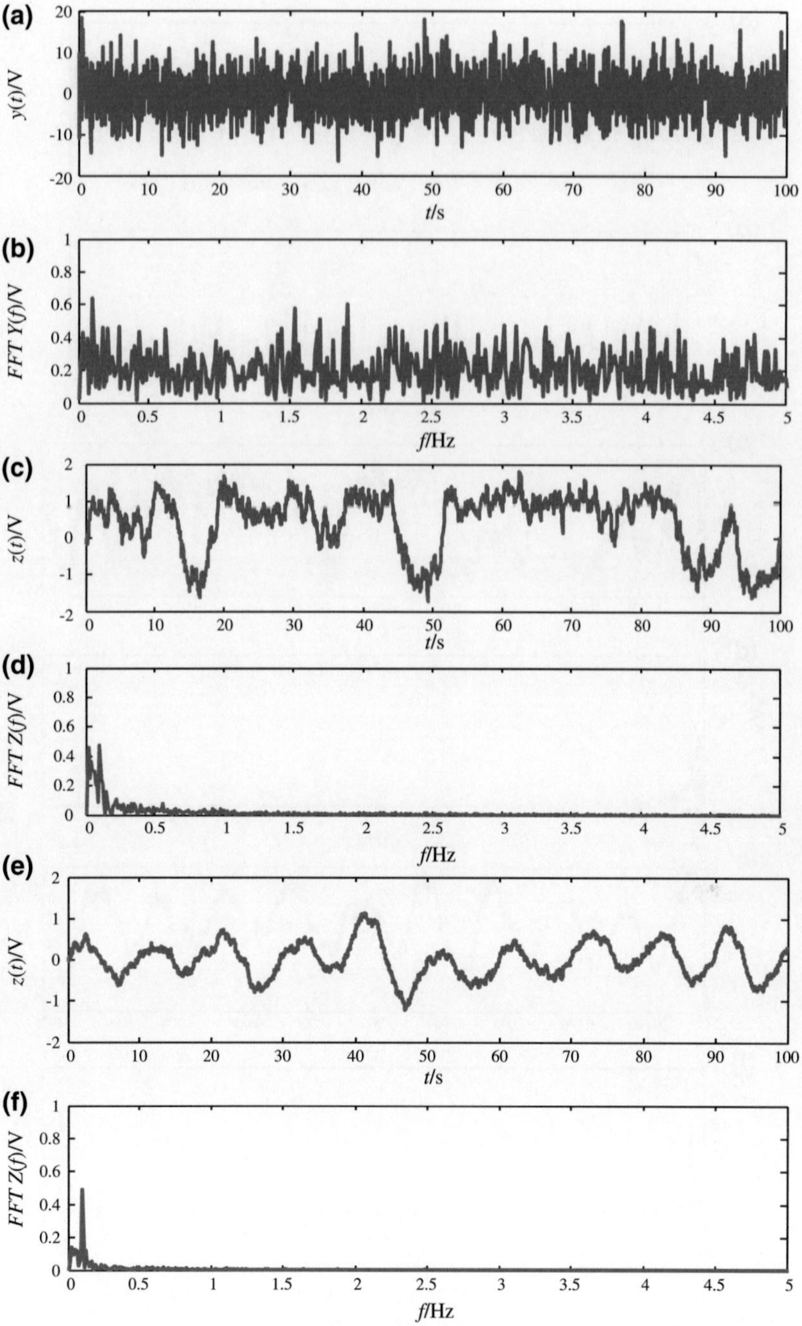

Fig. 7.1 Time-domain and its FFT of the input and output when $f = 0.1\,\text{Hz}$. **a** and **b** the input; **c** and **d** the output by one-time; **e** and **f** the output by averaged

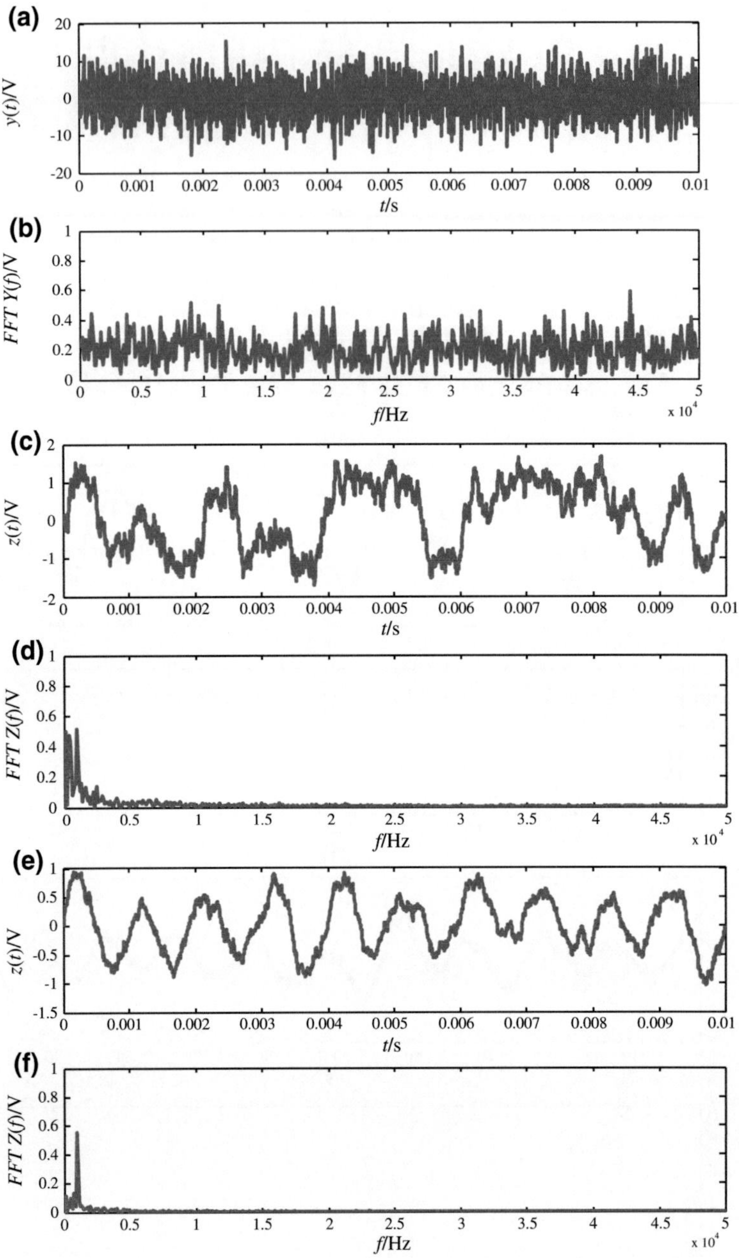

Fig. 7.2 Time-domain and its FFT of the input and output when $f = 1\,\text{kHz}$. **a** and **b** the input; **c** and **d** the output by one-time; **e** and **f** the output by averaged

detecting signal is standing out, which shows that the transform method is effective for adapting to the behavior of SR to enhance high frequency signals. Therefore, by adjusting parameter a of the bistable system, SR model can adapt to different frequency signal, while by adjusting parameter b, it can adapt to different noise intensity. In addition, compared (d) with (f) of Figs. 7.1 and 7.2, it is obvious that the averaged SR results are better than that of one-time SR result.

7.3 SR Model by Adding a Harmonic Excitation

7.3.1 *SR Interpretation via Melnikov Theory and Chaotic Dynamic Approach*

As stated in [10, 11], for a bistable system with noise and a periodic signal, the improvement of the signal-to-noise ratio (SNR) achieved by increasing the noise intensity is known as stochastic resonance (SR) (i.e., classical SR in these papers). Here, the signal to noise ratio (SNR) is expressed in dB as $SNR = 10\log10(S/N)$, where S and N are, respectively, the ordinate of the output power spectrum and the ordinate of the broadband output power spectrum at the signal frequency ω_0. As described in Sect. 2.1, the synchronization like phenomenon plays a key role in the SR mechanism.

Now we consider second-order dynamical systems described by the following equation [10]

$$\ddot{x}(t) = -\beta\dot{x}(t) - V'(x) + G(t) \tag{7.16}$$

where $V(x)$ is a potential function. The unperturbed counterpart of Eq. (7.1) is the Hamiltonian system expressed by $\ddot{x} = -V'(x)$. We assume that $V(x)$ is a double-well potential (Duffing-Holmes) as described in (7.3) with $a = b = 1$. $\ddot{x} = -V'(x)$ with the potential (7.3) and $a = b = 1$ has the homoclinic orbits [12].

Firstly, it is assumed that the excitation is only periodic, that is, in Eq. (7.16) $G(t) \equiv A_0 \sin(\omega_0 t)$. According to the Smale-Birkhoff theorem, the necessary condition for the occurrence of chaos is that the Melnikov function induced by the perturbation has simple zeroes. For Duffing system this condition is the Melnikov inequality

$$-(4/3)\beta + A_0 S_M(\omega_0) > 0 \tag{7.17}$$

where

$$S_M(\omega) = \sqrt{2}\pi\omega\text{sech}(\pi\omega/2) \tag{7.18}$$

is a system property known as the Melnikov scale factor [13].

Secondly, assume that the excitation consists of the quasiperiodic sum

$$G(t) \equiv A_0 \sin(\omega_0 t + \phi_0) + A_a \sin(\omega_a t) + \sum_{k=1}^{K} a_k \sin(\omega_k t + \varphi_k) \tag{7.19}$$

For this case a generalization of the Smale-Birkhoff theorem [13] yields the Melnikov inequality as the necessary condition for chaos

$$-\frac{4\beta}{3} + A_0 S_M(\omega_0) + A_a S_M(\omega_a) + \sum_{k=1}^{K} a_k S_M(\omega_k) > 0 \tag{7.20}$$

Finally, assume that the system's excitation is

$$G(t) \equiv A_0 \sin(\omega_0 t + \phi_0) + A_a \sin(\omega_a t) + \sqrt{2D\beta} R(t) \tag{7.21}$$

where $R(t)$ is a Gaussian process with unit variance and spectral density $g(\omega)$. Over any finite time interval, however large, each realization of the process $R(t)$ may be approximated as closely as desired by a sum [10]

$$R_N(t) = \sum_{k=1}^{K} b_k \sin(\omega_k t + \varphi_k) \tag{7.22}$$

so that the Melnikov inequality, that is, the necessary condition for chaos, can be written as in Eq. (7.20) where $a_k = \sqrt{2D\beta} b_k$. In formula (7.22), $b_k = \sqrt{g(\omega_k)\Delta\omega}$, φ_k are randomly chosen phases of uniform distribution on the interval $[0, 2\pi]$ and $\omega_k = k\Delta\omega$, $\Delta\omega = \omega_{\max}/K$, $\omega_{\max}$ is the frequency beyond which the spectrum vanishes (the cutoff frequency).

For the damped, forced system, the existence in a plane of section of a transverse point of intersection between the stable and unstable manifolds implies the existence of infinity of intersection points. Eventually, they may form a chaotic motion under a particular excitation. The strength of the chaotic transport, and therefore the characteristic rate α, increases as the left-hand side of in Eq. (7.20) becomes larger [13]. This is true regardless of whether the excitation is deterministic or stochastic. Moreover, again regardless of whether the excitation is deterministic or stochastic, a qualitative feature of the chaotic motions featuring escapes is that their spectral densities have a broadband portion with significant energy content at and near the system's characteristic rate α. Thus, we expect that we can build a bridge between chaos and stochastic resonance. That is to say, we can explain SR phenomena by chaotic dynamics approach.

Assume that the excitation is a sum of a harmonic signal and an additional harmonic component, that is, in Eq. (7.16), $G(t) \equiv A_0 \sin(\omega_0 t) + A_a \sin(\omega_a t)$. The system is therefore deterministic with, in general, quasiperiodic excitation. The necessary condition for chaos is given by in Eq. (7.20) in which $a_1 = a_2 = \cdots = a_K = 0$. We choose A_0 so that, for $A_a = 0$, the motion is confined to one well. In accordance with Melnikov theory this will be the case if the Melnikov inequality given by in Eq. (7.17) is not satisfied. We now add the excitation $A_a \sin(\omega_a t)$. For a certain region R_a of the parameter space $[A_a, \omega_a]$, the system can experience chaotic motion with jumps over the potential barrier. The Melnikov scale factor $S_M(\omega)$ provides the information needed to select frequencies ω_a such that the added excitation is effective in inducing chaotic behaviour. In general, ω_a should be equal or close to the frequency for which $S_M(\omega)$ is the largest.

Given the existence in the spectrum of a broadband portion qualitatively similar to that present in the case of classical SR, it is reasonable to expect that the synchronization like phenomenon that occurs in the classical SR case would similarly occur for the deterministically excited chaotic system. This was verified by a numerical simulation for a large number of cases. As a typical example in [10], the case for $\beta = 0.316$, $A_0 = 0.095$, $\omega_0 = 0.0632$ (for these values in Eq. (7.17) is not satisfied) and $\omega_a = 1.1$ is examined. Spectral densities of motions with these parameters and $A_a = 0.263$, 0.287, and 0.332, are shown in (a), (b) and (c) of Fig. 7.3, respectively. As it can be seen in Fig. 7.3a, when $\alpha = 0.0671$ is close to the signal frequency $\omega_0 = 0.0632$, The energy in the broadband portion of the spectrum is reduced clearly, while the energy at the signal's frequency is enhanced, compared with the respective counterparts in Fig. 7.3b, c, for which $\alpha = 0.0518$ and $\alpha = 0.1611$, respectively. The synchronization like phenomenon noted for classical SR is thus clearly evident in Fig. 7.3b. In addition, the motions in Fig. 7.3 of (a), (b), (c) are indeed chaotic. This shows that the additive harmonic signal plays the same effect as the noise in the enhancement of SNR.

7.3.2 Simulation for Enhancing the Detection of Weak Signal by Adding a Harmonic Excitation

7.3.2.1 Detecting Weak Signal at Low Frequency

Notice that the larger the left-hand side of Eq. (7.20), the stronger is the chaotic transport, and therefore the larger is the rate α. Let $A_a = 0$, $a_k = \sqrt{2D\beta}\sqrt{g(\omega_k)\Delta\omega}$ in Eq. (7.20). It is therefore clear from in Eq. (7.20) that for any given power of the stochastic excitation $2D\beta$, the left-hand side of Eq. (7.20) becomes larger and the rate α increases. We thus obtain the interesting qualitative result that, for a given Melnikov scale factor $S_M(\omega)$ and a given power of the stochastic excitation, the rate α increases as the spectral power of the excitation is distributed nearer to the frequency of $S_M(\omega)$'s peak, ω_{pk} (the greatest effectiveness being achieved by a single component with frequency equal or close to ω_{pk}).

We now illustrate the usefulness of this result for a system with classical SR (i.e., one for which in Eq. (7.21) $A_a = 0$, $D > 0$). We assume $R(t)$ has the Lorentzian spectral distribution $g(\omega) = \gamma(1 + \omega^2\tau^2)^{-1}$ cut off at the frequency $\omega_{\max}$; τ is the correlation time and γ is a normalization constant such that the variance of $R(t)$ is

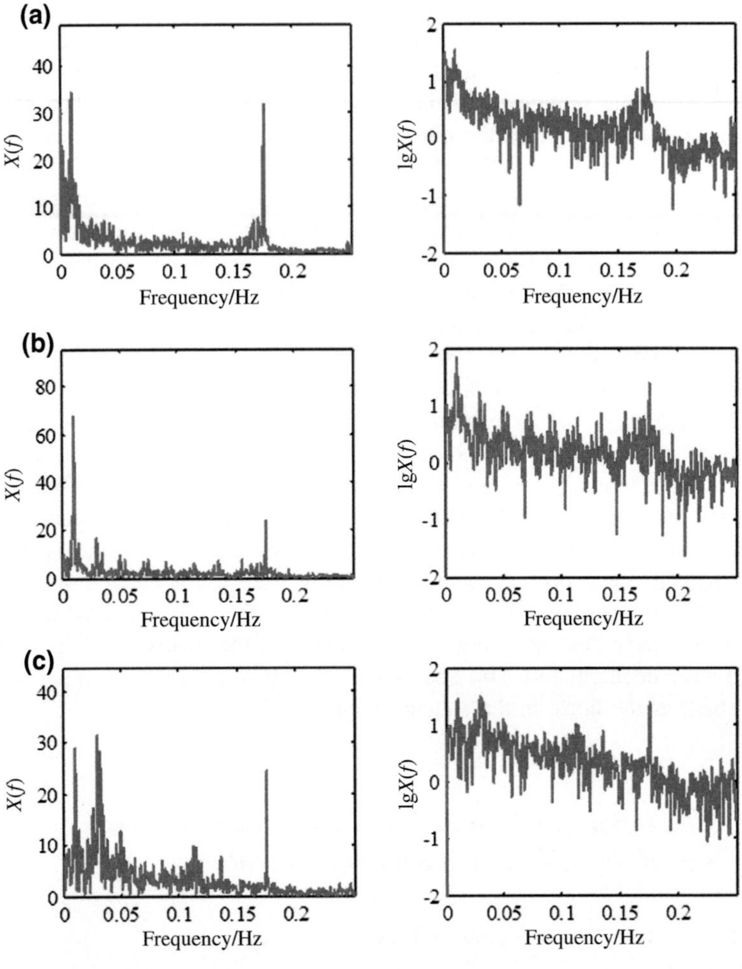

Fig. 7.3 Amplitude spectra of system with D = 0, A_0 and ω_0 keep constant. **a** The system is subjected to an additional harmonic excitation with $\omega_a = 1.1$ and $A_a = 0.263$. **b** All settings are the same as in, **a** except amplitude $A_a = 0.287$. **c** All settings are the same as in, **a** except amplitude $A_a = 0.332$

unity. The Melnikov scale factor $S_M(\omega)$ would in practice suppress contributions of components with frequencies $\omega > \omega_{\max}$.

Considering the case $\tau = \tau_1 = 0.2$, typical averaged output spectra $P(\omega)$ for $A_0 = 0.3, \omega_0 = 0.069, \omega_{\max} = 3.0, \beta = 0.25$ are shown in Fig. 7.4a–c for D = 0.1, 0.6, and 2.0, respectively (other parameters are the same as in [10]). The averaging was performed over 225 noise realizations approximated by in Eq. (7.20) with $100 < K < 500$. Note that $A_0 < 4\beta/3S_M(\omega_0)$, so that no chaotic behaviour can be induced by the periodic signal alone. However, it was verified that, for the noise realizations used to obtain the results of Fig. 7.4a–c, the Melnikov inequality given

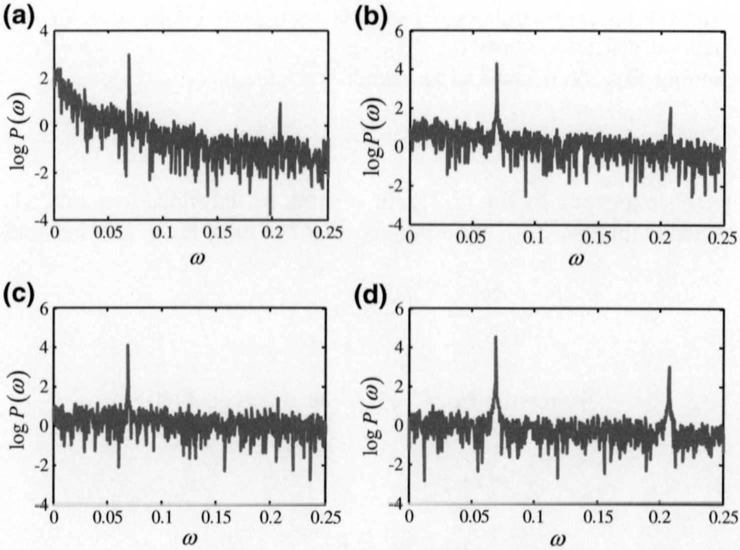

Fig. 7.4 Averaged power spectra of output for stochastically excited system: **a**–**c** increasing noise intensity D and $A_a = 0$. **d** The same noise intensity D as in **a**, and $A_a = 0.23$. Noise correlation time $\tau = 0.2$

by in Eq. (7.20) was satisfied, and that the respective motions were chaotic. Energy transfer to the signal frequency was found to be the highest when the rate α for the chaotic motion is close to the signal frequency ($\alpha = 0.0077$, $\alpha = 0.0667$, $\alpha = 0.1772$ for Fig. 7.4a–c, respectively).

As illustrated earlier, assume that $A_a = 0$, and that for a set of values A_0, ω_0, β and D the system has low SNR. We could improve the SNR by increasing D. However, it is more effective to increase the SNR by keeping D unchanged and adding an excitation $A_a \sin(\omega_a t)$ such that ω_a is equal or close to the frequency of $S_M(\omega)$'s peak and A_a is so chosen as to bring about a characteristic rate comparable to the signal frequency. In Fig. 7.4d, all parameters and the normalized spectrum $g(\omega)$ are the same as for Fig. 7.4a, except that the system is subjected to an additive sinusoidal excitation with amplitude $A_a = 0.23$ and frequency $\omega_a = 1.1$. This approach to increasing SNR is seen to be effective by comparing Fig. 7.4d with b (in Fig. 7.4d, $\alpha = 0.0706$ close to ω_0).

7.3.2.2 Detecting Weak Signal with Arbitrary Frequency

From Eq. (7.18), we can get that $S_M(\omega)$ achieves the maximum when $\omega \approx 0.76$. Once the frequency ω_a of additive harmonic excitation is equal or close to the frequency of $S_M(\omega)$'s peak, the SNR improvement is obvious. That is to say, the frequency of the detected characteristic signal only satisfies $\omega \leq \omega_{max} \approx 3.0$ and is very low. Now the problem is how to detect weak signal with arbitrary frequency by the method discussed above?

Combining Eqs. (7.3) and (7.16) obtains

$$\ddot{x} = -\beta\dot{x} + x - x^3 + G(t) \tag{7.23}$$

where $G(t)$ is expressed by Eq. (7.21), In general, let the added harmonic $\omega_a = 1.0$ and detected signal $\omega_0 < \omega_a$. By assuming that $t = \omega_1\tau$, Eq. (7.23) becomes

$$\frac{1}{\omega_1^2}\frac{\mathrm{d}x^2}{\mathrm{d}\tau^2} = -\frac{\beta\mathrm{d}x}{\mathrm{d}\tau} + x - x^3 + G(\tau) \tag{7.24}$$

Let $x_1 = x$, $x_2 = \frac{1}{\omega_1}\frac{\mathrm{d}x}{\mathrm{d}\tau}$, rewrite Eq. (7.24) to be state equation

$$\begin{aligned} \frac{\mathrm{d}x_1}{\mathrm{d}\tau} &= \omega_1 x_2 \\ \frac{\mathrm{d}x_2}{\mathrm{d}\tau} &= \omega_1\left(-\beta x_2 + x_1 - x_1^3 + G(\tau)\right) \end{aligned} \tag{7.25}$$

Thus, Eq. (7.25) can be applied to enhancement detection of weak characteristic signal with arbitrary frequency. It is important to emphasize that the normalized scale transform of SR described in Eq. (7.12) is used, instead of Eq. (7.25).

Now, as a typical example, assume that we want to detect a characteristic signal with amplitude $A_0 = 0.3$ and frequency $\omega_0 = 0.069 \times 2\pi \times 1{,}000$ (i.e., 69 Hz). In this case, $\omega_1 = 2\pi \times 1{,}000$, $\omega_a = 1.0 \times 2\pi \times 1{,}000$. Substituting these parameters into Eq. (7.25) obtains the solution as shown in Fig. 7.5. It can be seen that the SNR improvement is obvious from Fig. 7.6b.

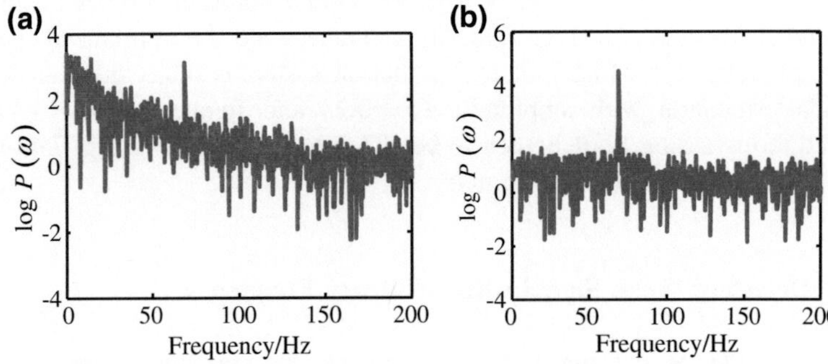

Fig. 7.5 Averaged power spectra of output for stochastically excited system: **a** increasing noise intensity $D = 0.1$ and $A_a = 0$ ($\alpha = 70.9466$, lower than $\omega_0 = 433.5398$ (i.e., 69 HZ)). **b** The same noise intensity D as in **a**, and $A_a = 0.23$, here $\alpha = 429.5146$ (i.e., 68.3594 Hz). Noise correlation time $\tau = 0.2$

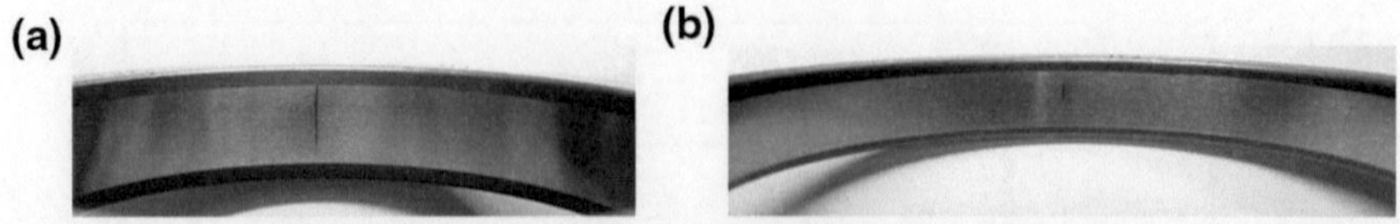

Fig. 7.6 Line defects seeded on the outer race of the test bearings. **a** MF (0.3 × 16 mm^2). **b** SF (0.11 × 6 mm^2)

7.4 Validation Using Experimental Data

The experimental data was from the Centre for Diagnostic Engineering at University of Huddersfield. The test rig consists of a three-phase electrical induction motor and a dynamic brake. The motor drives the brake by means of three shafts, which are joined by pairs of matched flexible couplings. The shafts are supported by two bearing housings, each containing one roller and one captive ball bearing. The bearing used in the experiments was a N406 roller bearing. The tested bearing was fitted in the bearing housing on the driven side.

In the experiments, the load applied to the test rig was 42.0 Nm and the rotational speed was 1,456 r/min (24.3 Hz). When the load was decreased, the rotational speed increased slightly. The three vibration signals were collected by accelerometers which were fixed on the cage near the tested bearings. Two different sizes of line defect were seeded on the outer race of bearings: a medium defect (approximately 0.3 × 16 mm^2), shown in Fig. 7.6a; and a small defect (approximately 0.11 × 6 mm^2), shown in Fig. 7.6b. Based on the geometric sizes and the rotational speed, the characteristic defect frequency (CDF) of the outer race was calculated to be 83.4 Hz. To get the finger print recordings, a defect free test was also conducted. All of the tests were repeated once during the experiments.

In the vibration data acquisition, the sampling rate was 64,938 Hz and the length of data was 810,439. For the convenience of analysis, the vertical radial acceleration signal was used to validate the SR-based algorithm for early detection of incipient fault. In the analysis, we extracted the data points by two times sampling interval. So the sampling rate was $f_s = 32{,}469$ Hz. The length of data was selected to be $2^{17} = 131{,}072$.

7.4.1 Envelop Analysis

The amplitude envelope of the raw vibration signals is computed using an algorithm based on the Hilbert transform, *H*, which is defined by

$$H[s(t)] = \frac{1}{\pi} \int_{-\infty}^{\infty} s(\tau) \frac{1}{t - \tau} \mathrm{d}\tau \qquad (7.26)$$

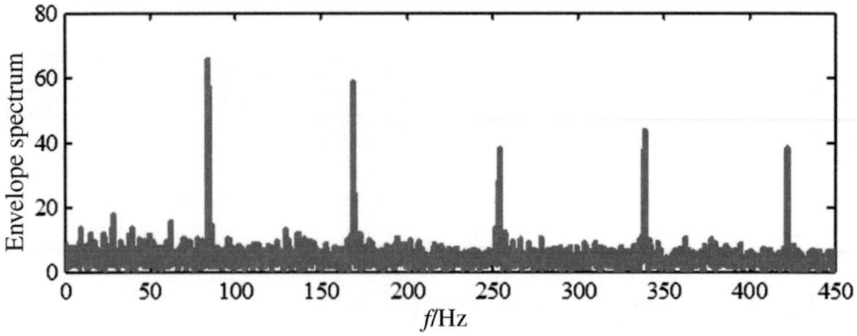

Fig. 7.7 A segment of FFT spectrum of envelope signal with medium defect

where $s(t)$ is a raw acceleration signal. Then the analytical signal, $\hat{s}(t)$, can be formed by

$$\hat{s}(t) = s(t) + jH[s(t)] \tag{7.27}$$

Thus the envelope of the raw acceleration signal can be computed by

$$En(t) = \sqrt{s^2(t) + H^2[s(t)]} \tag{7.28}$$

Finally, the signal computed by Eq. (7.28) is processed using the FFT to obtain the amplitude spectrum of the acceleration signal envelope. For the defect with certain degrees of severity, we can find that the component and its multiple components at the fault characteristic frequency are very obvious. For example, a classical result with medium defect of outer under middle load is shown in Fig. 7.7.

For the very small defect on the outer race, a classical results under the same load as that in medium defect are shown in Fig. 7.8. In this example, the selected pass band of filter is the same as that in medium defect. From Fig. 7.7, we can find that the signature frequency component and its multiple components are invisible and may be buried by the other unrelated components and random noise. That is to say, under the situation of incipient fault, the only envelope spectrum analysis cannot detect the small defect effectively and consistently.

7.4.2 SR Output of Driven by Envelope Signal

According to Eqs. (7.14) and (7.15) and the signature frequency component of outer defect, the benchmark frequency was selected to be f = 60 Hz for normalized scale transform of SR model. Some parameters are as follows. Added noise intensity $D = 0.0005$, $a = f/0.1 = 600$, $b = f/0.1 = 600$, model calculating time step $h = 1/f_s = 3.0799 \times 10^{-5}$, RMS of added noise $\sigma = \sqrt{2 \cdot D \cdot f_s} = 139.5752$.

The envelope signal of Fig. 7.7 plus added noise signal is considered to be the input signal and drives the model in Eq. (7.14). The corresponding results are shown in Fig. 7.9. Because the signature components in envelope spectrum are already obvious, the advantage of SR analysis cannot be exposed.

The envelope signal of Fig. 7.8 is considered to be the input signal and drives the model Eq. (7.14). The corresponding results are shown in Fig. 7.10. the signature

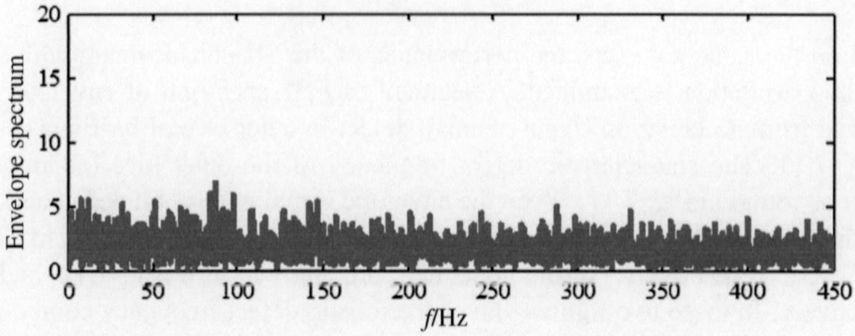

Fig. 7.8 A segment of FFT spectrum of envelope signal with small defect

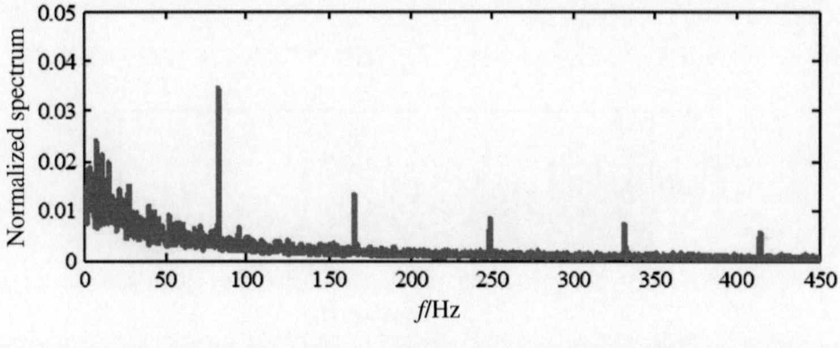

Fig. 7.9 The output results of SR model for vibration data of medium defect of outer

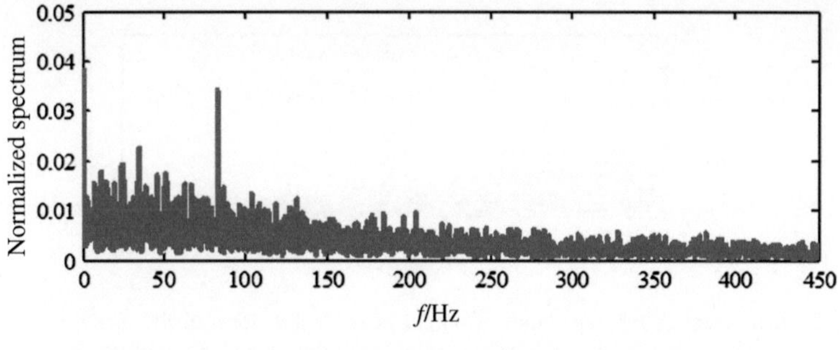

Fig. 7.10 The output results of SR model for vibration data of small defect of outer

component is very obvious in Fig. 7.10, in contrast, it is hard to be seen in the normal envelop spectrum in Fig. 7.8. This then clearly shows that SR is effective to enhance small signal component for incipient fault detection.

7.4.3 Weak Characteristic Signal Detection by SR of Adding a Harmonic Excitation

Based on the same data sets, the performance of the SR enhancement with adding harmonic excitation is examined. A segment of FFT spectrum of envelope signal analyzed from acceleration signal of small defect in outer race of bearings is shown in Fig. 7.11. The characteristic defect frequency of the outer race (about 83 Hz) cannot be found in Fig. 7.11. When the envelope signal with small defect and added harmonic signal drive the normalized scale transform of SR model in [12] or Eq. (7.25), respectively, results obtained are shown in Figs. 7.12 and 7.13 respectively. In these two figures, the characteristic defect frequency component of the outer race is relatively obvious (about 84 Hz), which shows that it is possible to use the approach of adding harmonic content for improving the performance of the SR based detection

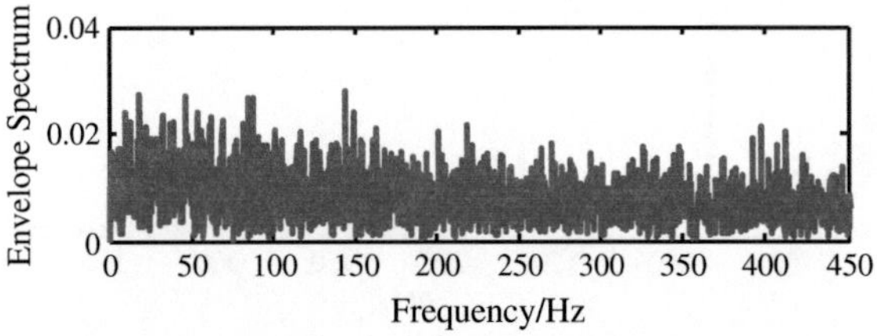

Fig. 7.11 A segment of FFT spectrum of envelope signal with small defect

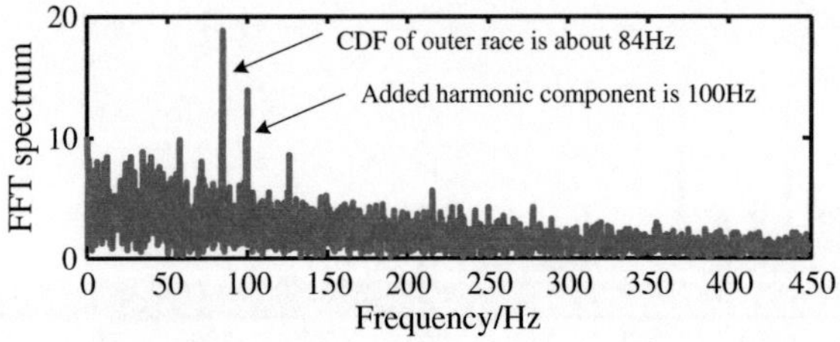

Fig. 7.12 A segment of FFT spectrum of output of scale transform of SR model in (7.12) was driven by envelope signal with small defect and added harmonic signal with $\omega_a = 2\pi \times 100$

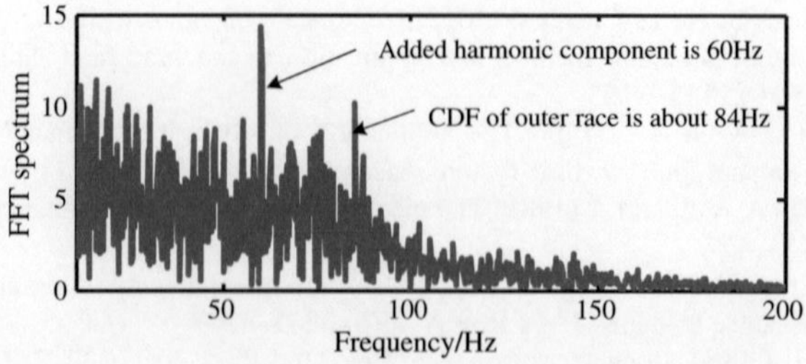

Fig. 7.13 A segment of FFT spectrum of output of Eq. (7.25) excited by envelope signal with small defect and added harmonic signal with $A_a = 0.07$ and $\omega_a = 2\pi \times 60$

7.5 Conclusion

According to the essence of classical stochastic resonance theory, the prominent role of the stochastic resonance phenomenon is that it can be used to boost weak signals embedded in a noisy environment. But for classical stochastic resonance, only weak signal with very low frequency can be detected in heavy noise. In this chapter, based on our recent studies [14–17], the signal enhancement via stochastic resonance is explained according to two main approaches. One is the scale transform of classical SR model for detecting relative higher frequency signal, thus the enhanced detection of periodic weak signal can be obtained via averaged approach. Another work is the investigation of the alternative mechanism for enhancing SNR, wherein the noise intensity is left unchanged but a harmonic excitation is added instead. This mechanism is shown more effective, allowing a better SNR to be obtained than the mechanism that of increasing the noise intensity. In addition, corresponding numerical simulations and case studies show that the proposed method for enhancing SNR is of more effectiveness.

Acknowledgments The authors would like to acknowledge the support of National Natural Science Foundation of China (Grant Nos. 51075391, 51105366, 51205401 and 51375484) and the Specialized Research Fund for the Doctoral Program of Higher Education of China under Grant No. 20114307110017.

References

1. Donald LB (1992) Chaotic oscillators and CMFFNS for signal detection in noise environment. IEEE Int Jt Conf Neural Netw 2:881–888
2. Hu NQ, Wen XS (2003) The application of duffing oscillator in characteristic signal detection of early fault. J Sound Vib 268(5):917–931
3. Qu LS, Lin J (1999) A difference resonator for detecting weak signals. J Measur 26(1):69–77
4. He ZJ et al (1996) Wavelet transform in tandem with autoregressive technique for monitoring and diagnosis of machinery. Chin J Mech Eng (Engl Ed) 9(4):311–317

5. Zhang XF, Hu NQ, Hu L, Cheng Z (2013) Multi-scale bistable stochastic resonance array: a novel weak signal detection method and application in machine fault diagnosis. Sci China Technol Sci 56(9):2115–2123
6. Hu NQ, Min C, Wen XS (2003) The application of stochastic resonance theory for early detecting rub-impact fault of rotor system. Mech Syst Signal Process 17(4):883–895
7. Benzi R, Sutera A, Vulpiani A (1981) The mechanism of stochastic resonance. J Phys A: Math Gen 14(11):453–457
8. De-chun G, Gang H, Xiao-dong W et al (1992) Experimental study of signal-to-noise ratio of stochastic resonance systems. Phys Rev A 46(6):3243–3249
9. Gammaitoni L (1998) Stochastic resonance. Rev Mod Phys 70(1):223–287
10. Franaszek M, Simiu E (1996) Stochastic resonance: a chaotic dynamics approach. Phys Rev E 54(2):1298–1304
11. McNamara B, Wiesenfeld K (1989) Theory of stochastic resonance. Phys Rev A 39 (9):4854–4869
12. Guckenheimer J, Holmes P (1990) Nonlinear oscillations, dynamical systems, and bifurcations of vector fields. Springer, New York
13. Wiggins S (1992) Chaotic transport in dynamical systems. Springer, New York
14. Zhang XF, Hu NQ, Cheng Z et al (2012) Enhanced detection of rolling element bearing fault based on stochastic resonance. Chin J Mech Eng 25(6):1287–1297
15. Zhang XF, Hu NQ, Hu L et al (2013) Stochastic resonance in multi-scale bistable array. Phys Lett A 377:981–984
16. Zhang XF, Hu NQ, Hu L et al (2012) Enhanced detection of bearing faults based on signal cepstrum pre-whitening and stochastic resonance (in Chinese). J Mech Eng 48(23):83–89
17. Hu N, Hu L, Zhang X, Gu F, Ball A (2012) Enhancement detection of characteristic signal using stochastic resonance by adding a harmonic excitation. In: The 25th international congress on condition monitoring and diagnostic engineering management (COMADEM2012), 17th–20th June 2012. Electronic application, Huddersfield, UK pp 34–43

Part III
Component Degradation and Design

Chapter 8
Active Thermography in Through-Life Engineering

Sri Addepalli and Lawrence Tinsley

Abstract With the increased use of high-value components in aerospace industries, there is huge emphasis on reliability as their failure in service could lead to catastrophic failure of the system. Thus, there continues to be increasing dependency of such high-value components to undergo critical maintenance routines in order to reduce the probability of failure. The prediction of Remaining Useful Life (RUL) is a critical factor in estimating the service cost. It has direct impact upon product and service pricing. With increasing maintenance costs, manufacturers have adopted a range of techniques such as non-destructive testing (NDT) to help assess the serviceability of these high-value components, where a component is inspected for quality without causing damage to the part. This allows for inspection of entire batches, instead of sample subsets. In service-focussed business models such as the aerospace sector, high-value components are required to perform for an optimum life cycle, balanced between maximising operation hours and a confidence in its safety. NDT has become a key process in determining the current state of degradation during the component's use, allowing estimation of remaining life and determination of repairs required. Detection of defects and anomalies is still a major challenge in the development of NDT practices in advanced manufacturing processes even more so with the introduction of new materials for higher reliability and performance. This research looks at expanding the NDT practices in maintenance by identifying the emerging challenges and suggesting areas of research for a robust development of NDT techniques and improved component degradation analysis. Active thermographic NDT is a recent technique that has become more widely included in NDT processes over recent years. However, due to its shallow depth and lower resolution limitations, it has not been exploited to its full potential in maintenance routines. Current challenges involve the further development of thermography as a quantitative technique as opposed to its traditionally qualitative implementations. This chapter focusses on the detection of damages caused due to component degradation using the pulsed active thermography technique. It also presents a novel approach on

S. Addepalli (✉) · L. Tinsley
EPSRC Centre for Through-Life Engineering Services, Cranfield University, Cranfield, Bedfordshire MK43 0AL, UK
e-mail: p.n.addepalli@cranfield.ac.uk

L. Redding and R. Roy (eds.), *Through-life Engineering Services*,
Decision Engineering, DOI 10.1007/978-3-319-12111-6_8

carrying out maintenance using an automated inspection system together with damage characterization of near and sub-surface damages on high-value components.

8.1 Introduction

High value components are a core focus in the aerospace industry. Over the last two decades, constant efforts have been put into continuous refinement of the design and manufacture of these components to improve the product performance and reliability, as failure of such components is a serious concern. Various design approaches have been adopted to manufacture components based on performance, product life, re-engineering and reverse engineering. This has led to the development of new and innovative products that demonstrate improved performance and reliability. With the introduction of these advanced, innovative products there is an ever increasing demand to understand the life of such components and the type of degradations such components are susceptible to. The last decade has seen the shift of product service responsibilities of end users back to the technology providers/ suppliers towards the maintenance of such high-value components leading to the establishment of Product Service Systems (PSS) [1]. This has forced manufacturers to adapt to innovative solutions to support maintenance through the life of the component. This simply means that the technology providers are obliged to incorporate maintenance costs as a part of their sales strategy.

With the constant development in advanced product design, it has become important to understand the various processes the component undergoes right from the design stage, to the manufacture, and through to the maintenance of such complex engineered products. There is now an ever increasing demand to monitor the quality of such complex innovative products which has led to the introduction of traditional inspection and monitoring techniques such as Non-Destructive Testing and Evaluation (NDT&E). The 20th century saw the rapid development of various analysis techniques and the associated instrumentation capable of determining the condition of the material or component or system without creating further damage to the inspected component. This science of monitoring and characterisation of damage to provide the structural health of the component non-destructively has been defined as Non-Destructive Testing or NDT [2]. As NDT does not create a permanent change in the characteristics and structure of the component, this method of inspection monitoring has been highly valued as it offers an effective defect and damage characterisation thereby preventing impending failure of component during service. These inspections thus play a crucial role in certifying the quality of the component right from as-produced stage to the in-service stage keeping control of the maintenance costs associated with it. With high-value complex technology products, the technology providers are under constant pressure to have an understanding of the various factors that affect the quality of their products. Now with the introduction of PSS, it is of utmost importance to provide the best possible monitoring and assessment, in addition to keeping the maintenance costs as low as possible.

NDT was developed primarily for weld inspections. Over decades, it found its way into most engineering industries including high capital industries where product quality and reliability are of utmost importance. The major NDT techniques currently in practice are

- Visual
- Dye penetrant testing
- Magnetic particle inspection
- X-Radiography and 3D Computed Tomography
- Ultrasonic testing
- Eddy current testing
- Acoustic emission
- Electromagnetic testing
- Vibration testing/analysis
- Infrared imaging

The performance of the NDT technique is dependent on the method used. For instance, surface cracks on metallic parts can be easily detected using dye penetrant testing; however, it will not detect any sub-surface defects or damages. In the world of NDT, no single technique can effectively characterize all defects found on the product and can only be complimentary when used together with alternate NDT methods, where possible [2].

Infrared Thermography has been increasingly used in defect and damage characterization over recent years as it is a non-contact, non-intrusive, fast, robust, cost effective NDT technique capable of rapidly inspecting large surfaces [3, 4]. The traditional NDT techniques were developed primarily for metal based structures and can be ineffective to characterize sub-surface features in advanced materials such as composites which can be non-conducting in nature. As thermography deals with surface temperature mapping, issues associated with signal attenuation due to fibre orientation in composites is a challenge for techniques such as Ultrasonic testing do not apply, and hence it is a perfect technique to locate near surface and sub-surface damages in a variety of materials [4, 5].

Some of the traditional NDT techniques use intrusive methods such as X-radiography, Laser shearography, or need surface contact as in the case of Ultrasonic testing. These techniques pose a challenge when applied to advanced, innovative, complex high-value components. Together with complexity associated with these advanced components, comes the maintenance cost which is a cost driver as in the case of the aerospace industry. Hence, the industry is in the constant process of improving their maintenance processes by introducing new, advanced methods to improve their component serviceability.

The current research focusses on the use of pulsed active thermography exploring its capabilities and opportunities to automate the inspection process.

8.2 Motivation

There have been significant developments in the manufacture of aerospace parts over the last decade. The introduction of innovative complex parts has made manufacturers review their design and maintenance processes. For instance, the manufacturers are constantly reducing the number of moving parts in the aero-engine. Engine components have had a complete redesign since the 1940s, indicating that the maintenance activities associated with these next generation parts need to be addressed thus indicating that the Maintenance, Repair and Overhaul (MRO) facilities also need to upgrade in order to keep pace with component developments.

A recent report suggested that the global fleet size will continue to grow over the next few years almost doubling the current fleet size even with the rise in oil prices [6]. It should also be noted that the new aircrafts are now redesigned and engineered in such a way that the maintenance requirements of these newer versions are far less than their preceding counterparts indicating that the maintenance calls for these advanced next generation aircrafts will reduce in number. Even though there will be a constant rise in the number of aircrafts being made, it is not necessary that the MRO sector will continue to grow equally. Thus it is of utmost importance that the MRO facilities adapt to newer technologies to support the related maintenance activities in a fast and efficient yet economical way with the highest level of confidence in carrying out their inspection routines.

A range of NDT techniques are currently in place within the aircraft industry. The most common techniques within engine maintenance are Visual, Dye penetrant testing, Magnetic particle inspection, X-Radiography and Ultrasonic testing. Infrared imaging or Thermography has now found its way into the aviation industry as an advanced NDT technique capable of providing fast inspection results [5, 7]. Maintenance of high value components is a challenge in itself and associated with it is the cost of assessing the condition of the component. Thermography, due to its ability to inspect the component without involving cleaning and preparation process, has become the highlight, and industries are trying to establish this as a quantitative technique as opposed to its traditionally qualitative deployment. The cost of the equipment, the time taken to inspect and the ability to inspect large areas have made thermography more lucrative with the high capital industry. The current research focuses mainly on this aspect and is currently developing an automated robotic thermographic inspection routine together with automated sentencing that will help the operator make quick meaningful decisions as to whether the components need repair or is beyond repair.

8.3 Thermography

Over the last decade, thermography has become a household technique both in the world of NDT and condition monitoring. The origins and science behind thermography has been cited elsewhere a number of times [8–12]. Thermography deals with the acquisition of a temperature map of a system or a component providing meaningful information of the component under study. The physical and mechanical properties control the heat flow in the material which in turn reflects as a surface temperature pattern. There are two major types of thermographic inspection techniques. They are:

(i) Passive Thermography
(ii) Active Thermography

8.3.1 Passive Thermography

In this method, an infrared radiometer, commonly referred to as an infrared camera captures the temperature pattern of the component in its natural working environment. The technique is widely used in engineering industry for condition monitoring purposes. For instance, when a bearing of the motor starts wearing, it is associated with a rise in temperature due to friction. The infrared camera is capable of recording this rise in temperature indicating the imminent failure of the bearing thereby helping the operator plan for an interim maintenance schedule as a part of their preventive maintenance strategy.

8.3.2 Active Thermography

In this method, the infrared camera records the temperature cooling profile of the component when external heat is applied. This excitation can be optical (e.g. flash lamp) or mechanical (e.g. shaker/sonic horn). It is an understanding that the heat flow through a homogeneous material will be constant. As soon as a discontinuity in the path of the heat flow, the diffusion characteristics change which form as a hot or a cold spot on the surface of the inspected component thereby indicating the presence or near-surface and sub-surface damage [10, 11].

Further we shall discuss two main types of active thermography techniques, the pulsed and lock-in thermography techniques.

8.3.3 Pulsed Thermography

Pulsed thermography has now become an established technique both in academic and industrial research over the last decade. It is known for its fast, robust, non-invasive, low-cost characteristics which permit the technique to be adapted to carry out large area inspections. In this technique, the surface temperature of the component is increased with the use of flash lamps. A high-speed infrared camera records the

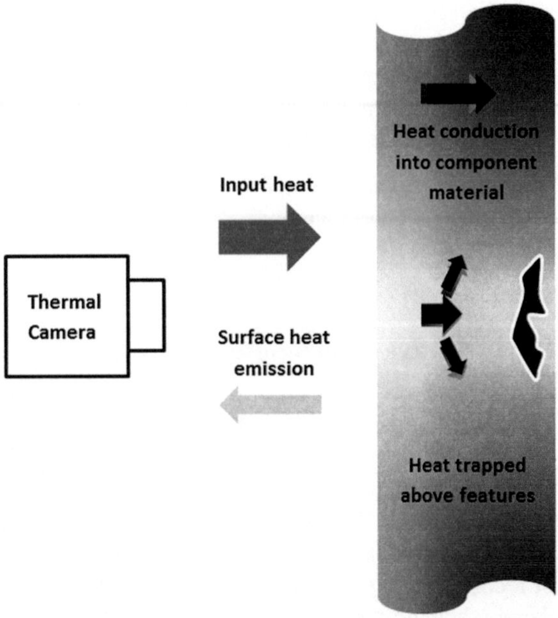

Fig. 8.1 Illustration of the working principle of pulsed active thermography

cooling characteristics of the component. It has been established that discontinuities occurring in the material result in the form of a hot or cold spot which is observed as a change in surface temperature characteristics (Fig. 8.1). In other words, for a homogeneous material, the transient heat characteristics are uninterrupted and as soon as a discontinuity appears in the material in the form of a damage or a defect, the transient heat characteristics change which is captured by the infrared camera.

Thus for a homogenous, semi-infinite sample, the heat equation becomes [13]

$$T_{surf}(t) - T_{surf}(0) = \frac{Q}{\kappa\rho c\sqrt{\pi t}} \tag{8.1}$$

where T_{surf}—Surface temperature, t—time, Q—input energy per unit area, κ—thermal conductivity, ρ—material density and c—specific heat capacity.

A natural logarithmic plot of the above equation reveals that the profile is linear with a slope of –0.5. When a discontinuity occurs, the profile characteristics deviate from the linearity indicating the presence of sub-surface features (Fig. 8.2).

8.3.4 Lock-in Thermography

Lock-in thermography has found applications in research and development organisations and is currently being explored in detail to look at damage characterisation. In this technique, the component to be inspected is subjected to periodic

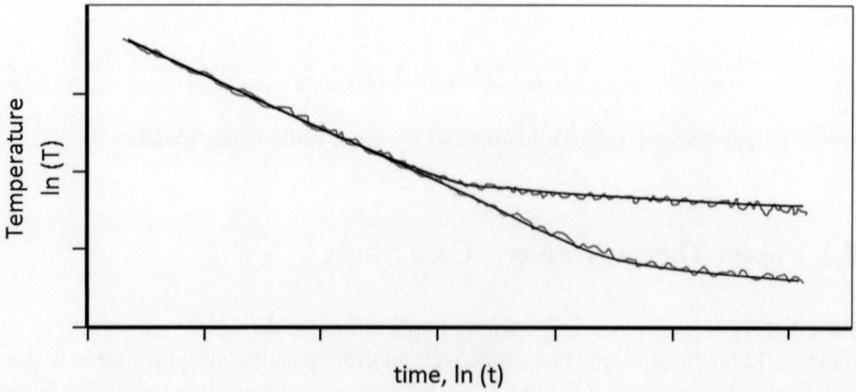

Fig. 8.2 Logarithmic plot of time versus temperature showing cooling characteristics of parent material (*bottom curve*) and damage (*top curve*) [13]

excitation using optical (e.g. halogen lamps) or mechanical (e.g. sonic horn) sources. This excitation results in developing stress patterns which produce changes in transient heat characteristics of the component synchronous with the excitation signal [14, 15]. This change in transient heat characteristics is recorded by the infrared camera, which is synchronous with the excitation signal to produce

(a) Phase image—a time dependent thermal image
(b) Amplitude image—thermal image dependent on diffusion characteristics of the component

This process of acquiring the thermal stress patterns by a synchronous infrared camera is referred to as 'Lock-in' thermography. The following is the equation which calculates the thermoelastic stresses generated in the component during excitation, [15]

$$\Delta T = -\frac{T}{\rho C_\varepsilon}\frac{\partial\sigma}{\partial T}\varepsilon + \frac{Q}{\rho C_\varepsilon} \tag{8.2}$$

where T—absolute temperature, C_ε—specific heat at constant strain, ρ—density, σ—stress, ε—strain and Q—heat input.

Partial differentiation of the above equation and deriving for principle stresses, we arrive at the equation

$$\Delta T = -KT\Delta(\sigma_x + \sigma_y) \tag{8.3}$$

where the thermoelastic constant is,

$$K = \frac{\alpha}{\rho\, C_p}, \tag{8.4}$$

and,

$$\Delta\sigma_x, \quad \Delta\sigma_y \tag{8.5}$$

— principle stresses, α — diffusion and C_p — specific heat capacity.

8.4 Pulsed Thermography—Case Study

Pulsed thermography has had various applications in the aerospace industry, the most well known being the inspection of composite panelling used for aircraft skin. This method is most ideal for the inspection of such advanced materials due to the sub-surface nature of their defects, and the planar direction of delaminations occurring during their service life. Also, when a composite panel is damaged, it permits the passage of water into the material. As water is more thermally inert than the surrounding sound areas of the composite panel, it will show up in the thermal image as trapping heat for longer, or taking longer to warm up, depending on the situation, thereby highlighting the damaged area. In aerospace this has been exploited by observing the skin of aircraft as soon as it has landed from the upper atmosphere, with water ingress in the parts showing up as cool spots against warm surrounding sound areas.

The inspection of composite panels post-manufacture with pulsed thermography has also had some popularity, due to the ease of inspection compared to ultrasound, and the wide area instant application, it has been used to detect sub-surface defects right from the manufacture stage.

This section focuses on the presentation of the research being carried out as a use-case study using automated pulsed thermography.

8.5 Automation of Thermographic NDT

Thermographic inspection carries numerous advantages for its non-invasive inspection technique to enable it to be used as a key inspection method for various degradation inspections such as overheating of degraded electrical components, thermal insulation, or steam pipe flow. In other areas where thermography is in competition with other well established NDT technologies, its advantages can be offset by its limitations, where addition of the technique to high volume inspection process may limit its attractiveness to the industry looking to implement it, who may not be interested in adding to their training and man-hour requirements in what is already a busy inspection routine. In order to address these issues, automation of the thermographic inspection is key to projecting the process into further use.

Automation of the inspection falls into two categories: the data capture stage and the data processing and analysis stage.

8.5.1 Data Capture Stage

Automation of data capture can be performed through use of an automated conveyor, a multi-axis machine or other robotic assisted system. In the current work, automation of data capture has been performed with a robotic arm. Unlike a conveyor system, this allows complex geometry components to be collected and presented to the inspection system with a repeatable attitude, with the additional functionality to permit multiple angles of inspection to allow for a full inspection.

8.5.2 Data Processing and Analysis Stage

After data capture, the processing of data can be time-consuming, requiring a trained technician to perform the analysis. To implement thermography as an additional process in high volume inspections could generate tens to hundreds of thousands of additional datasets for analysis each year, which is counter-productive when offering thermography as a rapid inspection technique. Offering thermography as a low-cost solution is attractive to reduce overall inspection burden, but must be balanced by automated data processing to assist in rapid decision making if the technique is to be used in more than the traditional qualitative approach.

To illustrate how this can be achieved, recent work is presented below. Initially, the inspection data is captured through the robotic assisted system, and was passed to a set of algorithms written in Matlab 2012b with image processing toolbox [16]. An upper-quartile threshold is applied to the data. Thresholding is a simple technique to rapidly identify hotspots, though selection of threshold requires a priori knowledge of a sound reference. Various techniques exist to offset this manual process, though their outputs still require trained analysis and were not the focus of these developments. An upper-quartile threshold of sample area was applied to offer an arbitrary data-simplification technique (Fig. 8.3).

The resulting image simplifies a monochrome gradient thermogram into a binary bitmap. This binary image encompasses structural features as well as damages, which need to be removed. The data is merged with a structure filter mask made from a sound reference of the same component, so that the remaining features are that of damage regions and do not contain spatial structural features. The process so

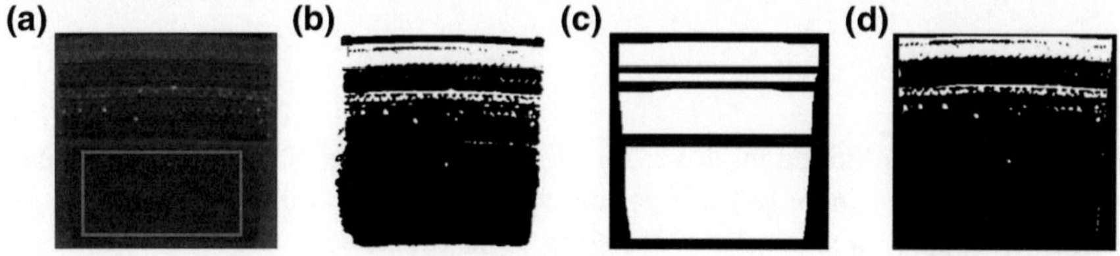

Fig. 8.3 Illustrates the data processing stage of threshold selection area (**a**), threshold result (**b**), filtermask (**c**), and filtered threshold result (**d**)

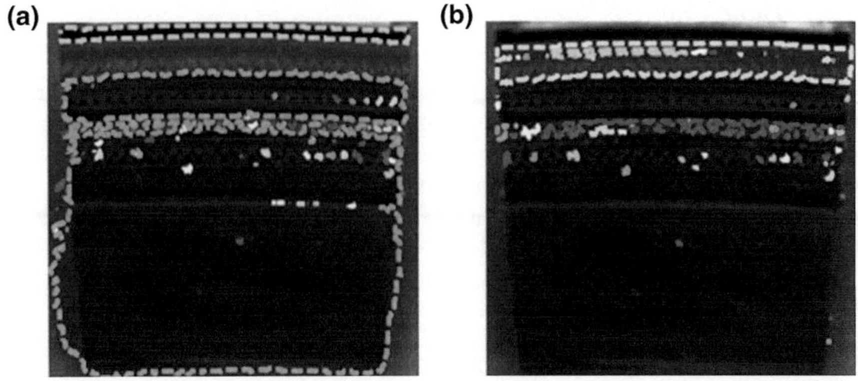

Fig. 8.4 Resulting damage detection **a** unfiltered, and **b** filtered displaying reduction in feature regions

far can be seen in Fig. 8.4. The boundaries of these remaining regions are then traced with a boundary tracing algorithm to provide region boundaries and their locations as segmented features within the data [17]. Traditionally, objects are sized with a calibrated scale stretched over the length and width of the feature—but to an algorithm, largest distance can be a simple task, minor distance may not be. The data enclosed by the boundary regions is segmented from the data and passed to a 2D Gaussian fitting algorithm [18]. This allows feature sizing in major and minor axis, with eccentricity providing a proxy measurement for shape. This allows for multi-axis sizing of features without manual placement of a calibrated scale over the features, effectively automating the detection and sizing of damage features. The reduction in feature detection can be seen in Fig. 8.4. In this instance, one of the filter mask regions were also traced as a damage feature collided with its boundary.

Future developments in this direction of automation could replace the thresholding process with adaptive thresholding or incorporate temperature decay gradients for a gradient that deviates from material cooling rate, or subtraction of reference benchmark data taken under identical conditions from the test data to provide an angle-specific, component-specific objective threshold of cooling profile deviations. Automation of the structure filter mask is also a key step in furthering automation.

8.6 Summary

NDT factors as an important driver for growth and competiveness in the aerospace and MRO sectors in order to keep pace with growing volume of demand and narrow the gap between cautious decisions to maximise component life. The use of thermographic NDT in industry and its advantages through its rapid non-intrusive application over traditional inspection methods have raised it as an important tool for driving speedy improvement and cost reduction in non-destructive inspection portfolios. In illustration of this, work in the automation of thermographic data capture and automation of the data processing stage have been presented, which will feature as a main route in current developments of the technique. The thermographic inspection method and such developments in automation of the technique are going to be significant in the drive to project the deployment of the method further into industry, and assist maintenance organisations in keeping pace with the increasing demand over coming years.

References

1. Meier H, Völker O, Funke B (2011) Industrial product-service systems (IPS2). Int J Adv Manuf Technol 52(9–12):1175–1191
2. Cartz L (1995) Nondestructive testing. In: ASM international, Ohio, p 229
3. Dudzik S (2009) A simple method for defect area detection using active thermography. Opto-Electron Rev 17(4):338–344
4. Adams RD, Cawley P (1988) A review of defect types and nondestructive testing techniques for composites and bonded joints. NDT Int 21(4):208–222
5. Kapadia A (2013) Non destructive testing of composite materials, pp 1–48 (Best Practice Guide)
6. Borkowski J (2007) Aircraft maintenance repair and overhaul market study: Glasgow international airport. In: Airport regions conference (ARC)
7. Menner P et al (2010) Aerospace applications of lock in thermography with optical, ultrasonic and inductive excitation. In: Thompson DO, Chimenti DE (eds) Review of quantitative nondestructive evaluation, vol 29. American Institute of Physics, College Park
8. Ring EFJ (2000) The discovery of infrared radiation in 1800. Imaging Sci J 48(1):1–8
9. Bagavathiappan S et al (2013) Infrared thermography for condition monitoring—a review. Infrared Phys Technol 60:35–55
10. Maldague X et al (2001) Fundamentals of infrared and thermal testing. In: Maldague X, Moore P (eds) ASNT nondestructive testing handbook: infrared and thermal testing. ASNT, Columbus, pp 31–52
11. Maldague XPV et al (2001) Techniques of infrared thermography. In: Maldague XPV, Moore PO (eds) ASNT nondestructive testing handbook: infrared and thermal testing. ASNT, Columbus, pp 307–357
12. Maldague XPV (2001) Theory and practice of infrared thermography for nondestructive testing. In: Chang K (ed) Wiley series in microwave and optical engineering. Wiley
13. Shepard S (2007) Flash thermography of aerospace composites. In: IV Conferencia Panamericana de END, Buenos Aires
14. Meola C et al (2006) Non-destructive evaluation of aerospace materials with lock-in thermography. Eng Fail Anal 13(3):380–388
15. Breitenstein O, Warta W, Langenkamp M (2010) Lock-in thermography: basics and use of evaluating electronic devices and materials. In: Itoh K et al (eds) Springer series in advanced microelectronics, 2nd edn. Springer, Heidelberg
16. NN MATLAB version 7.14.0.739 and Image Processing Toolbox Release 2012b (2012) The MathWorks Inc, Natick
17. Gonzalez R, Woods R, Eddins SL (2004) Digital image processing using MATLAB. Pearson Prentice Hall, Englewood Cliffs
18. Nootz G (2012) Fit 2D gaussian function to data. MATLAB Central File Exchange (cited 22 April 2013). Available from: http://www.mathworks.co.uk/matlabcentral/fileexchange/37087-fit-2d-gaussian-function-to-data/content/mainD2GaussFitRot.m

Chapter 9
Maintenance, Repair and Overhaul in Through-Life Engineering Services

E. Uhlmann, R. Stark, M. Rethmeier, J. Baumgarten, M. Bilz, C. Geisert, B. Graf, A. Gumenyuk, H. Grosser, F. Heitmüller, M. Manthei and S. Reinkober

Abstract Maintenance, Repair and Overhaul (MRO) is acquiring increasing commercial and socio-economic significance. For products and goods with high investment costs and a long lifespan, especially in the sectors of energy and transportation, a considerable portion of commercial profits are generated by after-sales

E. Uhlmann (✉) · R. Stark · M. Rethmeier · J. Baumgarten · M. Bilz · C. Geisert · B. Graf · A. Gumenyuk · H. Grosser · F. Heitmüller · M. Manthei · S. Reinkober
Fraunhofer IPK, Pascalstr. 8-9, 10587 Berlin, Germany
e-mail: eckart.uhlmann@ipk.fraunhofer.de

R. Stark
e-mail: rainer.stark@ipk.fraunhofer.de

M. Rethmeier
e-mail: michael.rethmeier@ipk.fraunhofer.de

J. Baumgarten
e-mail: jeannette.baumgarten@ipk.fraunhofer.de

M. Bilz
e-mail: martin.bilz@ipk.fraunhofer.de

C. Geisert
e-mail: claudio.geisert@ipk.fraunhofer.de

B. Graf
e-mail: benjamin.graf@ipk.fraunhofer.de

A. Gumenyuk
e-mail: andrej.gumenyuk@ipk.fraunhofer.de

H. Grosser
e-mail: hendrik.grosser@ipk.fraunhofer.de

F. Heitmüller
e-mail: florian.heitmueller@iwf.tu-berlin.de

M. Manthei
e-mail: marcel.manthei@ipk.fraunhofer.de

S. Reinkober
e-mail: sascha.reinkober@ipk.fraunhofer.de

L. Redding and R. Roy (eds.), *Through-life Engineering Services*,
Decision Engineering, DOI 10.1007/978-3-319-12111-6_9

services. In the field of research and development, not enough attention has been paid so far to tasks and approaches involving MRO. The field thus has a limited scientific background, despite a high potential in the business sector for technological and scientific optimization. The challenges and chances of MRO for sustainable enterprises will be explained with reference to the Fraunhofer Innovation Cluster Maintenance, Repair and Overhaul in Energy and Transport. The developments and project results of the four fields of innovation »Cleaning«, »Repair and Overhaul«, »Condition Monitoring and Diagnosis«, as well as »MRO Planning and Digital Assistance« will be explained.

9.1 Introduction on MRO

An increasingly sustainable handling of resources demands a longer lifespan and a higher usage intensity of products. As a result, their profitability is increasing. Combined with the growing complexity of machines and systems on the part of users as well as operators, this leads to a rising demand for services from the field of maintenance, repair and overhaul (MRO). MRO is acquiring increasing commercial and socio-economic significance. For products and goods with high investment costs and a long lifespan, especially in the sectors of energy and transportation, a considerable portion of commercial profits is generated by after-sales services. During the useful life of a product, not only continuous and scheduled servicing and maintenance work but also unpredictable repairs are necessary. An overhaul can not only restore the product to its original condition, but also incorporate improvements that bring it up to date with the latest technical and economic developments. MRO processes thereby make a significant contribution to resource protection and energy efficiency, while at the same time providing economic benefits. In the field of research and development, not enough attention has been paid so far to tasks and approaches involving MRO. The field thus has a limited scientific background, despite a high potential in the business sector for technological and scientific optimization. This development is certainly true for Germany as a location of research and industry. Changes in the field of MRO stem from technical and economical optimization potentials, achieved by the rising demand for MRO services and the concentration on after-sales services by providers of MRO activities.

9.2 Fraunhofer Innovation Cluster MRO

MRO—maintenance, repair and overhaul has a growing economical relevance. In the field of energy and transportation in particular, corrosion of materials, vandalism or material wear can cause high costs. Especially products and commodities with high investment costs and long lifespans have a great technological and economical optimization potential which has barely been considered in scientific research. The Fraunhofer Innovation Cluster »Maintenance, Repair and Overhaul in Energy and Transport« addresses this problem. It develops technologies for the optimization of MRO processes in the following four fields of innovation: Condition Monitoring and Diagnosis, MRO Planning and Digital Assistance, Cleaning, as well as Repair and Overhaul Technologies [1].

Status assessment and diagnosis encompasses the continuous, system-integrated recording of the current status of a plant and its structural components. Checks for fissures and flaws in wheel sets and shafts, increasingly with the help of high-performance electronic systems, are particularly important from a safety standpoint. Innovative inspection procedures should be developed to detect the condition of the plant, identify the ideal servicing time and assess the remaining useful life of individual structural components or of the entire plant. For instance, condition monitoring provides useful information on the state of the plant which helps to decide on future condition-based service times. Thus, it contributes towards a better exploitation of the useful lifespan, extends service intervals, avoids operational interruptions due to unscheduled maintenance work, and enhances the overall availability of the plant.

The efficiency and planning of MRO processes can be significantly improved by means of digital assistance. There is a need to develop new processes, methods and tools for MRO applications, in order to exploit the potential of virtual technologies for MRO optimization in practice. For instance, solutions for fast access to information on important lifecycle phases and MRO activities, solutions to interface conflicts between various multi-player systems, or to enable remote-servicing with mobile tele-cooperation devices via narrow-band connection are addressed. In addition, solutions to depict paper drawings, structural components and complex products and plants in 3D models at little cost and efforts are within the scope. Further topics are cost reduction by means of systematic and requirement-oriented MRO planning and assistance as well as fast reactions to changes in product conditions. Furthermore, new and smart global MRO component control systems for spare parts could help to reduce machine stoppages and downtimes. Finally, there is potential to shorten training times and to speed up MRO process implementations without failure.

The cleaning of machinery and plants plays a central role in the MRO process. Both the cleaning procedures and the field of applications are wide-ranging. The need for cleaning for optical reasons, as in the removal of graffiti, is highly pronounced in the railway transport business. Preventive cleaning geared to the preservation of functioning or of effectiveness is also essential. Expensive stoppages and repairs of machinery and plants can be avoided or at least reduced. Cleaning as a repair or manufacturing procedure is required particularly in connection with the removal of functional layers. In all these fields, flexible and ecologically efficient procedures will be developed and established.

In contrast to new production processes, service and maintenance procedures are far less foreseeable or predictable. This is why the repair of long-term machinery investments is usually undertaken on-site and at a time that is essentially non-specified in advance, whereas new manufactures are generally undertaken in series production in a production hall at a pre-specified time. The valuation and development of new repair technologies requires therefore flexibility and highly adaptive procedures.

9.2.1 Condition Monitoring and Diagnosis

In the field of innovation »Condition Monitoring and Diagnosis«, solutions for status assessments and prognoses of the remaining lifespan are developed in order to determine the ideal time for maintenance. They are the basis for MRO planning and its digital assistance. The condition of a machine or a plant can be acquired either in intervals or continuously. As opposed to former testing performed by the service staff within the framework of inspection tasks, the continuous and automated assessment of the system status as well as of consequential maintenance activities is prospectively aimed at. This continuous condition monitoring mainly requires robust, autonomous sensor and monitoring systems that are easy to install. The biggest challenge concerning the conversion to a condition-based maintenance strategy is the automated and continuous assessment of the actual status of a plant and its functionally relevant wear parts. To acquire status data, sensor systems or signals which are already integrated in the plant can sometimes be used. However, it is often necessary to install additional sensor technology in combination with corresponding intelligent monitoring systems. Since the progress of wear is mainly dependent on the method of operation and with that, on the workload of the machine, it is advisable to also gather the corresponding load data during operation in order to correctly evaluate the system's condition. This data particularly influences the prognosis of the residual lifespan or the remaining wear margin. In the contemplated field of plants for energy generation, special requirements regarding the location of the plants must be taken into account. Concerning stationary systems, this applies in particular to the not so readily accessible energy generation plants off-site or even off-shore for wind power. In order to realise the required features for data recording and evaluation in a condition monitoring system, it is therefore expedient to flexibly distribute single functions to decentralised or centralised systems as well as to provide a suitable infrastructure. Although condition monitoring of systems is demanded by different sides, there are only vague ideas about the fact which data or signals of the system or process must be metrologically monitored in order to gain condition-based parameters. Similarly, adequate wear models, which would provide a relation between those parameters and the current wear condition, are nonexistence. Even more difficult is the prognosis of the remaining useful life or the probable time of default.

9.2.1.1 Data Mining and Visualization of Diagnostic Messages for Condition Monitoring

Capital goods like production systems or systems used in the sectors of public transportation or energy are highly stressed and often operated near full capacity utilization. Therefore, strong efforts are made to ensure their technical availability. A well-established practice is the implementation of condition monitoring to be informed at all times about the health state of the system under consideration. According to Bertalanffy, "a system may be defined as a set of elements standing in interrelation among themselves and with environment" [2]. The complex interaction of subsystems and components needed for a correct overall system behavior makes condition monitoring extremely difficult, since a deep understanding of underlying physical correlations is the basic prerequisite for adequate diagnosis.

Another way to access the health status of a system is to use already existing internal data like system messages, which are generated by the subsystems and are distributed via communication bus. System messages can be divided into so-called protocol and diagnosis messages. Protocol messages provide information about the system's operating condition, while diagnosis messages comprise information about the health status of the overall system or its subsystems (refer to Fig. 9.1). Both types of messages include timestamps for the start and end of the event, as well as an ID that is uniquely linked to the event description.

A challenge concerning using system messages is that they are designed for specific tasks like system control as well as documentation of events and do not necessarily correlate to the underlying physics. Another disadvantage is that they are listed in chronological order of occurrence without any additional graphical representation. Nevertheless, they are widely used by service technicians to get a first impression of the system's current health status and to analyze previous events that may be associated with the current event.

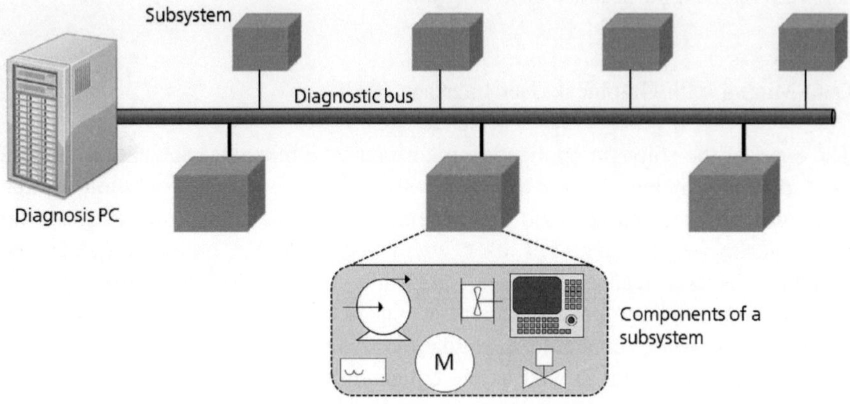

Fig. 9.1 Principle structure of an onboard diagnosis system

Approach for Data Mining

Since wear and tear and other degradation mechanisms are subtle processes, many serious failures are often announced by smaller intermittently occurring irregularities. These irregularities firstly disturb sub-functions without leading to an immediate breakdown of the overall system. Nevertheless, they may build specific patterns that become apparent within system messages. As mentioned before, weaknesses of today's possibilities of using logged system messages are:

- Message statement does not necessarily correlate to the underlying physics.
- Data is provided only as lists in tabular format.
- There is no graphical representation of the data.
- There is no support for an analysis by adequate software tools.

Therefore, manual data mining for condition monitoring purposes is very time-consuming and depends extremely on specific expert knowledge. To close this gap, the proposed approach deals with the development of a data mining software application to facilitate manual analysis and to (semi-)automatically detect emerging patterns, caused e.g. by an evolving malfunction of a component, at an early stage. This early detection can finally be used for the introduction of a condition-based maintenance strategy. The proposed solution consists of the following modules:

- On-board diagnosis system to collect system messages continuously (not part of the development),
- Database for the structured storage of system messages and additional information,
- Graphical user interface (GUI) to enable user interaction,
- Analysis toolbox including algorithms for data mining,
- Export functionality to share results for further analysis.

Figure 9.2 shows the interaction between the user and the proposed data mining software application in a use case diagram.

Data Mining with Graphical User Interface (GUI)

The basis of the software application is formed by a relational database scheme to store dynamic diagnosis data together with additional information about system configurations, as well as static descriptions of message codes (message list) and their causing subsystems. For user interaction, a JAVA-based graphical user interface (GUI) is realised to meet ergonomic analysis workflows following the interactive data mining process shown in Fig. 9.3.

On the one hand, sequential steps like the selection of interesting time intervals and system populations as well as the collection of messages to be used for data mining are presented separately to the user. On the other hand, it is possible to go back to the last step to realize necessary iterations. A central aspect is the

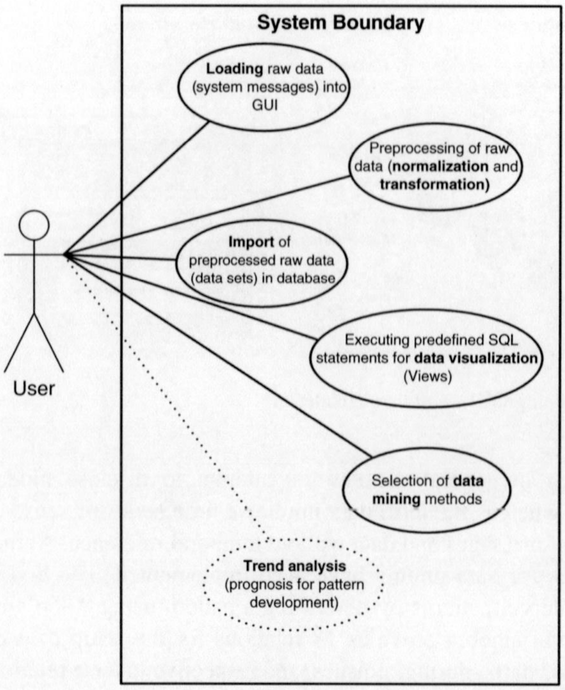

Fig. 9.2 Data mining GUI—use case diagram (at the current stage of work, trend analysis is not implemented)

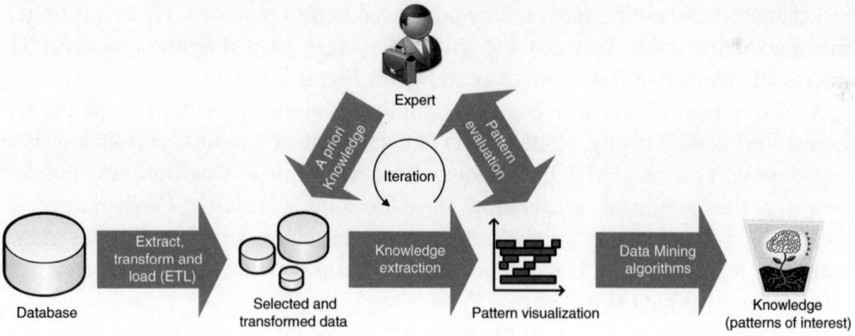

Fig. 9.3 Interactive data mining process

visualization of the dynamic condition data with different graphical diagrams (histogram and Gantt chart) to provide the user with compressed information. The overall software architecture is illustrated in Fig. 9.4. The module "generate report" includes the processing of data mining algorithms as well as graphical representations of the results.

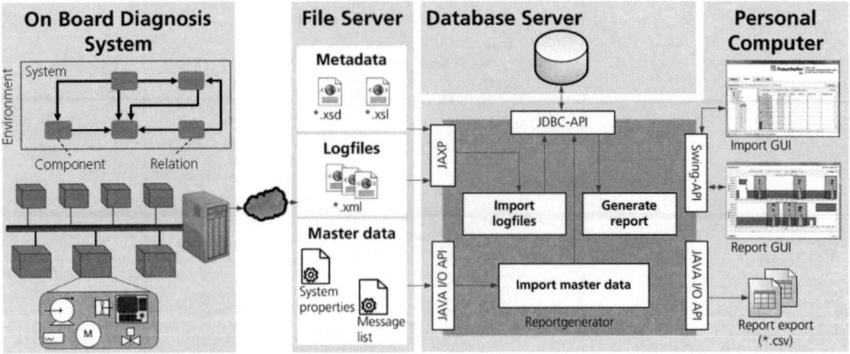

Fig. 9.4 Data mining GUI—software architecture

Data mining in general denotes the attempt to disclose hidden correlations between data, whereas the term data implies a time series of sensor measurements as well as event messages and data without temporal reference. At the current stage, two approaches for data mining have been implemented. The first approach finds similar time interval patterns by describing a pattern using Allen's interval algebra [3]. This interval algebra provides 13 relations for the setup of a pattern (before, meets, overlaps, starts, during, finishes, and respective inverse relations and equals). Any two intervals fulfill exactly one of the relations. Disadvantages of robustness due to describing the exact time boundaries within the relations were solved by using thresholds and fuzzy extensions.

The second approach finds frequent time point patterns by implementing the Apriori algorithm [4]. This algorithm is mostly applied to the domain of market basket analysis where frequent sets of purchased items are mined. Here, it is used to find association rules between the given messages of a diagnosis system. The process of interactive data mining is shown in Fig. 9.3.

As the human ability to recognise complex relationships within graphical representations is very strong, visualization of data and their temporal correlations is an important part of interactive data mining. This means that visualization is not only used for the graphical preparation of data mining results. Furthermore, the visualization is part of the overall interactive data mining process. Besides classic histogram representation, Gantt charts are used to illustrate temporal dependencies between interval data (Fig. 9.5).

The explained approach to support analysis tasks by graphical representation of condition data and the use of data mining algorithms to find reoccurring patterns in large data sets enables data analysts to recognise evolving technical problems and to launch appropriate counter-measures in terms of condition-based maintenance. Further research and development has to be done to fully automate the recognition of unknown patterns within large data sets and to forecast their future development with regard to trend analysis.

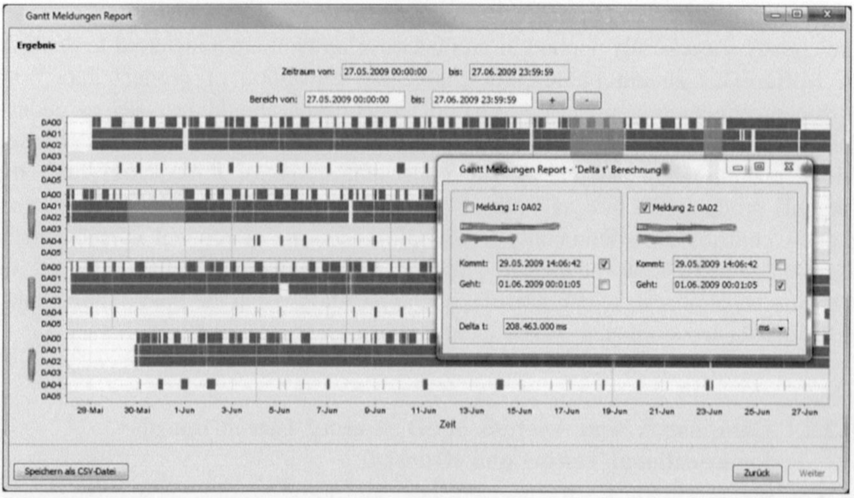

Fig. 9.5 Gantt chart of selected system messages

9.2.2 MRO Planning and Digital Assistance

The consistency of the digital process chain from product development to operation to MRO activities is an important foundation for efficient MRO planning and digital assistance. In the field of innovation »MRO Planning and Digital Assistance«, methods and digital tools for different fields of application are developed. Besides a higher level of automation in the MRO process, the focus is on adaptivity, for instance in the case of concurrent results, as well as on increased transparency, time saving, and error reduction. Equally, work scheduling should be supported by the availability of digital data and virtual models. The same is true of the engineering in the MRO process, the modular and spare part construction in particular. New solutions are needed to depict paper drawings, structural components and complex products in 3D models at little effort. Systematic and requirement-oriented MRO planning and assistance as well as faster reactions to changing product conditions are approaches towards lower costs. Moreover, in addition to a faster and flawless performance of MRO processes, machine downtimes and training times could be reduced with a global MRO inventory management. In the typical case of missing product information, these could be generated with technologies of reverse engineering. Virtual technologies and applications for mobile devices enable the provision and recording of all MRO-relevant information directly in the workshop or field. Furthermore, digital condition-based product and factory models facilitate an adaptive MRO process planning and operation. A consistent and user-friendly data management renders a continuous information management between MRO processes and systems engineering possible. Requirements for the supporting technologies are mobility, comprehensibility of processes, adaptivity, user-friendliness, and digital consistency. In order to improve efficiency, digital technologies which are already successfully applied in product development are transferred to the field of MRO. This should particularly enable the utilization of product data from upstream engineering processes. Overall, MRO data should be used to deduce insights for product and service improvements. The objective is a model-based MRO planning and assistance. The consistent usage of information from the (digital) product structure as well as the utilization of digital, geometric product models contribute predominantly to this aim. Systems of product lifecycle management (PLM), tools of factory planning and simulation, technologies of 3D measurements of reverse engineering, and applications from the field of augmented reality (AR) are employed as key technologies.

9.2.2.1 Computer Vision Analysis of 3D Scanned Circuit Boards for Functional Testing and Redesign

State-of-the-art reverse engineering services based on electrical or 2D optical measuring technologies for the reconstruction of schematics, layout plans and bill of materials have still a low degree of automation and are often destructive and error-prone. The aim of this research is to find single solutions for the mentioned challenges and to combine them in an automated process for the reconstruction of schematics, layout plans and bill of materials in order to enable repair and redesign tasks of long-living electronics, even when no spare parts or reference samples are available. Solutions comprise non-destructive measuring methods, PCB structure recognition, and the creation of error-proof netlists.

Printed Circuit Boards Specific 3D Digitization

The 3D digitization of printed circuit boards (PCB) is a non-trivial task which requires the use of advanced scanning technologies and a combination of different principles. To capture the entire geometry of a complex test PCB with dense parts placement, four layers and dimensions of approximately 300 mm × 180 mm, two scanning technologies were used to test digitisation quality. The PCB's surface was scanned with the structured light 3D scanner GOM ATOS III Triple scan. For the visualisation of inner structures, a computer tomography (CT) was provided by a service company with state-of-the-art equipment. The optical 3D scan was combined with photogrammetry and delivered information about the geometry, texture and labels of the visible surface, while the CT scan additionally showed inner layers. The resulting mesh by the optical 3D scan and the resulting voxel array by the CT scan were compared regarding the completeness of surface acquisition and accuracy (Fig. 9.6). Further, 3D measuring principles were taken into account: Shape from focus, gray-code principle, time-of-flight, white-light interferometry, and shape-from-shading.

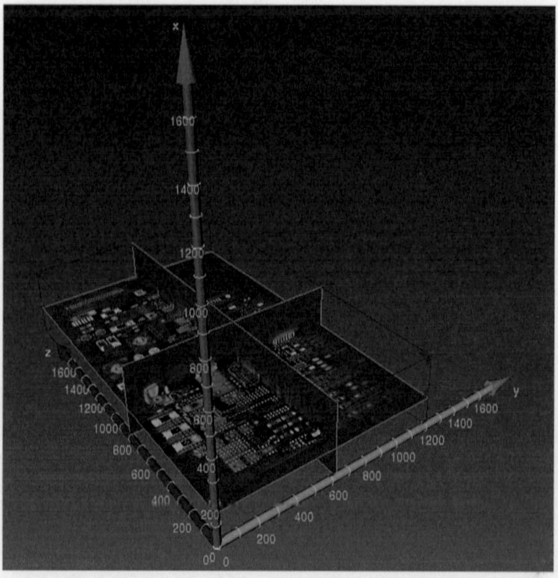

Fig. 9.6 Analysis of a PCB CT scan with software AVIZO® Standard

No scanning system was able to offer sufficient results concerning the depiction of needed elements such as parts, pins, and PCB tracks. A combination of approaches might be suitable and depends on whether a textured scan is required or not. 3D structured light and CT scanned data together allowed an overall analysis of the PCB and the recognition of individual components. Optical scans turned out to be highly vulnerable to light reflections and hidden structures. Thus, it was not possible to capture the whole PCB surface without having holes in the resulting mesh structure. A new 3D scanner based on the combined use of telecentric and pericentric optics as well as an illumination system for shadow-free and reflex-free optical acquisition is needed. To realise a sufficient resolution of 10 μm, a precise calibration and referencing system with constant environmental conditions regarding e.g. temperature or vibrations is necessary. CT scans enabled the visualization of inner layers and pins, whereas PCB tracks were not completely visible. Reasons can be found in an inclined voxel orientation and an insufficient resolution. There is no CT scanner on the market which can digitise circuit boards bigger than Eurocard format (160 mm × 100 mm) with a sufficient resolution. There is also need for a PCB specific measuring strategy. Parameters like current, voltage, measuring distance, prefiltering, amplification factor, integration time, projection number, and diverse software filter have to be chosen according to PCB characteristics.

Recognition of Printed Circuit Boards Parts

The optical recognition of single parts is still a challenge of the future. For the creation of schematics and layout plans, it is necessary to find out part type, identifying name and specification. Experiments showed that there is no optimal technical solution for tackling the problem of recognising PCB parts with computer vision approaches. The entire process is more a concatenation of algorithmic approaches which were decomposed into a post-processing pipeline.

With regard to optical detection, the application of basic image processing algorithms was tested to gain information about parts mounted on top of a PCB. Those included optical character recognition (OCR), Canny edge detection [5] and a simplistic watershed image segmentation [6]. Apart from the image segmentation, all algorithms were used directly from third party libraries like OpenCV.

The segmentation was tested on the voxel data set from the CT as well as the triangulated mesh from the structured light scanner. As mentioned above, the latter tended to produce holey scans when trying to capture small slits and indentations. That error in mesh completeness continued to complicate the segmentation process, as it gets significantly harder to separate single parts the more information is missing. In this case, the voxel data was, once again, the better option. An algorithm was used that encoded the entire voxel buffer as a graph and performed an adaptive broad search to separate mounted parts [7].

The market offers a variety of ready-to-use software for mechanic and electronic part databases with shape search functionality. CADENAS PARTsolutions was chosen as the host of the part database in this experiment. A sample database containing about 10 parts was set up as a basis to perform queries on.

Figure 9.7 shows a separated capacitor based on grey-scale segmentation, highlighting the complete capture of geometry. The segmentation was realised through the implementation of a filter pipeline consisting of a diffuser for elimination of image noise, a canny algorithm for edge detection and histogram

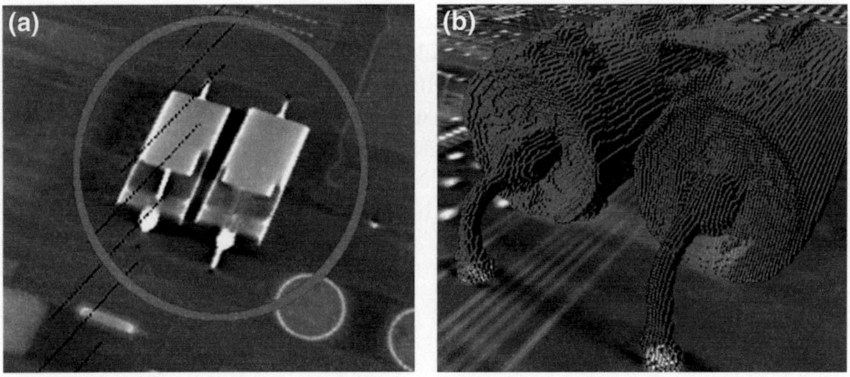

Fig. 9.7 **a** Planar view on a part; **b** Extracted segmented voxels

segmentation for the focus on relevant grey values. The resulting surface quality is not sufficient yet and has to be optimised in the future.

It turned out that the amount of similar looking parts in the electronic context is rather large, so that a shape-based database search of these segmented single parts is only sufficient to roughly categorise PCB parts. Improved segmentation principles would possibly be able to deliver highly-precise geometric part representations which are needed for better results of geometric database searches. However, based on the current segmentation results, parts identification does not work robustly so that alternatives were analysed. 2D image analysis is error-prone and highly-dependent on a proper camera angle. The character recognition method was able to retrieve a few parts in a test database that was set up beforehand. The biggest potential lies in the combination of optical character recognition and geometric detection algorithms. Future research aims at performing plausibility checks through redundant searches for geometry and labelling. Databases should include both IDs and part geometries.

Segmentation of Pins and Printed Circuit Board Tracks

In order to get an enhanced data set for later calculations, the segmentation of pins and PCB tracks was split into different tasks to separate the data acquisition and pre-processing from the actual pattern recognition. A major problem during the pre-processing step was the erroneous rotation angle of the PCB within the voxel buffer. This problem cannot be avoided as a PCB cannot be placed in a perfect planar position inside a CT scanner. The solution was the implementation of an algorithm which detected the PCB's edges and computed the deviation against the slices of the voxel array. Subsequently, the PCB scan was aligned inside the voxel buffer using the transformation matrix gained from the previous calculation. After alignment, greyscale colour correction and noise reduction were applied to maximize the contrast throughout the subsequent creation of a master view. This master view is a specific region of interest in the voxel buffer of which it can be assumed that it contains all pins and PCB tracks (respectively of a single layer), but no mounted parts. All voxel slices belonging to that region were inspected in a loop and merged together pixel-wise with the lighter pixel winning in comparison. The resulting 2D image was digitally scaled and interpolated showing pins at a very high contrast. By applying thresholding and binarisation, all pins were detected and the few remaining false positives could be eliminated with a semi-manual user interface. While pin detection works robustly with a good image source, the entire process is very dependent on the quality of the CT scan. Especially noisy scans resulting from deflected rays complicate the following computer vision process. Also, if the rotation angle that needs to be corrected beforehand is too big, the resulting data set can be slightly distorted.

9.2.3 Cleaning

The cleaning of machines and plants plays a central role in the MRO process. In the context of maintenance and repair, cleaning procedures are applied in almost all industrial sectors. Of particular importance are the fields of transport (automobile manufacture, aviation, rail transport) and energy (mechanical engineering, turbine manufacturing, power plants, energy systems). The goals and requirements of the cleaning of machines and plants can be divided into visual, preventive and functionally relevant. Besides the high processing, conditioning and disposal costs of aqueous, chemical and mechanical cleaning processes which are oftentimes manually performed, impairments of the environment, of the health of workers and of the cleaned components are likely to occur. In recent years, conventional cleaning procedures have been criticised by the public and the increasingly stricter case laws. On that account, the application of flexible and eco-efficient cleaning processes has taken on greater significance. In addition, newly developed and adjusted cleaning technologies are able to reduce stoppages and downtimes. Cleaning is part of the maintenance process and thus mandatory for ensuring a flawless operation. Nevertheless, a flexible, mobile and appropriate industrial manufacturing equipment as well as knowledge of the processes and solutions for automation are lacking for numerous of these important tasks.

9.2.3.1 Development of a Miniaturised Cleaning Nozzle for Dry Ice Blasting

Requirements concerning the availability of plants or vehicles for maintenance increase constantly. Contamination or wear in operation call for regular maintenance work. Due to temporal and economic reasons, it should mainly be performed on-side. To that end, adequate and appropriate industrial manufacturing equipment and processes are essential to perform this kind of work.

Dry ice pellets are used primarily as an abrasive for blasting processes during cleaning tasks. Due to its special physical properties—dry ice as blasting abrasive is solid at ambient conditions with a temperature of −78.5 °C and changes directly into the gaseous state during blasting—as well as its low hardness, it is suitable for gentle cleaning and processing of sensitive surfaces. As a result of the sublimation the process is dry and residue-free. Transport and acceleration of the abrasive effected is pneumatically. However, the low hardness makes the pellets also very sensitive to external forces such as impacts or friction.

Until now, dry ice blasting has been used mainly to clean easily accessible surfaces. For areas with limited accessibility, for example back-tapers, various blasting nozzles are available, which, however, exclusively use rebound effects for the pellet distraction. That effect leads to an increased crushing and a partial sublimation of the blasting abrasive inside the nozzle. This in turn leads to a loss of mass and velocity, which reduces the kinetic energy and consequently the removal rate. For that reason, a new deflecting nozzle was developed for MRO tasks. The new approach is based on the well-known injector principle, which uses a secondary flow of compressed air to deflect the jet of dry ice pellets.

As the operating principle in Figs. 9.8 and 9.9 shows, the dry ice jet will be deflected by an air jet by 90°. The air jet throttles the dry ice flow in the hose and at the nozzle inlet by a jet contraction. Therefore, the pellet speed in the deflection area is so small that the dry ice will not be crushed by impacts on the rear wall of the nozzle. Another advantage of this concept is the possibility of influencing the material removal rate and intensity of the pellet deflection by varying pressure settings on the used two-hose dry ice blasting system. The theoretical proof of

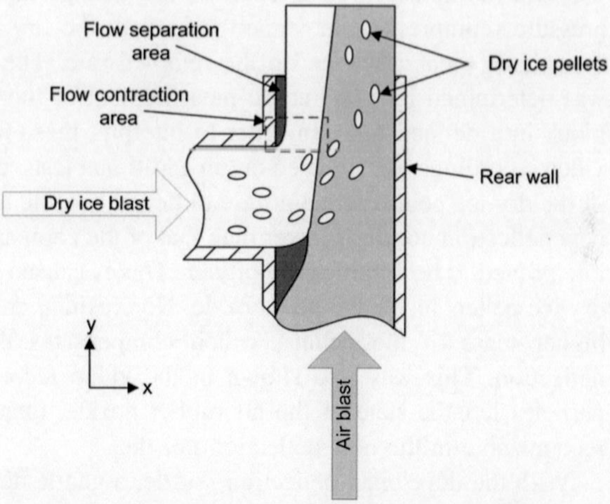

Fig. 9.8 Injector principle

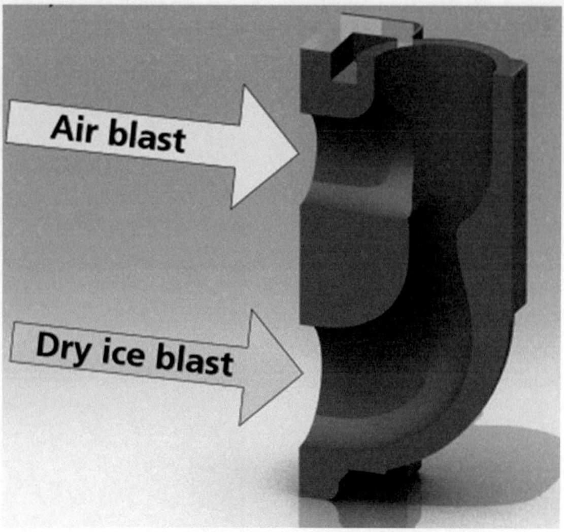

Fig. 9.9 Crosscut through the deflection nozzle

concept was verified by flow simulations with Computational Fluid Dynamics (CFD) software. In the result plots of the simulations the influences of the particle trajectories were clearly visible.

Due to the simulation results and its complex inner geometry, the deflection nozzle was manufactured with the Selective Laser Melting (SLM) technology in order to perform blasting tests. For technical reasons, separate adaptors for the connection to the blasting system had to be made, which are shown together with the deflection nozzle in Fig. 9.10.

During the blasting tests, a state of the art rubber nozzle was used to compare the results. The nozzle outlet of the rubber nozzle was bent by 90°. The dry ice mass flow was kept constant for both nozzles. Various configurations of the transport pressure (compressed air stream for the transportation of dry ice pellets) and jet pressure (compressed air stream to deflect the dry ice pellets) were specified to investigate their influence on the removal rate. The removal rate or performance was determined by gravimetric measurement of the material removal on a plastic block in a defined time. In order to interpret these results graphically, high speed video recordings were carried out in additional tests, evaluating the speed and shape of the dry ice pellets at the nozzle outlet. While the average particle velocity of the new deflection nozzle is lower than that of the compared rubber nozzle, the removal rate proved to be significantly higher. This is caused by the gentle deflection of the dry ice pellets inside the new nozzle. No crushing and sublimation leads to a much higher mass of the pellets, which compensates the loss of speed during the deflection. This was also shown in the video recordings. Whereas mainly snow particles left the state of the art rubber nozzle, intact pellets were predominantly recognizable in the new deflection nozzle.

With the developed deflection nozzle, a gentle deflection of dry ice pellets was evaluated by use of a secondary air stream. The nozzle concept allows for a much higher material removal in cleaning processes through a more effective use of the sensitive blasting abrasive, as compared to conventional deflection nozzles. The

Fig. 9.10 Deflection nozzle with adaptors

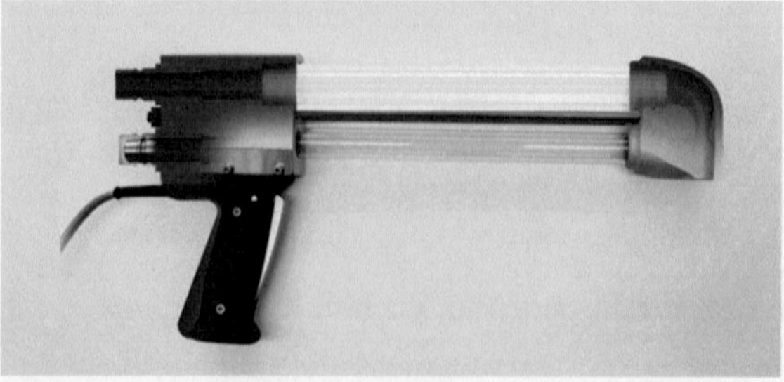

Fig. 9.11 Developed deflection nozzle, fully equipped with blasting gun for the use with a two-hose blasting system

maximum removal rate reached in the tests of the new deflection nozzle is 16 times higher than that of the investigated state of the art rubber nozzle using identical process parameters. In addition, the size ratio of the new nozzle is significantly lower and thus facilitates easier cleaning of areas with difficult accessibility. Figure 9.11 shows the final deflection nozzle with blasting gun for use in MRO tasks.

9.2.4 Repair and Overhaul Technologies

Cost-intensive machines and plants usually do not yield a profit until they are used over a long period of operation. To guarantee this lifespan, worn components have to be repaired or spare parts have to be manufactured. In terms of high-cost components, a repair offers considerable potential savings as opposed to a complete replacement. The requirements of new repair processes are predetermined by high demands concerning the flexibility and adaptivity of the procedure. At the moment, those requirements are usually met by manual activities. This method, however, is very cost- and time-intensive and it always contains the risk that the final quality is mainly influenced by the employee. An automation of the process will contribute to an improvement of productivity as well as of quality. The Fraunhofer Innovation Cluster »Maintenance, Repair and Overhaul (MRO) in Energy and Transportation« intends to introduce and establish permanent, energy-efficient MRO processes and technologies based on different areas of research. One research focus of the cluster is on repair technologies for high value components such as gas turbines blades which have a long life time, but are regularly exposed to high loads. In general, the field of innovation »Repair Technologies« pursues the following goals:

- Developing strategies to increase the level of automation in the fields of repair and maintenance,
- Increasing the flexibility of repair processes,

- Developing concepts for rapid manufacturing of spare parts and structural components and
- Increasing the useful life of structural components by employing materials which have higher wear resistance, and protective layers [8].

9.2.4.1 Repair Cell for Engine and Turbine Components

Aviation engines and industrial gas turbines are subject to an operation-related wear. In this industrial context, so-called »patch processes« have been established for the repair of engine and turbine components. Damaged component parts are identified and separated, and replacements are attached. Subsequently, the contour is re-established with mechanical procedures. Due to the multitude of necessary work stages and all kinds of different damages to component parts, the manufacturing process chain of such repairs is understandably quite complex.

Repair offers a considerable saving of costs as opposed to the replacement with spare parts, but it also consists of a high proportion of manual processes with a low reliability of individual repair stages and a low reproducibility of intended results. Thus, the overall goal of this project is to improve the profitability and planning predictability of repairs in the field of turbines. Furthermore, analyzing the mechanisms of action and correlations between manufacturing technologies should lead to an increase of reproducibility and automation. As an innovative approach, the Fraunhofer Innovation Cluster MRO establishes a partly-automated, robot-supported repair cell, which integrates all necessary work stages. Essential individual technologies will be further developed and provided for the repair cell. In the sector of mechanical processing, a power-controlled, iterative machining strategy will be striven for which will help to achieve the requested component quality in view of the comparatively low stiffness of the robot. Naturally, certain challenges concerning the employment of an industrial robot as opposed to machine tools have to be taken into account. There is a lower absolute accuracy up to a factor of 1,000 and a lower mechanical stiffness, as well as lower path accuracy around the factors of 100–1,000. The factors are based on technical data provided by the manufacturer. In comparison to CAD-CAM systems for CNC machines, existing path planning systems can only be applied limitedly and the dynamic behaviour is strongly dependent on the position of the robot. Nevertheless, concerning individual technologies, robot-supported test stands are built which enable the examination of different processing strategies. For grinding and milling procedures in particular, comparisons can be drawn in order to determine which procedure stresses the specifics of the variants of power- and path-controlled processing. For grinding processes, an iterative machining strategy with a passively-controlled processing head for a basic convex workpiece surface is developed, which facilitates the creation of imperfections in form of indentations and notches. The adaptation of control and regulating algorithms for the force control of milling and grinding processes has been completed to a great extent. In consequence of procedure-specific attributes, different requirement profiles concerning the disintegration of process forces, active attenuation as well as path accuracy are essential. It is recognised that particularly economies of scale in the sector of bearing friction represent a special challenge for active force control.

Concept of a Robot-based Repair Tool Kit

Operations in maintenance and repair are less predictable and projectable than manufacturing processes for new parts. Each component has to be inspected and rated with regard to the benefit and feasibility of a possible repair process. This results in small batches down to a lot size of 1, since every defect is different while the required repair technologies are similar or even identical. Hence, a very flexible concept is necessary to apply the different technologies. This is one of the reasons why repair is mainly carried out manually. Nevertheless, the aero engine industry aims to automate the repair process in order to achieve reproducible results with a constant quality and to reduce labor costs [9–11]. Machine tools are mostly single-purpose machines, designed to give the best performance for a specific machining task. Industrial robots, on the other hand, are designed as flexible machines for handling or measuring, but usually not for machining processes. Using standard industrial robots as a flexible multi-purpose machine tool by equipping them with tools and enhancing their accuracy was the initial idea of the two-year research project »Repair tool kit for engine and turbine components«. A robot cell was set up to demonstrate the application of the repair process chain described below.

Process Chain

For the repair of turbine blades the following process chain is applied [9]:

- Cleaning
- Inspection
- Cutting (defect area)
- Laser cladding
- Adaptive machining
- Shot peening and
- Quality control

Within the repair tool kit, all of these steps, except for the laser cladding, have been carried out using industrial robots.

Adaptive Robot-guided Grinding and Polishing

An increase in flexibility and effectiveness in robot-guided grinding and polishing processes can be achieved by using adaptive machining strategies [12]. In order to provide an adaptive machining strategy for force-controlled robotic abrasive belt grinding, explicit technical knowledge about the abrasive belts used and the exact geometry of the workpiece to be repaired is essential. The process parameters need to be altered to the desired local depth of cut which depends on the failure mode closely linked to the actual geometry of the workpiece. A geometry and process-dependent robot trajectory with adapted technology parameters for each supporting point can be provided for the belt grinding process by a software tool. The tool allows automated trajectory planning of the industrial robot on the basis of geometry and process data, with a defined number of supporting points dependent on the surfaces curvature. The program starts with an import and preparation of the actual and set geometry, provided from a 3D measuring device and CAD data, respectively. A set-actual comparison is conducted in the second module of the program, giving information about the required local depth of cut for every position of the workpiece. Based on the geometrical data, in the third module a robot trajectory with a specific number of supporting points dependent on the workpiece curvature is calculated. Knowing the tool-dependent depth of cut function at each supporting point of the programmed robot trajectory, the technological parameters feed rate v_f, grinding normal force F_N, and cutting speed v_c can precisely be adapted to the required depth of cut. Due to the depth of cut function and restrictions in feasible feed rates coming from the dynamics of the robot, the achievable depth of cut for each process step and supporting point is limited.

The fourth module allows for an evaluation of a possible discrepancy between calculated and real process steps, in order to optimise the experimentally determined depth of cut function. Every re-iterated process step takes the previous depth of cut into account, so that the technical parameters are continuously adapted in order to minimise tool wear, imperfections of the robots trajectory, inaccuracies due to vibrations, etc. This compensation data is considered in the following process steps. This process can be re-iterated for as many cycles as needed until the required part accuracy has sufficiently been achieved. Thus, an iterative approximation from actual to set geometry is achieved. The actual number of required iterations depends on the geometry and material of the workpiece, the tool specification used, and the kinematic restrictions of the robot and is calculated for each supporting point in advance. Consequently, a continuous transition from roughing to finishing is achieved. The intention is to perform the last process step with the lowest possible depth of cut with high feed rate, thus minimizing the relative error of the experimentally determined depth of cut function and the risk of damaging the workpiece. Figure 9.12 shows exemplarily the calculated process steps with depth of cut along the workpiece geometry.

Determining Technological Parameters for the Used Abrasive Belts

In order to set defined depths of cut by adapting process parameters at each supporting point, the exact knowledge of the tool-dependent depth of cut function is necessary. The result of a force-controlled abrasive belt grinding process is a typical

Fig. 9.12 Calculated processing steps

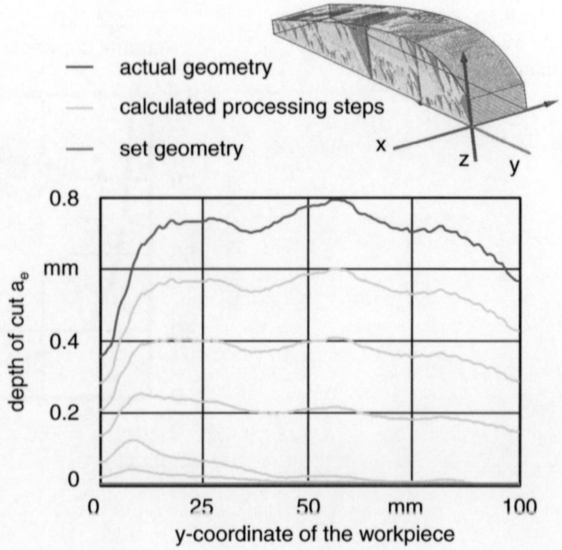

degressive curve characteristic of the depth of cut function with increasing feed rates. Based on experimental data, the feed rate can be adapted for each supporting point to achieve the required material removal or rather depth of cut at every position of the workpiece. Furthermore, choosing minimal feed rates and therefore high material removal rates or depths of cut could lead to thermal damages at the workpiece surface. For every tool-workpiece combination, a threshold must be carefully determined, giving the minimum possible feed rate and accordingly the maximum depth of cut per processing step. For the abrasive belts used, CS420 X-Waterproof, the standard specification of the determined minimum feed rate amounted to 300 mm/min. At lower levels of feed rate, thermal damage was observed. Thus, a range of applicable feed rates for each supporting point depending on the geometrical and kinematical boundary conditions can be defined. Figure 9.13 schematically displays the depth of cut function with suitable range of feed rates for a single convex workpiece made of Inconel 718 and machined with corundum abrasive belts.

Conclusion

Using robots for machining and repair processes can provide a flexible and cost-effective solution for some machining tasks. Robot-based adaptive abrasive belt grinding has a great potential to replace previously manually performed repair processes in the area of high-performance materials and thus increase the degree of automation [13]. For robot-based re-contouring with abrasive belts, a grinding strategy was developed that enables efficient and semi-automated machining of Inconel 718.

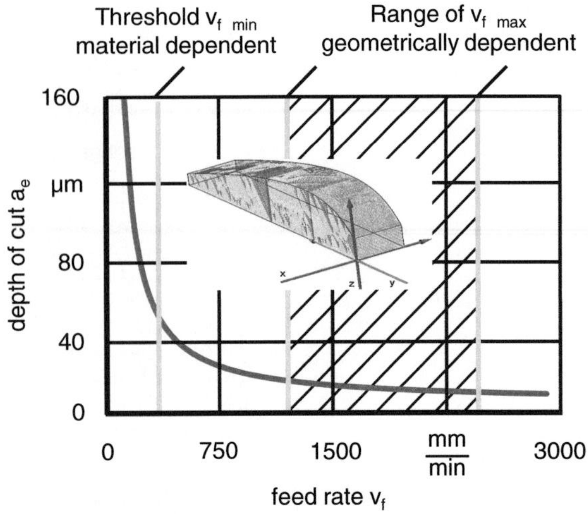

Fig. 9.13 Restrictions for choosing the feed rate as influencing parameter

The robotic industry and research institutes have been conducting research on handling processes for about half a century, whereas using robots for machining has only just gained attention within the last couple of years. This is one reason why robots are used for machining primarily in the academic world to this day. Another reason is the comparatively poor operability concerning the use of robots and the nonexistence of standard solutions, e.g. for workpiece and tool calibration. The robot-based machining processes will be developed more and more in the future in order to provide OEMs and SMEs with ready-to-use robot cells for machining as there is an increasing demand for such systems. Besides R&D for technological processes, it will also be necessary to provide solutions for data management, programming and documentation, as well as to increase the ease of use of industrial robots.

9.2.4.2 Design of Experiments for Laser Metal Deposition in Maintenance, Repair and Overhaul Applications

Nomenclature

- P laser power (w)
- d laser spot diameter (mm)
- v welding velocity (mm/min)
- $\dot{m}$ powder mass flow (g/min)

Laser metal deposition is a technology to create a metallurgically bonded material deposition on a substrate. The technology is shown in Fig. 9.14. A laser beam is used to melt the surface of a specimen. A powdery filler material is injected

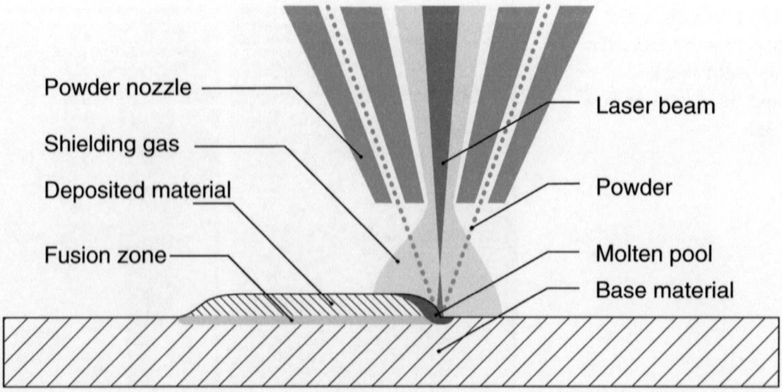

Fig. 9.14 Laser metal deposition

in the molten pool. After solidification, the filler material forms single weld beads. The process is characterised by a low heat input which leads to low distortion and low thermal damage in the base material. The degree of dilution is below 5 % and a fine microstructure can be achieved. Low metallurgical impact is particularly important for modern materials like high-strength steels (e.g. dual-phase or martensitic steel) or nickel-based superalloys because their microstructure is crucial for their material properties. Because of these advantages, laser metal deposition is increasingly used in MRO applications. Examples are found in the mould and die industry. Laser metal deposition can be used for the repair of sintered tools [14] or with vanadium-carbide tool steel for die repair [15]. The state-of-the-art of laser metal deposition is described in [16]. The application as repair technology for stainless steel and titanium alloys is shown in [17].

The design of experiments (DoE) is an important tool in engineering. Different DoE methods have been developed in the past and have been used for laser metal deposition. For example, a central composite design was used by Sun to determine the effect of process parameters on the degree of dilution for titanium alloy [18]. Paul et al. [19] and Lee [20] used a Taguchi design to determine the influence on the deposition rate. Comparative studies between different DoE methods have been done by Youssef et al. [21] for a lathe turning operation.

Laser metal deposition can be used in different repair applications with specific requirements regarding weld bead geometry. In order to repair worn surfaces or to produce a hard facing layer, weld beads with a flat and wide geometry are advantageous. They can be deposited next to each other in order to cover a large surface. The repair of damaged volumes, e.g. tip repair of compressor blades, requires high and narrow weld beads. They are deposited on top of each other to rebuild the damaged volume, as shown in Fig. 9.15. In order to adjust the welding process for a specific repair task, knowledge about process parameters and their influence on weld bead geometry is necessary.

Fig. 9.15 Different weld bead shapes; **a** Surface: flat and wide weld bead; **b** Volume: high and narrow weld bead

In addition, this knowledge is needed to automate the process. Based on 3D scanning to obtain a model of the damaged area, an adequate bead geometry needs to be chosen. With statistical methods, the corresponding welding parameters can be determined.

Materials and Experimental Procedure

A TRUMPF 2.0 kW Yb:Yag laser was used for metal deposition. For all experiments, 4 L/min Helium 5.0 was used as carrier gas for the powdery filler material and 10 L/min Argon 5.0 was used as shielding gas. Nickel-based superalloy René 80 was used as filler material with a powder grain size between 45–125 µm.

Single weld beads were deposited using a DoE full factorial design with factor levels according to Table 9.1. All factor levels are combined with each other, resulting in 2^4 combinations. A randomised schedule was used. To improve statistical reliability, each factor combination was used twice in the experiment. The effect of single factors is analysed by comparing mean responses.

Effects on Bead Width and Height

Bead width was measured in the range of 1.7–3.0 mm and height in the range of 0.4–1.8 mm. Consequently, the varied welding parameters have a substantial impact on the weld bead geometry. The effect of welding parameters is shown in

Table 1 Factor levels for DoE

	−1 (min)	+1 (max)
Laser power P in W	800	1,700
Spot diameter d in mm	1.2	1.8
Welding velocity v in mm/min	320	680
Powder mass flow $\dot{m}$ in g/min	5.0	11.0

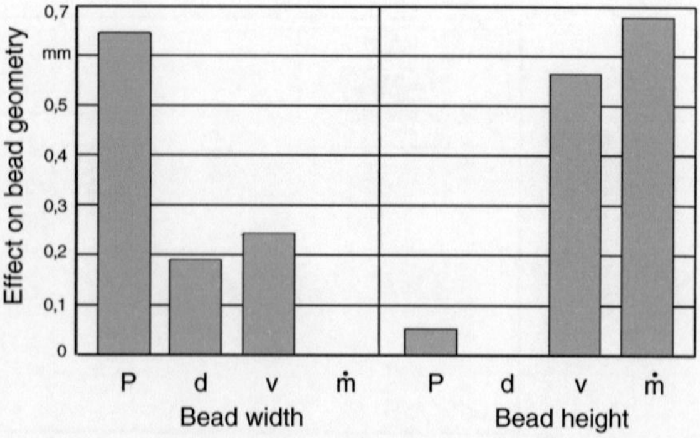

Fig. 9.16 Effect of welding parameters on bead width and height

Fig. 9.16. For bead width, laser power has the biggest effect. The bead height is mainly influenced by welding velocity and powder mass flow.

The large effects allow adjusting the bead geometry according to varying parameters. Using a linear combination of all effects and their interactions, an expected geometry can be calculated for normalised welding parameters according to function (9.1) and (9.2). Compared to the main factors, the effect of interactions between process parameters is small.

$$\text{width} = \left(2.32 + \frac{0.64}{2}P + \frac{0.2}{2}d - \frac{0.26}{2}v - \frac{0.14}{2}v\dot{m}\right)\text{mm} \tag{9.1}$$

$$\begin{aligned}\text{height} = \Bigg(0.96 + \frac{0.06}{2}P - \frac{0.57}{2}v + \frac{0.69}{2}\dot{m} - \frac{0.06}{2}Pd \\ - \frac{0.17}{2}v\dot{m}\Bigg)\text{mm}\end{aligned} \tag{9.2}$$

Figures 9.17 and 9.18 show a comparison between predicted values according to (9.1) and (9.2) and measurements. The straight line demonstrates the ideal case where predicted values and measurements are the same. This is only theoretically possible when there is no random influence and no measurement inaccuracy.

Homoscedasticity can be assumed. The distance between measurement and predicted value is not dependent on the value of the measurement. It is noticeable that the residuals for bead height are grouped together in clusters.

Summary and Outlook

In MRO applications, it is necessary to have flexible repair technologies which can be adjusted quickly to the specific repair task. For laser metal deposition, this adjustment is possible with the knowledge about the effect of process parameters on the weld bead geometry. With a full factorial design, this effect can be determined.

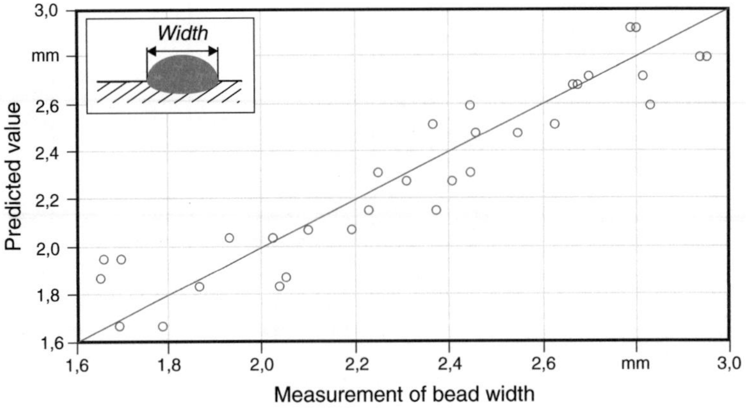

Fig. 9.17 Predicted values according to (9.1) and measurement; bead width

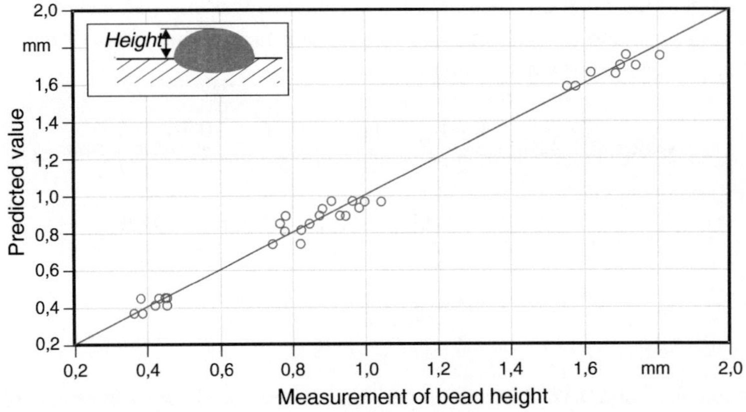

Fig. 9.18 Predicted values according to (9.2) and measurement; bead height

A good agreement between the statistical model and experimental results is obtained. It is of practical relevance that the main factors with effect on width and height are different. This allows for the independent adjustment of different bead dimensions. Current research activities deal with the automation of laser metal deposition. In a process chain with a three dimensional scanning process, material deposition can be used to rebuild damaged areas. Based on a 3D model of the damaged area, the welding bead shape has to be chosen correspondingly. Statistical methods offer an important contribution to this process step and therefore to an automated repair process.

9.3 Outlook to MRO in the Future

After the success of the Fraunhofer Innovation Cluster MRO, the Fraunhofer Innovation Cluster »Life Cycle Engineering for Turbomachines« has been launched in December 2012. It concentrates on the holistic consideration of materials and technologies for turbomachines in the aviation industry and for gas turbines in the energy generation industry. Prospectively, the cluster region Berlin-Brandenburg in Germany will play a special role in the new cluster. Turbomachinery producers are obliged to meet future challenges concerning resource conservation, profitability and aviation safety. Due to the climate change, the German energy policy framework requires the reduction of CO_2 emissions as well as a higher proportion of renewable energies. The latter calls for more flexible load changes of fossil power plants and of turbomachines. Furthermore, the increasing scarcity of resources concerning materials and fuels constitutes another important challenge. In addition to the reduction of emissions, the aviation programs ACARE 2020 [22] and FLIGHTPATH 2050 [23] demand an increase of profitability regarding the operation of airplanes as well as an improvement in aviation management and safety. In order to meet these challenges, industry and science have united in the Fraunhofer Innovation Cluster »Life Cycle Engineering for Turbomachines«. The established concept of »Life Cycle Engineering« (LCE) looks at all life phases of a product: In terms of construction, production, operation, as well as reuse and recycling of the primary materials it is always necessary to include economic, environmental and technical framework conditions. The comprehensive life cycle approach is especially relevant for products with a high investment volume and a long lifespan—as, for example, turbomachines. This approach is able to answer the claim for higher efficiency and lower costs with the simultaneous improvement of environmental sustainability. The aim of the new Fraunhofer Innovation Cluster LCE is therefore to transfer the concept of Life Cycle Engineering to turbomachines: energy-efficient and resource-conserving technologies shall be provided for all life cycles of turbomachines.

Acknowledgments The Fraunhofer Innovation Cluster MRO in Energy and Transport and the projects within the cluster have been co-financed by the EU within the Europäischer Fond für regionale Entwicklung (EFRE) program, by the German industry and by the Fraunhofer-Gesellschaft.

References

1. Besinger F, Lichtfuß H-J, Röhner M, Uhlmann E (2011) Turbomaschinen in Berlin-Brandenburg. Regiverlag, Berlin
2. Bertalanffy Lv (2003) General system theory—foundations, development, applications, 14th edn. George Braziller Inc., New York, p 252
3. Allen JF (1983) Maintaining knowledge about temporal intervals. In: Communications of the ACM, 26 Nov 1983. ACM Press, pp 832–843

4. Agrawal R, Srikant R (1994) Fast algorithms for mining association rules in large databases. In: Proceedings of the 20th international conference on very large data bases, VLDB. Santiago, Chile, Sept 1994, pp 487–499
5. Canny J (1986) A computational approach to edge detection. IEEE Trans Pattern Anal Mach Intell 8(6):679–698
6. Tingquan D, Yanchao L (2013) An improved watershed image segmentation algorithm combining with a new entropy evaluation criterion. In: Proceedings of SPIE 8784, 5th international conference on machine vision (ICMV 2012): algorithms, pattern recognition, and basic technologies, 13 Mar 2013, p 87841V
7. Politz F (2011) Entwicklung einer Lösung zur automatisierten Bauteilidentifikation in der MRO-Diagnostik. Bachelor thesis, Beuth Hochschule für Technik, Berlin
8. Uhlmann E, Behrendt J (2011) International conference MRO, maintenance, repair and overhaul. Verlag Fraunhofer IRB, Stuttgart
9. Bremer C (2006) AROSATEC publishable final report. FP6-502937, 30 Nov 2006
10. Huang A, Gong M, Chen XQ, Zhou L (2003) SMART robotic system for 3D profile turbine vane airfoil repair. Int J Adv Manufact Technol 21:275–283
11. Heitmüller F (2012) Turbinenschaufeln robotergestützt reparieren. Futur 14, Nr. 1, pp 24–26
12. Ren X, Cabaravdic M, Zhang X, Kuhlenkötter B (2007) A local process model for simulation of robotic belt grinding. Int J Mach Tools Manuf 47:962–970
13. Sun Y, Giblin D, Kazerounian K (2009) Accurate robotic belt grinding of workpieces with complex geometries using relative calibration techniques. Robot Comput Integr Manuf 25:204–210
14. Capello E, Colombo D, Previtali B (2005) Repairing of sintered tools using laser cladding by wire. J Mater Process Technol 164–165:990–1000
15. Leunda J, Soriano C, Sanz C, Garcia Navas V (2011) Laser cladding of vanadium-carbide tool steels for die repair. Physics Procedia 12:345–352
16. Birger EM, Moskvitin GV, Polyakov AN, Arkhipov VE (2011) Industrial laser cladding: current state and future. Weld Int 25(3):234–243
17. Graf B, Gumenyuk A, Rethmeier M (2012) Laser metal deposition as repair technology for stainless steel and titanium alloys. Phys Procedia 39:376–381
18. Sun Y, Hao M (2012) Statistical analysis and optimization of process parameters in Ti6Al4V laser cladding using Nd:YAG laser. Opt Lasers Eng 50:985–995
19. Paul CP, Ganesh P, Mishra SK, Bhargava P, Negi J, Nath AK (2007) Investigating laser rapid manufacturing for Inconel-625 components. Opt Laser Technol 39:800–805
20. Lee HK (2008) Effects of the cladding parameters on the deposition efficiency in pulsed Nd: YAG laser cladding. J Mater Process Technol 202:321–327
21. Youssef YA, Beauchamp Y, Thomas M (1994) Comparison of a full factorial experiment to fractional and Taguchi designs in a lathe dry turning operation. Comput Ind Eng 27:59–62
22. European Commission (2001) European aeronautics. A vision for 2020. European Communities, Luxembourg
23. European Commission (2011) Flightpath 2050. European Union, Luxembourg

Chapter 10
Modeling and Sequential Repairs of Systems Considering Aging and Repair Effects

Haitao Liao and Huairui Guo

Abstract The reliability of a repairable system depends on the system age and the number of repairs it experienced. When these effects are considered, predicting the system reliability metrics, such as the cumulative number of failures and failure intensity in the future, becomes a challenging problem. Many existing models utilize Monte Carlo simulations to do prediction but this entails significant computational efforts. This chapter presents a modified Proportional Failure Intensity model to analyze repairable systems. By further modification (approximation) to the model, the system reliability metrics and the associated confidence bounds can be effectively predicted without conducting time-consuming simulations. Moreover, to make repair/replacement decisions, most research assumes the repair model of the system is available beforehand. In practice, however, the model needs to be estimated based upon failures and sequential repair/replacement decisions must be made based on the predicted system reliability metrics. The proposed model is utilized in this decision-making paradigm considering a *short-run cost rate* criterion. Unlike the widely used *long-run cost rate*, this criterion emphasizes the economic impact of a repair/replacement decision on the next fail-and-fix cycle of the system. Two benchmark data sets are analyzed to demonstrate the model in practical use.

Notation

Cdf	Cumulative distribution function
pdf	Probability density function
$N(t)$	Cumulative number of failures up to time t
$m(t)$	$E[N(t)]$
$\lambda(t)$	Failure intensity at time t

H. Liao (✉)
Systems and Industrial Engineering Department, University of Arizona, Tucson, AZ 85721, USA
e-mail: hliao@email.arizona.edu

H. Guo
ReliaSoft Corporation, 1450 S. Eastside Loop, Tucson, AZ 85721, USA

L. Redding and R. Roy (eds.), *Through-life Engineering Services*, Decision Engineering, DOI 10.1007/978-3-319-12111-6_10

T_i, t_i	Random time of the ith failure and its observation
$\lambda_0, \alpha, \beta, \gamma$	Model parameters
$f(t_{i+1}\vert t_i)$	Conditional pdf of the (i + 1)th failure time given the ith failure is, observed at t_i
$F(t\vert t_i)$, $R(t\vert t_i)$	Conditional Cdf and reliability given the ith failure is observed at t_i
η	Repair time with pdf $h_\eta(\eta) = \lambda_\eta e^{-\lambda_\eta \eta}$
ξ	Replacement time with pdf $h_\xi(\xi) = \lambda_\xi e^{-\lambda_\xi \xi}$
C_R	Replacement cost per unit time
C_r	Repair cost per unit time
$\psi(\cdot)$	Moment generating function of a random variable

10.1 Introduction

There are many probabilistic models that address the reliability of repairable systems. Among these models, the nonhomogeneous Poisson Process (NHPP) has been widely used for modeling minimal repairs. One of the most important NHPP models is the Crow-AMSAA model [3], given by:

$$\lambda(t) = \lambda_0 \beta t^{\beta-1}, \tag{10.1}$$

where $\lambda_0 > 0$ and $\beta > 0$. When $0<\beta<1$, the failure intensity is decreasing, reflecting the reliability improvement; when $\beta = 1$, the failure process reduces to the homogeneous Poisson Process (HPP) with a constant failure intensity; when $\beta > 1$, the system is deteriorating.

This model assumes that the system after each repair is as bad as old. However, repairs often restore a failed system to somewhere between the as-good-as new and the as-bad-as old states. Because of this, much attention has been paid to such general repair effects [5–7, 9, 11, 18, 19]. In particular, the Proportional Hazards (PH) model proposed by Cox [1, 2] has been used to describe the failure intensity of a repairable system. For instance, Lawless and Thiagarajah [9] presented a Proportional Failure Intensity model as:

$$\lambda(t) = e^{\alpha+\beta t} e^{\gamma(t-t_{N(t-)})}, \tag{10.2}$$

where α, β and γ are the model parameters, $t_{N(t-)}$ is the most recent failure time before time t, and the log-linear function $e^{\alpha+\beta t}$ is used as the baseline failure intensity. Essentially, this model assumes that repairs do not change the baseline failure intensity but shift it vertically along the intensity axis. Although this model considers both repair and aging effects, it has no closed-form solutions for the future failure intensity, number of failures, and mean time between failures because $t_{N(t-)}$

in the future is unknown. To obtain $t_{N(t-)}$, simulation needs to be conducted. The Kijima virtual age models [6, 7] encounter the same problem. To facilitate engineering applications, a model that takes into account aging and repair effects and is easy to use for reliability prediction will be of significant practical value.

Another important issue related to repairable systems is maintenance decision-making. Usually, a maintenance strategy is suggested according to the reliability of the system under its nominal operating conditions, and repair/replacement decisions are made based on an assumed repair model with known parameters. In reality, however, systems of the same type may be operated and maintained differently, and the system-specific repair models may not be available until several repairs to those systems have been performed. As a result, making sequential repair/replacement decisions for individual systems based on their own failure data is more appropriate.

This chapter presents a modified Proportional Failure Intensity model to analyze repairable systems considering both aging and repair effects and illustrates an approximation approach to predict the reliability of such systems without conducting simulation. The model integrates the Crow-AMSAA and Proportional Failure Intensity models to assist in making sequential repair/replacement decisions.

The remainder of this chapter is organized as follows. Section 10.2 gives the proposed model. Section 10.3 provides maximum likelihood estimates of the unknown parameters and tests for aging and repair effects, and presents an approach to obtain approximate confidence bounds for predicted reliability metrics. Section 10.4 addresses a new maintenance decision-making criterion and formulates the sequential repair/replacement problem. In Sect. 10.5, two numerical examples are provided to demonstrate the implementation of the proposed model in such decision-making problems. Finally, Sect. 10.6 concludes this chapter.

10.2 A Modified Proportional Failure Intensity Model

It is quite common that the chronological age of a system may not accurately reflect the reliability of the system. Instead, both usage and repair history of the system need to be considered. To some extent, the cumulative number of failures can be utilized as an aggregate measure of usage and repair history. Motivated by this, a modified Proportional Failure Intensity model is proposed as:

$$\lambda(t) = \lambda_0 \beta t^{\beta-1} e^{\gamma N(t-)}, \qquad (10.3)$$

where $\lambda_0 > 0$, $\beta > 0$, and γ are the model parameters, and $N(t-)$ is the cumulative number of failures before time t. This model uses the Crow-AMSAA model $\lambda_0 \beta t^{\beta-1}$ as the baseline function, thus inheriting all of its advantages. Moreover, the cumulative repair effect is reflected by the term $\gamma N(t-)$. If $\gamma > 0$, the repair action has a negative effect that increases the failure intensity; if $\gamma < 0$, the repair effect is positive (i.e., the repair makes the system better than the state just prior to the failure); and when $\gamma = 0$, the model becomes the Crow-AMSAA model.

10.3 Statistical Inference and Approximation

10.3.1 Maximum Likelihood Estimates

To estimate the model parameters, the Maximal Likelihood Estimation (MLE) approach can be utilized. Let $0 < t_1 < t_2 < \cdots < t_n$ be n consecutive failure times of the system. From Eq. (10.3), the conditional reliability of the system given that the ith failure occurs at $t_i (i = 1, 2, \ldots)$ is:

$$R(t|t_i) = e^{-e^{i\gamma} \lambda_0 \left(t^\beta - t_i^\beta\right)}, \tag{10.4}$$

and the corresponding conditional pdf of the $(i + 1)$th failure time t_{i+1} is:

$$f(t_{i+1}|t_i) = \lambda_0 \beta t_{i+1}^{\beta-1} e^{i\gamma} e^{-e^{i\gamma} \lambda_0 \left(t_{i+1}^\beta - t_i^\beta\right)}. \tag{10.5}$$

For time-truncated data, the likelihood function is:

$$L(\lambda_0, \beta, \gamma | Data) = f(t_n|t_{n-1}) f(t_{n-1}|t_{n-2}) \ldots f(t_1) R(T|t_n), \tag{10.6}$$

where T is the truncation time. Then, the log-likelihood function can be expressed as:

$$\begin{aligned} \text{In}(L) = & n \ln \lambda_0 + n \ln \beta + (\beta - 1) \sum_{i=1}^{n} \ln t_i + \frac{n(n-1)}{2} \gamma \\ & - \lambda_0 \sum_{i=1}^{n} \left(e^{(i-1)\gamma} t_i^\beta \right) + \lambda_0 \sum_{i=1}^{n+1} \left(e^{(i-1)\gamma} t_{i-1}^\beta \right) - \lambda_0 T^\beta e^{n\gamma}. \end{aligned} \tag{10.7}$$

and the MLE's $\hat{\lambda}_0$, $\hat{\beta}$, and $\hat{\gamma}$ can be obtained by maximizing ln(L). Note that for failure-truncated data, the corresponding likelihood function is given by:

$$L(\lambda_0, \beta, \gamma | Data) = f(t_n|t_{n-1}) f(t_{n-1}|t_{n-2}) \ldots f(t_1). \tag{10.8}$$

10.3.2 Tests for Aging and Repair Effects

The likelihood ratio (LR) test is a statistical tool for comparing two hierarchically nested models. Let $\underline{\theta}_k$ be a vector consisting of k additional model parameters. Suppose the null hypothesis is $H_0 : \underline{\theta}_k = \underline{\theta}_{0k}$, where $\underline{\theta}_{0k}$ consists of the given values of $\underline{\theta}_k$. The LR statistic given by $\Lambda = -2 \ln \frac{\sup\left\{L_{\underline{0}_k} : \underline{\theta}_k = \underline{\theta}_{0k}\right\}}{\sup\left\{L_{\underline{\theta}_k} : \underline{\theta}_k \neq \underline{\theta}_{0k}\right\}}$ follows the chi-squared (χ_k^2) distribution with k degrees of freedom, where $\sup\left\{L_{\underline{0}_k} : \underline{\theta}_k = \underline{\theta}_{0k}\right\}$ and

$\sup\{L_{\underline{\theta}_k} : \underline{\theta}_k \neq \underline{\theta}_{0k}\}$ are the maximum likelihood values obtained when $\underline{\theta}_k = \underline{\theta}_{0k}$ and $\underline{\theta}_k \neq \underline{\theta}_{0k}$, respectively. The null hypothesis is rejected if $\Lambda > \chi^2_{k,\alpha}$, in which $\Lambda > \chi^2_{k,\alpha}$, is the upper α quantile of the χ^2 distribution with k degrees of freedom. Since the significance of a specific effect is reflected by its associated parameter in the repair model, checking the significance of an effect is equivalent to check the significance of its parameter.

10.3.3 Fisher Information Matrix

It is convenient to obtain the confidence bounds of λ_0, β, and γ using the asymptotic normality properties of MLE. From the Fisher information matrix evaluated at the MLE, the variance-covariance matrix of parameter estimates is:

$$\begin{pmatrix} \mathrm{var}(\hat{\lambda}_0) & \mathrm{cov}(\hat{\lambda}_0,\hat{\beta}) & \mathrm{cov}(\hat{\lambda}_0,\hat{\gamma}) \\ \mathrm{cov}(\hat{\lambda}_0,\hat{\beta}) & \mathrm{var}(\hat{\beta}) & \mathrm{cov}(\hat{\beta},\hat{\gamma}) \\ \mathrm{cov}(\hat{\lambda}_0,\hat{\gamma}) & \mathrm{cov}(\hat{\beta},\hat{\gamma}) & \mathrm{var}(\hat{\gamma}) \end{pmatrix} = \begin{pmatrix} -E\left[\frac{\partial^2 \ln L}{\partial \lambda_0^2}\right] & -E\left[\frac{\partial^2 \ln L}{\partial \lambda_0 \partial \beta}\right] & -E\left[\frac{\partial^2 \ln L}{\partial \lambda_0 \partial \gamma}\right] \\ -E\left[\frac{\partial^2 \ln L}{\partial \lambda_0 \partial \beta}\right] & -E\left[\frac{\partial^2 \ln L}{\partial \beta^2}\right] & -E\left[\frac{\partial^2 \ln L}{\partial \gamma \partial \beta}\right] \\ -E\left[\frac{\partial^2 \ln L}{\partial \lambda_0 \partial \gamma}\right] & -E\left[\frac{\partial^2 \ln L}{\partial \gamma \partial \beta}\right] & -E\left[\frac{\partial^2 \ln L}{\partial \gamma^2}\right] \end{pmatrix}^{-1}, \tag{10.9}$$

where the entries are:

$$E\left[\frac{\partial^2 \ln(L)}{\partial \lambda_0^2}\right] = -\frac{n}{\lambda_0^2};$$

$$E\left[\frac{\partial^2 \ln(L)}{\partial \beta^2}\right] = -\frac{n}{\beta^2} - \lambda_0 \sum_{i=1}^{n} \left[e^{(i-1)\gamma} t_i^{\beta} (\ln t_i)^2\right] + \lambda_0 \sum_{i=1}^{n} \left[e^{(i-1)\gamma} t_{i-1}^{\beta} (\ln t_{i-1})^2\right];$$

$$E\left[\frac{\partial^2 \ln(L)}{\partial \gamma^2}\right] = -\lambda_0 \sum_{i=1}^{n} \left[e^{(i-1)\gamma} t_i^{\beta} (i-1)^2\right] + \lambda_0 \sum_{i=1}^{n} \left[e^{(i-1)\gamma} t_{i-1}^{\beta} (i-1)^2\right];$$

$$E\left[\frac{\partial^2 \ln(L)}{\partial \lambda_0 \partial \beta}\right] = -\sum_{i=1}^{n} \left[e^{(i-1)\gamma} \left(t_i^{\beta} \ln t_i - t_{i-1}^{\beta} \ln t_{i-1}\right)\right];$$

$$E\left[\frac{\partial^2 \ln(L)}{\partial \lambda_0 \partial \gamma}\right] = -\sum_{i=1}^{n} \left[e^{(i-1)\gamma} \left(t_i^{\beta} - t_{i-1}^{\beta}\right)(i-1)\right];$$

$$E\left[\frac{\partial^2 \ln(L)}{\partial \gamma \partial \beta}\right] = -\lambda_0 \sum_{i=1}^{n} \left[e^{(i-1)\gamma} t_i^{\beta} \ln t_i \cdot (i-1)\right] + \lambda_0 \sum_{i=1}^{n} \left[e^{(i-1)\gamma} t_{i-1}^{\beta} \ln t_{i-1} \cdot (i-1)\right].$$

From Eq. (10.9), the asymptotic confidence bounds of model parameters can be calculated [13]. For example, based on the asymptotic distribution (lognormal) of $\hat{\lambda}_0$ the two-sided confidence bounds of λ_0 are:

$$\left[\hat{\lambda}_0 \exp(z_{\alpha/2}\sqrt{\mathrm{var}\left(\hat{\lambda}_0\right)} \Big/ \hat{\lambda}_0),\ \hat{\lambda}_0 \exp(-z_{\alpha/2}\sqrt{\mathrm{var}\left(\hat{\lambda}_0\right)} \Big/ \hat{\lambda}_0)\right], \tag{10.10}$$

where $z_{\alpha/2}$ is the $\alpha/2$ quantile of the standard normal distribution. The confidence bounds of other model parameters can also be obtained using this method.

10.3.4 Approximation and Confidence Bounds for m(t) *and* λ(t)

Equation (10.3) coupled with the MLE approach, is able to provide the estimates of the reliability metrics both at and between failure times; however, to do prediction and calculate the associated confidence bounds, simulation is required as the value of $N(t)$ in the future is unknown. To facilitate prediction, we approximate Eq. (10.3) by:

$$\lambda(t) = \lambda_0 \beta t^{\beta-1} e^{\gamma m(t)} \tag{10.11}$$

where $m(t) = E[N(t)]$.

Essentially, this approximation is carried out by substituting the counting process $N(t)$ with its mean $m(t)$. Although the substitution leads to an NHPP model, Fig. 10.1 as well as the numerical examples in Sect. 10.5 show that Eq. (10.11) can provide satisfactory predictions compared with actual observations.

If $\gamma = 0$, Eq. (10.11) becomes the Crow-AMSAA model; otherwise, the model satisfies the following differential equation:

$$\frac{dm(t)}{dt} = \lambda_0 \beta t^{\beta-1} e^{\gamma m(t)}. \tag{10.12}$$

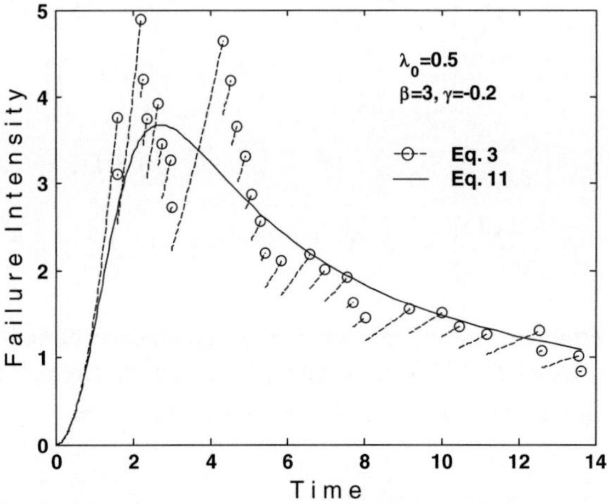

Fig. 10.1 Approximation of system failure intensity using (Eq. 10.11)

Its general solution can be expressed as:

$$m(t) = -\frac{1}{\gamma}\ln\left(C - \gamma\lambda_0 t^\beta\right). \tag{10.13}$$

From the initial condition $m(0) = 0$, we have $C = 1$, and then $m(t)$ becomes:

$$m(t) = -\frac{1}{\gamma}\ln\left(1 - \gamma\lambda_0 t^\beta\right). \tag{10.14}$$

Therefore, $\lambda(t)$ can be expressed as:

$$\lambda(t) = \frac{dm(t)}{dt} = \frac{\lambda_0 \beta t^{\beta-1}}{1 - \gamma\lambda_0 t^\beta} \tag{10.15}$$

It can be seen that, when $\beta = 1$ this model becomes the Musa-Okumoto software reliability model [14]. Note that for Eq. (10.14) to be valid the following inequality must hold:

$$\gamma < \frac{1}{\lambda_0 t^\beta}. \tag{10.16}$$

Moreover, since $\lim_{t\to\infty} \frac{1}{\lambda_0 t^\beta} = 0$ as $\beta > 0$, $\gamma \leq 0$ is assumed throughout this chapter, which excludes the case that repairs deteriorate the system.

It can be seen that Eq. (10.11) has the flexibility of modeling various repairable systems. Taking the first derivative of $\lambda(t)$ with respect to t yields:

$$\frac{d\lambda(t)}{dt} = \lambda_0 \beta t^{\beta-2} e^{\gamma m(t)}[(\beta - 1) + \gamma\lambda(t)t]. \tag{10.17}$$

The sign of the derivative depends on the function $(\beta - 1) + \gamma\lambda(t)t$ because the other terms are positive. If $\lambda(t) < \frac{1-\beta}{\gamma t}$, $\lambda(t)$ is increasing (i.e., the system is deteriorating); if $\lambda(t) > \frac{1-\beta}{\gamma t}$, $\lambda(t)$ is decreasing (i.e., the system's health is improving).

Note that the approximation technique significantly reduces computational efforts in predicting the system reliability metrics. Without conducting simulation, the approximate solutions to $m(t)$ and $\lambda(t)$ can be obtained by substituting the MLE $\hat{\lambda}_0$, $\hat{\beta}$ and $\hat{\gamma}$ into Eqs. (10.14) and (10.15), respectively. In addition, the approximate confidence bounds for $m(t)$ and $\lambda(t)$ can be calculated based on the variance-covariance matrix of the parameter estimates. From Eq. (10.14), the first partial derivatives of $m(t)$ with respect to the model parameters are:

$$\frac{\partial m(t)}{\partial \lambda_0} = \frac{t^\beta}{1-\gamma\lambda_0 t^\beta}; \qquad \frac{\partial m(t)}{\partial \beta} = \frac{\lambda_0 t^\beta \ln t}{1-\gamma\lambda_0 t^\beta};$$

$$\frac{\partial m(t)}{\partial \gamma} = \frac{1}{\gamma^2}\ln\left(1-\gamma\lambda_0 t^\beta\right) + \frac{1}{\gamma}\frac{\lambda_0 t^\beta}{1-\gamma\lambda_0 t^\beta}.$$

Using the delta method, the variance of $\hat{m}(t)$ can be expressed as:

$$\begin{aligned}\operatorname{var}(\hat{m}(t)) = {} & \left(\frac{\partial m(t)}{\partial \beta}\right)^2 \operatorname{var}\left(\hat{\beta}\right) + \left(\frac{\partial m(t)}{\partial \lambda_0}\right)^2 \operatorname{var}\left(\hat{\lambda}_0\right) + \left(\frac{\partial m(t)}{\partial \gamma}\right)^2 \operatorname{var}(\hat{\gamma}) \\ & + 2\left(\frac{\partial m(t)}{\partial \beta}\right)\left(\frac{\partial m(t)}{\partial \lambda_0}\right)\operatorname{cov}\left(\hat{\beta},\hat{\lambda}_0\right) + 2\left(\frac{\partial m(t)}{\partial \gamma}\right)\left(\frac{\partial m(t)}{\partial \lambda_0}\right)\operatorname{cov}\left(\hat{\gamma},\hat{\lambda}_0\right) \\ & + 2\left(\frac{\partial m(t)}{\partial \beta}\right)\left(\frac{\partial m(t)}{\partial \gamma}\right)\operatorname{cov}\left(\hat{\beta},\hat{\gamma}\right). \end{aligned} \tag{10.18}$$

Assuming $\hat{m}(t)$ follows the lognormal distribution, the two-sided confidence bounds of $m(t)$ can be approximately expressed as:

$$\left[e^{z_{\alpha/2}\sigma_{\hat{m}(t)}/\hat{m}(t)}, e^{-z_{\alpha/2}\sigma_{\hat{m}(t)}/\hat{m}(t)}\right], \tag{10.19}$$

where $\sigma_{\hat{m}(t)} = \sqrt{\operatorname{var}(\hat{m}(t))}$. A similar procedure can be used to calculate the approximate confidence bounds for $\lambda(t)$ by substituting the associated terms into the above equations.

10.4 Application to Sequential Repair/Replacement Decision-Making

Many repair, failure replacement and preventive maintenance strategies have been reported in the literature (e.g., [4, 10, 12, 15, 18, 20]). In this chapter, it is assumed that the system is either repaired or replaced only upon failures and the system is as good as new after each replacement.

10.4.1 Mathematical Formulation

The proposed model will be utilized in a sequential repair/replacement decision-making strategy considering a *short-run cost rate* criterion. Specifically, the *short-run cost rate* Ψ_i associated with the repair/replacement decision upon the ith observed failure at t_i is a random variable and is given by:

$$\Psi_i = \frac{C_R \xi}{T_1 + \xi} I + \frac{C_r \eta}{T_{i+1} - t_i + \eta}(1 - I), \tag{10.20}$$

where $I = \{0,$ if the system is repaired; 1, if the system is replaced, T_1 is the next random operating duration if the system is replaced (takes a random time, ξ), $T_{i+1} - t_i$ is the next random operating duration after the repair (takes a random time, η), and C_R and C_r are the replacement cost and repair cost per unit time, respectively.

Let $X = \frac{C_R \xi}{T_1 + \xi}$ and $Y_i = \frac{C_r \eta}{T_{i+1} - t_i + \eta}, i \in \{1, 2, \ldots\}$. It can be shown that X is in the range of (0, C_R) as

$$\lim_{\substack{T_1 \to 0 \\ (\text{or } \xi \to \infty)}} \frac{C_R \xi}{T_1 + \xi} = C_R \quad \text{and} \quad \lim_{\substack{T_1 \to \infty \\ (\text{or } \xi \to 0)}} \frac{C_R \xi}{T_1 + \xi} = 0.$$

Similarly, Y_i is in the range of $(0, C_r)$ as

$$\lim_{\substack{T_{i+1} \to t_i \\ (\text{or } \eta \to \infty)}} \frac{C_r \eta}{T_{i+1} - t_i + \eta} = C_r \quad \text{and} \quad \lim_{\substack{T_{i+1} \to \infty \\ (\text{or } \eta \to 0)}} \frac{C_r \eta}{T_{i+1} - t_i + \eta} = 0.$$

Considering the expected *short-run cost rate*, a repair or replacement decision will be made in favor of the one that achieves $\min(E[Y_i], E[X])$. In other words, the following sequential repair/replacement decision rule is utilized:

$$\text{Decision} = \begin{cases} \text{repair} & \text{if } E[Y_i] < E[X] \\ \text{replace} & \text{if } E[Y_i] \geq E[X] \end{cases}. \tag{10.21}$$

In reality, the *short-run cost rate* criterion would be more realistic than the *long-run cost rate* alternative in repair/replacement decision making due to product obsolescence.

10.4.2 Distribution of Short-Run Cost Rate

To calculate the expected *short-run cost rates* in Eq. (10.20), it is necessary to derive the distributions of X and Y_i. Let $G_X(x)$ and $g_X(x)$ be the Cdf and pdf of X, respectively. By definition, $G_X(x)$ is given by:

$$\begin{aligned} G_X(x) &= P_r(X < x) = P_r\left(T_1 > \frac{(C_R - x)\xi}{x}\right) \\ &= \int_{\frac{(C_R - x)\xi}{x}}^{\infty} \int_0^{\infty} f(t_1) h_\xi(\xi) d\xi \, dt_1 \quad x \in (0, C_R). \end{aligned} \tag{10.22}$$

The pdf of X can be expressed as:

$$g_X(x) = \frac{dG_X(x)}{dx} = \int_0^{\infty} f\left(\frac{(C_R - x)\xi}{x}\right) \frac{C_R \xi}{x^2} h_\xi(\xi) d\xi, \quad x \in (0, C_R). \tag{10.23}$$

The following theorems give the exact *short-run cost rate* distributions when the system is replaced (see Appendices 1 and 2 for the proofs).

Theorem 1 *If the system is replaced and both the replacement time and the first failure time after the replacement follow the exponential distributions, then the short-run cost rate X follows the Truncated Cauchy distribution with pdf* $\frac{C_R \lambda_0 \lambda_\xi}{((\lambda_\xi - \lambda_0)x + \lambda_0 C_R)^2}$ *with the support of* $(0, C_R)$.

Theorem 2 *If the system is replaced with an exponentially distributed replacement time (with parameter* λ_ξ*) and the hazard rate of the first failure time after the replacement is in the form of* $\lambda(t) = \lambda_0 \beta t^{\beta-1}$, *then the* short-run cost rate X *has the pdf* $\frac{\lambda_\xi C_R}{x(C_R - x)} \psi'(-\lambda_\xi)$ *with the compact support of* $[0, C_R,]$ *where* $\psi'(-\lambda_\xi)$ *is the first derivative, with respect to* $-\lambda_\xi$, *of the moment generating function* [17] $\psi(-\lambda_\xi)$ *of the two-parameter Weibull random variable with scale parameter* $\frac{\sqrt[\beta]{\lambda_0 (C_R - x)}}{x}$ *and shape parameter* β.

Note that when $\beta = 1$ (without the aging effect) the two theorems give the same distribution. In addition, the regular Cauchy distribution with the support of $(-\infty, +\infty)$ does not have moments; however, the moments (e.g., $E[X]$) of the random variable X following the Truncated Cauchy distribution exist due to the truncation. Moreover, since the moment generating function $\psi(\cdot)$ of the Weibull random variable has no closed form, numerical integration is needed to calculate $\psi(\cdot)$ and $\psi'(\cdot)$.

As for Y_i, its conditional pdf given that $T_i = t_i$ is:

$$g_{Y_i}(y_i) = \int_0^{\infty} f\left(\frac{(C_r - y_i)\eta}{y_i} + t_i \middle| t_i\right) \frac{C_r \eta}{y_i^2} h_\eta(\eta) d\eta, \quad y_i \in (0, C_r), \tag{10.24}$$

where $f(\cdot|\cdot)$ is given by Eq. (10.5). Obviously, when both the repair time and the next operating duration $(t_{i+1} - t_i)$ after the ith repair follow the exponential distributions (i.e., $\beta = 1$ and $\gamma = 1$), Y_i follows the Truncated Cauchy distribution with the support of $(0, C_r)$. The following theorem gives the probabilistic property of the *short-run cost rate* when both the aging and repair effects are considered (see Appendix 3 for the proof).

Theorem 3 *If the system is repaired with an exponentially distributed repair time, and the failure intensity of the system can be described by the modified*

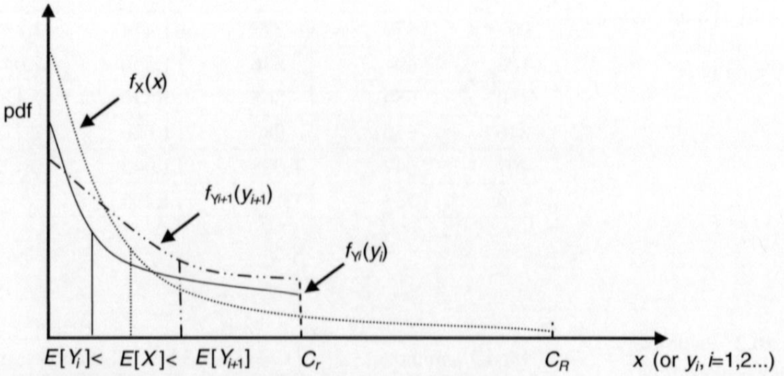

Fig. 10.2 Probability density functions of X and $Y_i, i = 1, 2, \ldots$

Proportional Failure Intensity model in the form of Eq. (10.3), *then the* short-run cost rate ($Y_i, i \in \{1, 2, \ldots\}$) *has the pdf* $\frac{\lambda_\eta C_r e^{\lambda_0 e^{i\gamma} t_i^\beta}}{y_i(C_r - y_i)}\psi'(-\lambda_\eta)$ *with the support of* $(0, C_r)$, *where* $\psi'(-\lambda_\eta)$ *is the first derivative, with respect to* $-\lambda_\eta$, *of the moment generating function* $\psi(-\lambda_\eta)$ *of the three-parameter Weibull random variable with location parameter* $\frac{-y_i t_i}{C_r - y_i}$, *scale parameter* $\frac{\sqrt[\beta]{\lambda_0 e^{i\gamma}(C_r - y_i)}}{y_i}$ *and shape parameter* β.

Note that the distribution of Y_i varies as the number of failures and the age of the system increase. Based on the above theorems, the repair/replacement decision can be made upon each failure. Figure 10.2 depicts the decision-making strategy. Upon the *i*th failure (at time t_i), a repair will be performed since $E[Y_i] < E[X]$, while the system will be replaced upon the (i + 1)th failure (at time t_{i+1}) as $E[Y_{i+1}] > E[X]$.

10.5 Numerical Examples

10.5.1 Aircraft Air-Conditioning Equipment Data

A subset of the failure data of 13 Boeing aircrafts given by Proschan [16] is considered first. The subset noted as Aircraft #3, consisting of 29 failure times, is reported in Table 10.1. The results from the LR test presented in Sect. 10.3.2 indicate that there is no aging effect but the repair effect is significant. Therefore, the proposed model (3) with $\beta = 1$ is used, and the Musa-Okumoto model can be utilized for prediction.

The estimation results of the Crow-AMSAA model and the proposed model are given in Table 10.2 and shown in Figs. 10.3 and 10.4. Both figures show that the failure intensity of the system is decreasing over time. Therefore, the repair/replacement policy is always in favor of repairs, at least based on the current observations.

Table 10.1 Failure times (in hours)—aircraft air-conditioning data

90	470	777	1,474	1,835
100	494	836	1,550	2,043
160	550	865	1,576	2,113
346	570	983	1,620	2,214
407	649	1,008	1,643	2,422
456	733	1,164	1,705	

Table 10.2 Parameters estimates—aircraft air-conditioning data

Model parameter	Crow-AMSAA	Proposed model
$\hat{\beta}$	0.90073	1
$\hat{\lambda}_0$	0.0260	0.018733
$\hat{\gamma}$	0	−0.02956
Log-likelihood value	−157.1624	−156.3486

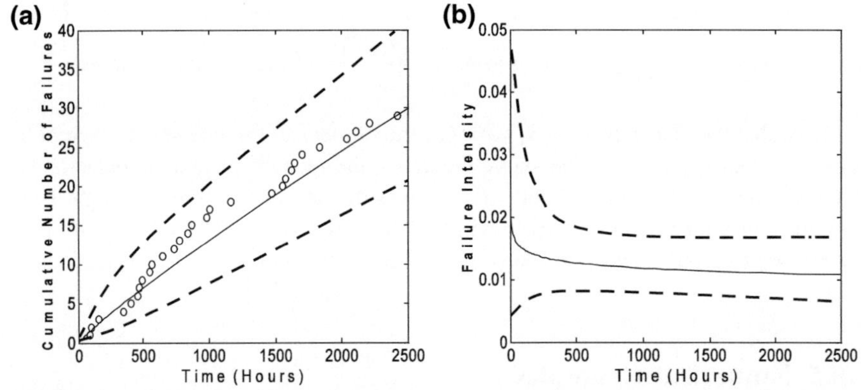

Fig. 10.3 Crow-AMSAA model (with 95 % confidence bounds)—air-conditioning data. **a** Expected failure number. **b** Failure intensity

10.5.2 Hydraulic Systems Data

The data set presented by Kumar and Klefjsö [8] is utilized for another demonstration. The failure data given in Table 10.3 is related to the hydraulic system of a load-haul-dump (LHD) machine (LHD09). The data set consists of 27 failure times. The times between failures can be obtained by subtracting the previous failure time from the current one. Table 10.4 gives the estimates of model parameters considering various cases.

The LR approach is utilized to test for the aging and the repair effects. First, to study the aging effect, the following test is conducted:

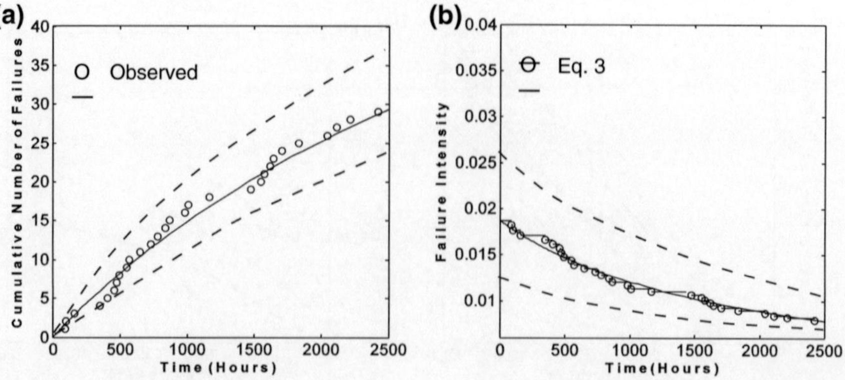

Fig. 10.4 Proposed model (with 95 % confidence bounds)—air-conditioning data. **a** Expected failure number. **b** Failure intensity

Table 10.3 Failure times (in hours)—hydraulic system LHD 09

278	1827	2052	3094	3735	4267	4371
539	1859	2228	3236	3939	4291	4682
1529	1910	2475	3274	4121	4323	4743
1720	1920	2640	3523	4237	4361	

Table 10.4 Estimates of model parameters—hydraulic system LHD 09

Model parameter	Proposed full model	Reduced model (γ = 0, Crow-AMSAA)	Reduced model (β = 1, no aging effect)
$\hat{\beta}$	1.6743	1.6539	1
$\hat{\lambda}_0$	1.9375E−5	2.2466E−5	5.6926E−3
$\hat{\gamma}$	−1.8654E−3	0	0
Log-likelihood	−163.6416	−163.6421	−166.5519

$$\mathrm{H_0}: \text{No time trend}(\beta = 1); \quad \mathrm{H_1}: \text{Time trend exists}(\beta \neq 1).$$

As shown in Table 10.4, by setting β = 1, the maximum log-likelihood value is obtained as −166.5519. The test statistic is $\Lambda = 5.8206 > \chi^2_{1,\,0.05} = 3.8415$. Therefore, $\mathrm{H_0}$ is rejected, implying that the time trend is significant. Similarly, the second test can be implemented to check the significance of the repair effect by:

$$\mathrm{H_0}: \text{No repair effect}(\gamma = 0) \quad \mathrm{H_1}: \text{Repair effect exists}(\gamma \neq 0).$$

In this test, the maximum log-likelihood value equals −163.6421. The test statistic is $\Lambda = 0.0011 < \chi^2_{1,\,0.05} = 3.8415$, implying that there is no clear repair effect.

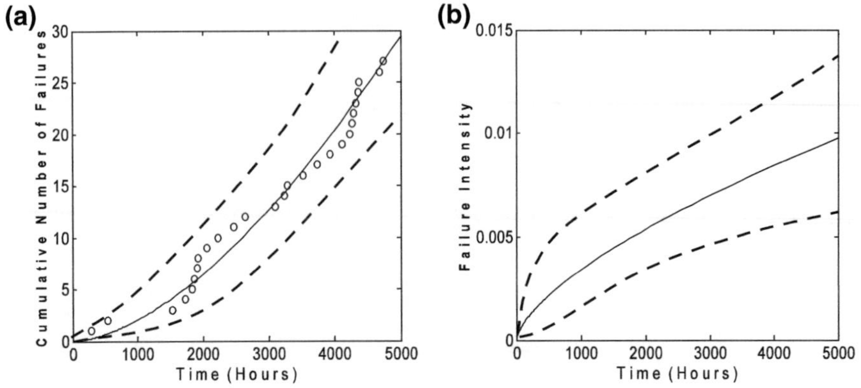

Fig. 10.5 Crow-AMSAA model (a special case of the proposed model, with 95 % confidence bounds)—hydraulic failure data

These tests show clear evidence of the aging effect and the insignificant repair effect, therefore the Crow-AMSAA model that is the special case of the proposed model is sufficient. The expected failure number and failure intensity (point estimates) are given in Fig. 10.5. The results show that the failure intensity is increasing over time, and the repair action belongs to minimal repair.

We next demonstrate the sequential repair/replacement strategy addressed in Sect. 10.4. Since it is necessary to estimate the repair model first, we assume that several repairs have been performed before making the repair/replacement decision. Specifically, the failure data up to the decision-making time are utilized.

We assume $C_r = 3.8$/unit time, $\lambda_n = 0.05$/unit time, $C_R = 10$/unit time, and $\lambda_\xi = 0.03$/unit time. Table 10.5 shows a series of parameter estimates as new failures occur, as well as the corresponding *short-run cost rate* estimates and repair/replacement decisions. Figure 10.6 also presents the *short-run cost rates* and

Table 10.5 Repair/replacement decisions—hydraulic system LHD 09

Number of failures i (failure time in hours)	Parameters estimates $\hat{\beta}, \hat{\lambda}_0, \hat{\gamma}$	$E[\Psi_i]$ (Repair)	$E[\Psi_i]$ (Replace)	Repair	Replace
17 (3,735)	1.4982, 7.5590E−5, 0	0.7868/UT	0.9155/UT	√	
18 (3,939)	1.4693, 9.3908E−5, 0	0.7800/UT	0.9441/UT	√	
19 (4,121)	1.4544, 1.0500E−4, 0	0.7793/UT	0.9594/UT	√	
20 (4,237)	1.4715, 9.2000E−5, 0	0.7944/UT	0.9399/UT	√	
21 (4,267)	1.5292, 5.9022E−5, 0	0.8294/UT	0.8800/UT	√	
22(4,291)	1.5884, 3.7365E−5, 0	0.8655/UT	0.8246/UT		√
23 (4,323)	1.6412, 2.4806E−5, 0	0.8982/UT	0.7797/UT		
24 (4,361)	1.6883, 1.7193E−5, 0	0.9279/UT	0.7428/UT		

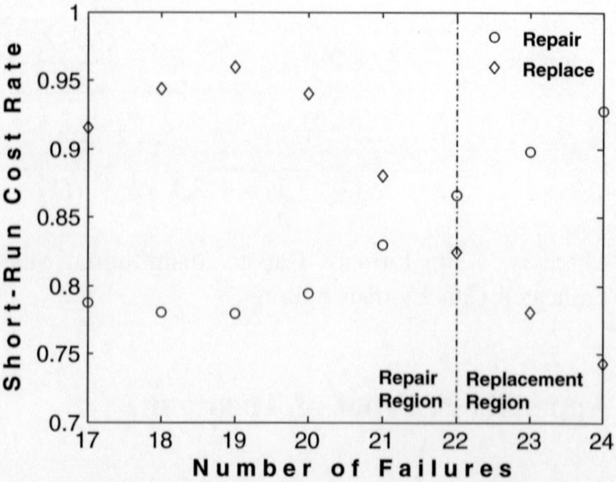

Fig. 10.6 Decision regions for sequential repair/replacement—hydraulic system LHD 09

indicates the repair and replacement decision regions. It can be seen that the differences in *short-run cost rates* between repair and replacement are increasing when the number of failures falls into the replacement region (upon and after the 22nd failure); therefore, if the replacement decision is not made upon the 22nd failure event, we will have stronger evidence against estimation uncertainty to make the right replacement decision later.

10.6 Conclusion

In this chapter, the Crow-AMSAA model and the Proportional Failure Intensity model are reviewed. To overcome the inconvenience of these models, a new model is proposed, which considers both aging and repair effects. After modification, the model is convenient to use, and the confidence bounds of the associated reliability metrics can be obtained without conducting time-consuming simulation. In particular, the model enables making sequential repair/replacement decisions driven by repair data over time.

Appendix 1: Proof of Theorem 1

Suppose both the replacement time and the first failure time after the replacement follow exponential distributions with parameters λ_ξ and λ_0, respectively. Then, the pdf of X is:

$$g_X(x) = \int_0^\infty \frac{C_R}{x^2} \xi \lambda_\xi e^{-\lambda_\xi \xi} \lambda_0 e^{-\frac{\lambda_0 (C_R - x)\xi}{x}} d\xi = \frac{C_R \lambda_0 \lambda_\xi}{x^2} \int_0^\infty \xi e^{-\left(\lambda_\xi + \frac{\lambda_0 C_R - \lambda_0 x}{x}\right)\xi} d\xi$$
$$= \frac{C_R \lambda_0 \lambda_\xi}{x^2} \left(\frac{x}{(\lambda_\xi - \lambda_0)x + \lambda_0 C_R} \right)^2 = \frac{C_R \lambda_0 \lambda_\xi}{((\lambda_\xi - \lambda_0)x + \lambda_0 C_R)^2}, \tag{10.25}$$

which is in the form of Cauchy distribution. Since $X \in (0, C_R)$, X follows the Truncated Cauchy distribution.

Appendix 2: Proof of Theorem 2

Suppose the replacement time has the exponential distribution with parameter λ_ξ and the hazard rate of the first failure time after the replacement is in the form of $\lambda(t) = \lambda_0 \beta t^{\beta - 1}$. Then, the pdf of X is:

$$g_X(x) = \int_0^\infty \frac{C_R}{x^2} \xi \lambda_\xi e^{-\lambda_\xi \xi} \lambda_0 \beta \left(\frac{(C_R - x)\xi}{x} \right)^{\beta - 1} e^{-\lambda_0 \left(\frac{(C_R - x)\xi}{x} \right)^\beta} d\xi$$
$$= \frac{\lambda_0 \beta C_R (C_R - x)^{\beta - 1}}{x^{\beta + 1}} \int_0^\infty \xi^\beta e^{-\left(\frac{\lambda_0 (C_R - x)^\beta}{x^\beta} \right) \xi^\beta} \lambda_\xi e^{-\lambda_\xi \xi} d\xi$$
$$= \frac{\lambda_\xi C_R}{x(C_R - x)} \int_0^\infty (\xi e^{-\lambda_\xi \xi}) d\left(1 - \exp\left(- \left(\frac{\sqrt{[\beta]}\lambda_0 (C_R - x)}{x} \right)^\beta \right) \right). \tag{10.26}$$

Considering $\psi(a) = E\left[e^{a\xi}\right]$ is the moment generating function of a random variable ξ and $\psi'(a) = E\left[\xi e^{a\xi}\right]$ is the first derivative of the function with respect to a, the pdf of X becomes $g_X(x) = \frac{\lambda_\xi C_R}{x(C_R - x)} \psi'(-\lambda_\xi)$, where $\psi'(-\lambda_\xi)$ is the first derivative, with respect to $-\lambda_\xi$, of the moment generating function $\psi(-\lambda_\xi)$ of the two-parameter Weibull random variable with scale parameter $\frac{\sqrt[\beta]{\lambda_0 (C_R - x)}}{x}$ and shape parameter β. Furthermore, the integral in Eq. (10.26) exists because:

$$\xi \geq 0, \quad \lambda_\xi > 0 \Rightarrow e^{-\lambda_\xi \xi} \leq 1 \Rightarrow \xi e^{-\lambda_\xi \xi} \leq \xi \Rightarrow$$
$$\int_0^\infty (\xi e^{-\lambda_\xi \xi} d) \left(1 - \exp\left(- \left(\frac{\sqrt[\beta]{\lambda_0 (C_R - x)}}{x} \xi \right)^\beta \right) \right) < \int_0^\infty \xi d \left(1 - \exp\left(- \left(\frac{\sqrt[\beta]{\lambda_0 (C_R - x)}}{x} \xi \right)^\beta \right) \right),$$

where the term on the right hand side is the expectation $(<\infty)$ of the Weibull random variable.

Appendix 3: Proof of Theorem 3

Suppose the repair time has the exponential distribution with parameter λ_η and the failure intensity of the system follows the modified Proportional Failure Intensity model with the failure intensity $\lambda(t) = \lambda_0 \beta t^{\beta-1} e^{\gamma N(t-)}$. Then, the pdf of Y_i is:

$$\begin{aligned} g_{Y_i}(y_i) &= \int_0^\infty \frac{C_r}{y_i^2} \eta \lambda_\eta e^{-\lambda_\eta \eta} \lambda_0 \beta \left(\frac{(C_r - y_i)\eta}{y_i} + t_i \right)^{\beta-1} e^{i\gamma} e^{-e^{i\gamma}\lambda_0 \left(\left(\frac{(C_r-y_i)\eta}{y_i} + t_i \right)^\beta - t_i^\beta \right)} d\eta \\ &= \frac{\lambda_0 \beta C_r e^{i\gamma}}{y_i^2} e^{\lambda_0 e^{i\gamma} t_i^\beta} \int_0^\infty \left(\frac{(C_r - y_i)\eta}{y_i} + t_i \right)^{\beta-1} e^{-\lambda_0 e^{i\gamma} \left(\frac{(c_r-y_i)\eta}{y_i} + t_i \right)^\beta} \lambda_\eta \eta e^{-\lambda_\eta \eta} d\eta \\ &= \frac{C_r \lambda_\eta e^{\lambda_0 e^{i\gamma} t_i^\beta}}{y_i (C_r - y_i)} \int_0^\infty \eta e^{-\lambda_n \eta} d \left(1 - \exp \left(- \left(\sqrt[\beta]{\lambda_0 e^{i\gamma}} \frac{(C_r - y_i)}{y_i} \left(\eta - \frac{-y_i t_i}{C_r - y_i} \right) \right)^\beta \right) \right), \end{aligned} \tag{10.27}$$

which is in the form of $\frac{\lambda_\eta C_r e^{\lambda_0 e^{i\gamma} t_i^\beta}}{y_i (C_r - y_i)} \psi'(-\lambda_\eta)$, where $\psi'(-\lambda_\eta)$ is the first derivative, with respect to $-\lambda_\eta$, of the moment generating function $\psi(-\lambda_n)$ of the three-parameter Weibull random variable with location parameter $\frac{\sqrt[\beta]{\lambda_0 e^{i\gamma}}(C_r - y_i)}{y_i}$ scale parameter $\frac{\sqrt[\beta]{\lambda_0 e^{i\gamma}}(C_r - y_i)}{y_i}$ and shape parameter β.

References

1. Cox DR (1972) The statistical analysis of dependencies in point processes. In: Lewis PAW (ed) Stochastic point processes. Wiley, New York, pp 55–66
2. Cox DR, Lewis PAW (1966) The statistical analysis of series of events. Methuen, London
3. Crow LH (1974) Reliability analysis for complex, repairable systems. In: Proschan F, Serfling RJ (eds) Reliability and biometry. SIAM, Philadelphia, pp 379–410
4. Elsayed EA (1996) Reliability engineering. Addison Wesley Longman Inc, New York
5. Kaminskiy M, Krivtsov V (1998) A Monte Carlo approach to repairable system reliability analysis. Probabilistic safety assessment and management. Springer, New York, pp 1063–1068
6. Kijima M (1989) Some results for repairable systems with general repair. J Appl Probab 26:89–102
7. Kijima M, Sumita N (1986) A useful generalization of renewal theory: counting process governed by non-negative Markovian increments. J Appl Probab 23:71–88
8. Kumar U, Klefjsö B (1992) Reliability analysis of hydraulic systems of LHD machines using the power law process model. Reliab Eng Syst Saf 35(3):217–224
9. Lawless JF, Thiagarajah K (1996) A point-process model incorporating renewals and time trends, with application to repairable systems. Technometics 38:131–138
10. Lie HC, Chun HY (1986) An algorithm for preventive maintenance policy. IEEE Trans Reliab 35:71–75

11. Lindqvist BH, Elvebakk G, Heggland K (2003) The trend-renewal process for statistical analysis of repairable systems. Technometrics 45(1):31–44
12. Malik MAK (1979) Reliability preventive maintenance policy. AIIE Trans 11:221–228
13. Meeker WQ, Escobar LA (1998) Statistical methods for reliability data. Wiley, New York
14. Musa JD, Okumoto K (1984) A logarithmic poisson execution time model for software reliability measurement. In: Proceedings of seventh international conference on software engineering, pp 230–238
15. Park KS (1979) Optimal number of minimal repairs before replacement. IEEE Trans Reliab 28:137–140
16. Proschan F (1963) Theoretical explanation of observed decreasing failure rate. Technometrics 5:375–383
17. Ross S (1997) Introduction to probability models. Academic Press, New York
18. Wang H, Pham H (1996) Optimal age-dependent preventive maintenance policies with imperfect maintenance. Int J Reliab Qual Saf Eng 3:119–135
19. Yanez M, Joglar F, Modarres M (2002) Generalized renewal process for analysis of repairable systems with limited failure experience. Reliab Eng Syst Saf 77(2):167–180
20. Zhang YL, Yam RCM, Zuo MJ (2002) Optimal replacement policy for a multistate repairable system. J Oper Res Soc 53:336–341

Chapter 11
Cold Spray Coating Technology for Metallic Components Repairing

Pasquale Cavaliere

Abstract Cold spraying is a coating technology on the basis of aerodynamics and high-speed impact dynamics. In this process, spray particles (usually 1–50 mm in diameter) are accelerated to a high velocity (typically 300–1,200 m/s) by a high-speed gas flow that is generated through a convergent-divergent de Laval type nozzle. A coating is formed through the intensive plastic deformation of particles impacting on a substrate at a temperature below the melting point of the spray material. It can be considered a safe and green technology because of the absence of a high-temperature gas jet, radiation, and explosive gases. The coatings formed using the cold flow deposition processes are dense and oxide-free. The cold flow deposition process has emerged as an important alternative to other thermal spraying processes. An example of a key application of the cold spray process is the recovery of costly aircraft parts during overhaul and repair. Cold spray also can be used in the development of unique materials and for the production of actual parts. Cold spray can be used to produce a new class of materials that could not be achieved by conventional ingot metallurgy. The cold spray process represents leading edge technology and provides superior performance over conventional technologies. Even if it has great application potentials in aerospace, automobile manufacture, chemical industry, etc., there are still many fundamental aspects to be uncovered. Because adhesion of the metal powder to the substrate and deposited material is achieved in the solid state, the characteristics of cold spray deposits are quite unique. Cold spray is suitable for depositing a wide range of traditional and advanced materials on many types of substrate materials, especially in non-traditional applications that are sensitive to the temperature of the process. Cold spray is capable of potentially providing restoration, sealing, surface modification, wear resistance, thermal barriers, heat dissipation, rapid prototyping, aesthetic coatings, fatigue resistance and many other applications without the undesirable effects of process temperatures or metallurgical incompatibilities among materials. It can also be used to increase the heat resistance of a material. Research into improving the cold spraying technology is still being conducted worldwide today.

P. Cavaliere (✉)
Department of Innovation Engineering,
University of Salento, Via Per Arnesano, 73100 Lecce, Italy
e-mail: pasquale.cavaliere@unisalento.it

L. Redding and R. Roy (eds.), *Through-life Engineering Services*,
Decision Engineering, DOI 10.1007/978-3-319-12111-6_11

11.1 Introduction

Cold spray is a solid state coating technology based on the acceleration of powder particles impacting on a substrate through the employment of an high speed gas jet. Cold Spray was discovered during the testing of a supersonic wind tunnel in the mid 1980s at the Institute for Theoretical and Applied Mechanics of the Siberian Division of the Russian Academy of Science in Novosibirsk and first reported by the Russian scientists. This is a unique form of spraying in that it relies solely on kinetic energy to form the coating. It can be considered a safe and green technology because of the absence of a high-temperature gas jet, radiation, and explosive gases. The coatings formed using the cold flow deposition processes are dense and oxide-free. The cold flow deposition process has emerged as an important alternative to other thermal spraying processes. An example of a key application of the cold spray process is the recovery of costly aircraft parts (Champagne and Leyman [1]) during overhaul and repair (Ogawa and Niki [2]). Cold spray also can be used in the development of unique materials and for the production of actual parts (Nooririnah et al. [3]). Cold spray can be used to produce a new class of materials that could not be achieved by conventional ingot metallurgy (Villafuerte [4]). The cold spray process represents leading edge technology and provides superior performance over conventional technologies (Liu et al. [5]). Even if it has great application potentials in aerospace, automobile manufacture, chemical industry, etc., there are still many fundamental aspects to be uncovered (Birtch [6]). Because adhesion of the metal powder to the substrate and deposited material is achieved in the solid state, the characteristics of cold spray deposits are quite unique. Cold spray is suitable for depositing a wide range of traditional and advanced materials on many types of substrate materials, especially in non-traditional applications that are sensitive to the temperature of the process (Cavaliere and Silvello [7]). Cold spray is capable of potentially providing restoration, sealing, surface modification, wear resistance, thermal barriers, heat dissipation, rapid prototyping, aesthetic coatings, fatigue resistance and many other applications without the undesirable effects of process temperatures or metallurgical incompatibilities among materials (Sova et al. [8]). Mechanical properties of cold spray coatings strongly depend on particles bonding on the substrate (Assadi et al. [9]); such bonding is dependent on mechanical properties of the substrate and particles and on the processing parameters employed during spray (Schmidt et al. [10]). The enhancement of shear instability represents an instrument to improve bonding; such non-equilibrium phenomena lead to the deformation inducing melting during impact which increases the bonding effect during spray, clearly the phenomenon is enhanced as improving processing parameters as a function of the different materials of the substrate and particles (Klinkov et al. [11]). The surface finishing strongly influences the substrate deposit interface microstructure and consequently the microstructural and mechanical behavior; depending on the difference between the hardness of the impacting particles and substrate, the processing parameters can lead to different aspect of the interface: interfacial instability and adiabatic shear instability. The conclusion is that with the same material, the processing conditions can lead to a shift between the behaviors. From an experimental point of view it seems that the matter of fact is the incoming of partial melting due to particular localized conditions, such conclusions (supported by numerical investigations) have been demonstrated for some specific materials and conditions but the results seem not conclusive at the present. Flow pressure governing particles speed and gas temperature is fundamental. Flow pressure and particle speed are directly related to the impact energy. If such energy is too low the particles do not bond to the substrate and consequently erode it. Temperature is fundamental because it regulates the particles ductility leading them to correctly flatten for bonding.

11.2 Basic Features and Equipment

With respect to other thermal spray processes, cold spray is characterized by lower operating temperatures and higher spray speed (Fig. 11.1).

Such aspects have a strong effect on the improvement of microstructural and mechanical properties.

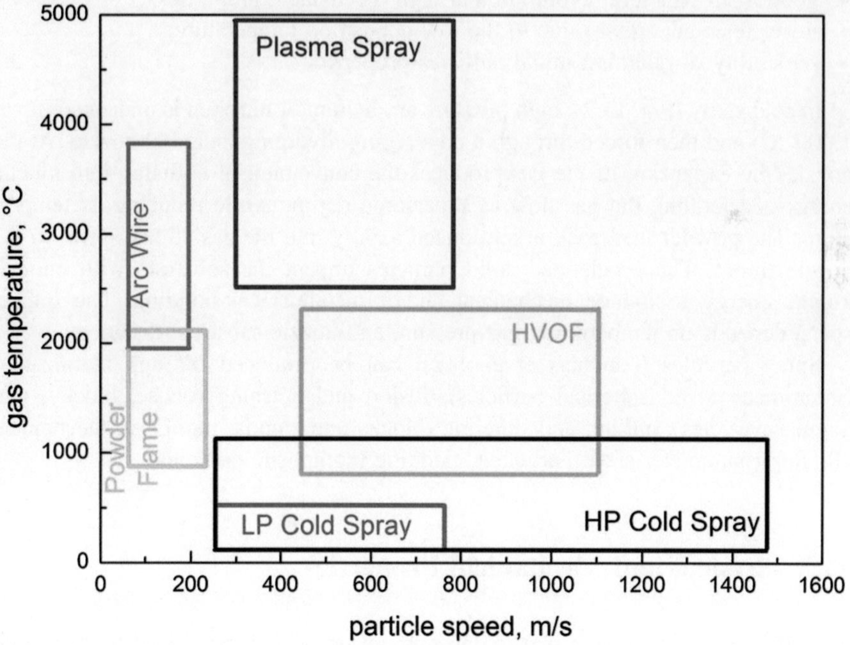

Fig. 11.1 Particle speed versus gas temperature for different thermal spray processes

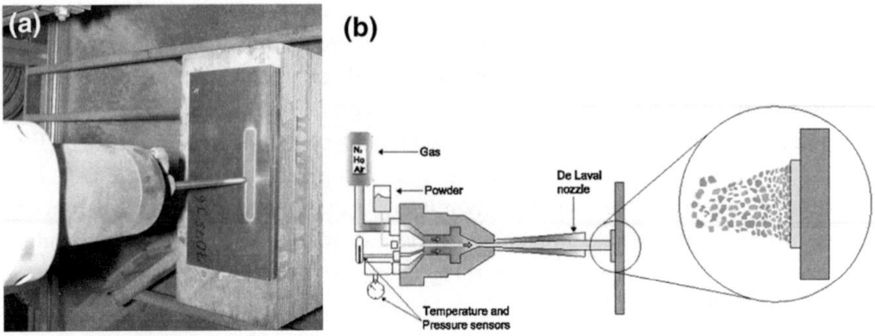

Fig. 11.2 Cold spray gun (**a**) and machine scheme (**b**)

The main advantages of employing such technology as repairing and restoring ones can be summarized as:

- safety due to the absence of high temperature or explosive gases and radiation;
- high deposition efficiency for all metals and alloys;
- high productivity due to high feed rate;
- high possible combination of dissimilar materials;
- simple or no-preparation of the substrate;
- good microstructural evolution and high mechanical properties;
- lower residual stresses due to the low deposition temperature;
- possibility of retaining initial particles properties.

In cold spray (Fig. 11.2), high pressure air, helium or nitrogen is preheated (up to 1,000 °C) and then forced through a converging-diverging DeLaval nozzle. At the nozzle, the expansion of the gas produces the conversion of enthalpy into kinetic energy accelerating the gas flow to supersonic regime while reducing its temperature. The powder feedstock is introduced axially into the gas stream, prior to the nozzle throat. The accelerated solid particles impact the substrate with enough kinetic energy to induce mechanical and/or metallurgical bonding. The impact speed depends on temperature, gas pressure and nozzle-substrate distance.

Spray particles (Fauchais et al. [12]) can be produced through atomization (micrometer sized spherical particles), fusion and crashing (dense, blocky, and angular particles), milling and sintering (blocky and angular particles), mechanical alloying (nanometer sized particles), cladding (composite particles).

11.3 Physical and Mechanical Properties

Particles impact is a severe plastic deformation process characterized by very high strain rates (10^{-5} s^{-1}). In such contest, material behavior is strongly dependent on the material type. Qualities of cold spray deposits are evaluated measuring the following microstructural and mechanical properties:

- deposit microhardness;
- deposit grain size;
- substrate-coating adhesion strength;
- deposit porosity;
- residual stresses.

This properties depend on particles size and physical-mechanical properties of the sprayed material, mechanical properties of the substrate and processing parameters. The processing parameters have a strong effect on coating microstructure from submicrometer to nanocrystalline range with large improvement of adhesion and mechanical properties (Ajdelsztajn et al. [13]). Particle speed governs the material deformation mode which depends on particle composition and substrate material. For each kind of particle-substrate combination just a narrow range of processing parameters is allowable in order to obtain good efficiency and optimal microstructural and mechanical properties of the sprayed structures (Table 11.1).

The main parameters, affecting the final properties of the deposited samples, are working gas pressure and temperature. The working gas can be Air, He or N_2, normally if He is employed, more types of particles can be sprayed. Particle speed is related to gas temperature and starting particle dimensions (Fig. 11.3). The higher is the particle velocity and the smaller is the starting particles dimensions the higher

Table 11.1 Main parameters of cold spray technology

Spray material	Substrate	Particle speed (m/s)	Temperature (°C)
Ti, Ti6Al4 V	Ti, Ti6Al4 V	600–1,000	25–800
Ti, Ti6Al4 V	Steel	650–1,100	100–800
Al, AA7075	AZ91	650–1,300	25–400
Al, AA7075	AA7075	800–1,300	250–400
Ni, Ni–Cr	Ni, IN718	500–1,000	25–600
AISI316	Steel	600–1,300	350–800

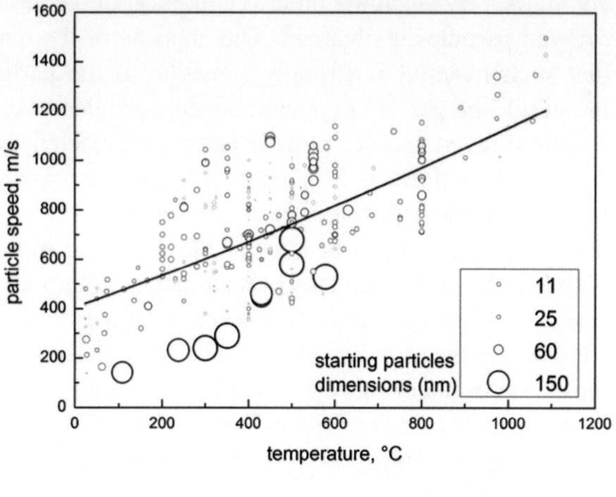

Fig. 11.3 Particle speed as a function of gas temperature and particle dimension

will result deposit hardness and grain refinement. Processing parameters result fundamental in the variation of adhesion and porosity of the deposits. The higher the pre-heat temperature is, the higher is the thermal energy retained by the particles impacting on the substrate. The deposition of the coatings is normally carried out with a single step at a high transverse velocity. Thus, the gas interaction with the substrate material is limited. In terms of cold spray coatings, it is important to note that the gas temperature will increase the substrate temperature. Normally the efficiency of the deposits in terms of porosity reduction is related to high particles deformation (related to the particles speed and pressure); such porosity is inversely proportional to the gas temperature. In fact, the higher is the temperature the larger results the melting due to deformation that conduces to an increase in porosity also in the case of high levels of adhesion strength. In the case of pure aluminum particles, sprayed on magnesium alloys substrates, adhesion strength normally shows maximum values for high temperature and pressure; in such conditions deposit porosity shows minimum levels. A similar behavior is observed for titanium particles sprayed on pure titanium substrate; in such cases high adhesion strength is shown for high pressure and intermediate temperature. A more energetic impact of the particles on the substrate leads to more plastic deformation; this part of the coating formation process leads to metallurgical bonding as well as porosity reduction.

In such case the processing conditions are optimal to reduce melting (Ti has a melting point very higher with respect to aluminum alloys) and to have a non-melting mediated deformation in order to reach low levels of porosity with very high levels of adhesion strength. Copper particles deposited on AA2024 and AA7075 show maximum values of adhesion strength in correspondence to high temperature and high pressure; in the same conditions the porosity of the deposits shows minimum levels. In the case of copper deposited on steel substrate the maximum values in adhesion strength and the minimum levels of porosity are reached for intermediate values of pressure and temperature. The same behavior has been observed in the case of pure titanium deposited on steel substrate. Such behavior is due to the very similar hardness between Cu-steel and Ti-steel, resulting in a strong jetting of the particles appearing just in a very narrow range of spraying conditions. By changing process conditions, changes in the in-flight velocity of cold sprayed particles is obtained. The analysis of the deposition of Ni particles shows that such material is strongly sensitive to the particles velocity (function of the pressure), the gas temperature and the particles size.

The sprayed metallic powder particles impacted onto the substrate in a solid state are heavily deformed in a very short period, typically less than 10 ns, and well bonded to the substrate. Furthermore, the severe and nearly adiabatic plastic deformation generates heating of the impacted region and forms very fine grains with the size of several tens of nanometers within the particles (Gang et al. [14]). It is suggested that the grain refinement results from in situ dynamic recrystallization. Actually, it seems that a higher particle velocity is necessary to make much larger plastic deformation. In the case of Ni particles sprayed on steel substrate, the best performances are in correspondence to intermediate values of pressure and velocity. The ability of a cold sprayed particles to undergo plastic deformation and adiabatic shear is a function of material properties, such as yield strength and melting point as well as process conditions, such as deposition temperature, particle velocity and particle size. When a particle reaches the substrate at its critical velocity, adiabatic shear takes place contributing to the formation of a metallurgical bonding. For each condition of substrate, particle pressure and temperature, it is possible to individuate a critical velocity of the particles. For speeds higher than such critical velocity, bonding is not achieved. The substrate deformation contributed to the formation of an intimate, possibly metallurgical, contact between the materials. In the case of substrate much more harder than the cold sprayed particles, it is possible to obtain good quality deposits with lower temperatures and particles velocity levels. As the difference between the hardness of the substrate and the particles decreases, it is necessary to employ more severe conditions in terms of high pressure and high temperature. In general, the yield strength of the material affects the critical velocity required to induce adiabatic shear and therefore the ability of the material to form a strong, metallurgical bond. An interesting aspect is represented by the grain size variation as a function of processing parameters. The deposit grain size decreases with increasing pressure and increasing gas temperature (directly related to the increase of the particle speed).

By focusing the attention on fatigue properties, increase in strength is possible just if a strong compressive residual stress state is induced on the surface. Such compressive state is possible by correctly couple coating-substrate materials and processing parameters. In aluminium alloys the increase in strength is strongly dependent on processing parameters; in magnesium alloys an increase in fatigue life is recorded for many conditions of materials coated with pure aluminium. In titanium alloys, Nickel alloys and Steel the increase in fatigue life is related to the particle hardness, if the hardness of impacting particles is higher with respect to the hardness of the substrate, higher levels of compressive stress can be reached with good effect on fatigue life.

An example of fatigue behaviour, showing the broad potential of the technology, of cold spray coatings is shown in Fig. 11.4.

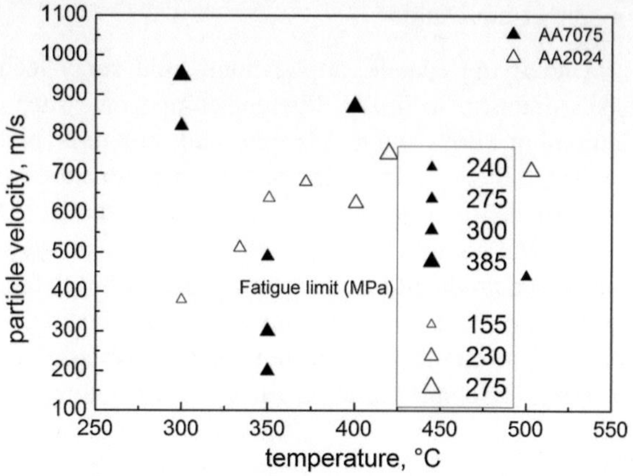

Fig. 11.4 Fatigue properties of AA7075 and AA2024 coated with AA7075

Such behavior can be observed just in cold spray high pressure equipment, normally portable low pressure machines induce lower levels of superficial residual stresses leading to a non-detectable improvement of fatigue properties. Residual stresses are strongly dependent on bulk properties. In the case of softer bulk, the deformation process strongly changes, a larger amount of deformation is transferred to the bulk with lower deformation levels and consequently lower residual stresses in the coating. The bonding mechanism in cold spray is mainly metallurgical. As a matter of fact, the deposition of particles with hardness very similar to the bulk one results very difficult. In the case of higher bulk hardness, an high deformation of particles is recorded leading to metallurgical bonding. Residual stresses depend on the strain effect due to the impact and the thermal misfit due to the temperature difference between the particles and the substrate. In the case of softer bulk, the effect of thermal misfit is much more less pronounced with respect to the strain effect. As additional demonstration, some experiments, performed on pre-heated substrates, show that over some temperature range, pre-heating leads to a stronger effect of thermal misfit on residual stresses also in the case of very high pressures employed during spray. Such aspect was also confirmed by numerical simulations performed in different processing conditions.

11.4 Applications

This technology will not replace any other thermal spray application but it will enlarge the range of possible applications in terms of capability of spraying different materials (Vlcek et al. [15]) and of fields of application (Gärtner et al. [16]).

The principal cold spray applications are (Maev and Leshchynsky [17]):

- corrosion-protective coatings production;
- wear resistance coatings;
- non-equilibrium alloys production;
- defects and damages repairing;
- direct fabrication.

Underlying specific applications, cold spray technology is employed in aerospace industry to repair different components such as aluminum alloys boosters, aluminum alloys wing, fuselage and structural parts (Pattison et al. [18]). It is finding a broad field of application in production and restoration of high temperature oxidation resistant coatings based on the NiCrAl and NiCrAlY systems. The application to the formation of aluminum alloys coatings on Mg-based alloys allows the production and repairing of gear boxes for the helicopter industries. Such technology open the possibility to produce coatings for Mg alloys not available at the present due to the high temperature oxidation behavior of magnesium. Ni–Cr coatings are employed to repair high temperature components of aerospace rockets.

Table 11.2 Cold spray fields of application

Particles	Substrate	Application
Pure Al	Al, and Mg alloys	Corrosion
Zn, Zn alloys	Steel	Corrosion
Ni, Ni alloys	Steel, Ti alloys	Corrosion, wear
Al-based nanocomposites	Al alloys	Wear
Ni-based nanocomposites	Ni alloys	Wear, corrosion
7XXX Al alloys	Al alloys	Fatigue
Ni, Ti alloys	Ni, Ti alloys	Restoration and repair
Steel	Steel	Restoration and repair

Cold spray has a large appeal in the modern industry because of the eliminate or reduce fabrication steps. Because of the stringent becoming of health and safety regulation its application is crucial because cold spray is a good gran alternative to other similar production techniques. By evolving, cold spray applicability will occupy a broad field of non-traditional production technologies. A schematic view of the possible applications as a function of different particle-metal combinations is given in Table 11.2.

11.5 Conclusions

Cold spray is the emerging technology in the broader family of thermal spray. Such technology do not replace any one else but enlarge the rate of temperature and particle speed that can be employed during coatings or bulk production. In the present chapter a detailed description of cold spray advantages with respect to other coating technologies is described. The possibility to work at lower temperature and manipulate the material in the solid state allows the obtaining of nanostructured materials with superior mechanical and physical-chemical properties. In the present dissertation, many examples of coatings production optimization are presented. In particular, the effect of processing parameters on coatings microstructural and mechanical behavior is described. Large emphasis is given to the improvement of fatigue properties through cold spray coatings for aluminum alloys. Further extensive research are needed to optimize processing conditions, nozzle design, substrates and particles properties, substrate-particle combinations and post deposition treatments. The implementation of finite element modelling of process and in service behavior is wide open.

References

1. Champagne VK, Leyman PF (2008) Repair of apache mast support on AH64 helicopter using cold spray. In: Failure prevention for system availability—proceedings of the 62nd meeting of the society for machinery failure prevention technology
2. Ogawa K, Niki T (2010) Repairing of degraded hot section parts of gas turbines by cold spraying. Key Eng Mater 417–418:545–548
3. Nooririnah O, Khalil AY, Aludin MS, Rohana A, Zuraidah RR (2013) Overview potential applications of cold spray process for aviation industry in Malaysia. Adv Mater Res 701:735–738
4. Villafuerte J (2010) Recent trends in cold spray technology: looking at the future. Surf Eng 26 (6):393–394
5. Liu Q et al (2010) Surface modification and repair for aircraft life enhancement and structural restoration. Mater Sci Forum 654–656:763–766
6. Birtch W (2012) Advances on the repair and dimensional restoration of magnesium and aluminum components by cold spray. In: Source of the document national association for surface finishing annual conference and trade show 2012, SUR/FIN 2012
7. Cavaliere P, Silvello A (2014) Processing parameters affecting cold spray coatings performances. Int J Adv Man Tech 71:263–277. doi:10.1007/s00170-013-5465-0
8. Sova A, Grigoriev S, Okunkova A, Smurov I (2013) Potential of cold gas dynamic spray as additive manufacturing technology. Int J Adv Man Tech 69:2269–2278
9. Assadi H, Gärtner F, Stoltenhoff T, Kreye H (2003) Bonding mechanism in cold gas spraying. Acta Mater 51:4379–4394
10. Schmidt T, Gärtner F, Assadi H, Kreye H (2006) Development of a generalized parameter window for cold spray deposition. Acta Mater 54:729–742
11. Klinkov SV, Kosarev VF, Rein M (2005) Cold spray deposition: significance of particle impact phenomena. Aerosp Sci and Tech 9:582–591
12. Fauchais P, Montavon G, Bertrand G (2010) From powders to thermally sprayed coatings. J Ther Spray Tech 19(1–2):56–80
13. Ajdelsztajn L, Jodoin B, Schoenung JM (2006) Synthesis and mechanical properties of nanocrystalline Ni coatings produced by cold gas dynamic spraying. Surf Coat Technol 201:1166–1172
14. Gang J, Morniroli JP, Grosdidier T (2003) Nanostructures in thermal spray coatings. Scr Mater 48:1599–1604
15. Vlcek J, Gimeno L, Huber H, Lugscheider E (2005) A systematic approach to material eligibility for the cold-spray process. J Ther Spray Tech 14(1):125–133
16. Gärtner F, Stoltenhoff T, Schmidt T, Kreye H (2006) The cold spray process and its potential for industrial applications. J Ther Spray Tech 15(2):223–232
17. Maev R, Leshchynsky V (2006) Air gas dynamic spraying of powder mixtures: theory and application. J Ther Spray Tech 15(2):198–205
18. Pattison J, Celotto S, Morgan R, Bray M, O'Neill W (2007) Cold gas dynamic manufacturing: a non-thermal approach to freeform fabrication. Int J Mach Tool and Manuf 47:627–634

Part IV
System Degradation and Design in TES

Chapter 12
Through-Lifecycle Aspects for Functional Products to Consider During Development and Operation: A Literature Review

John Lindström

Abstract The paper, which is based on a literature review, concerns which potential through-lifecycle aspects are relevant to consider for Functional Products during development and later operation until end-of-life. The aspects which are already proposed as part of the current definition of Functional Products are corroborated. Additionally, the additional new potential aspects found should be further verified prior to being proposed to extend the Functional Products definition—and in particular the service-support system and management of operation constituents. An additional seven potential new aspects have been found, whereof some may be relevant for the concepts of Through-life Engineering Services, Product-Service Systems and Industrial Product-Service Systems as well.

12.1 Introduction

This paper, based on a literature review, concerns potential through-lifecycle aspects for Functional Products (FP[1]) to consider during development and later operation until end-of-life. The main objective for FP is to provide a function to customers with an agreed upon level of availability. Availability can be seen as a function of reliability and maintainability. The duration of FP contracts can range

[1] The concept of FP has similarities with, for instance, Functional Sales [1], Extended Products [2], Total Care Products [3], Product-Service System (PSS) and Industrial Product-Service Systems (IPS[2]) [4], Servitizing [5], Service Engineering [6] or Through-life Engineering Services (TES) [7] in the sense of increasing the focus on soft parts such as services, knowledge and know-how, etc., additionally offered. Tukker and Tischner [8] have identified three main PSS categories: product-oriented, use-oriented and result-oriented. The FP, originating from hardware aspects, has most commonalities with PSS/IPS[2], Total Care Products, TES and Functional Sales, however adding additional complexity development-wise.

J. Lindström (✉)
ProcessIT Innovations, Luleå University of Technology, 971 87 Luleå, Sweden
e-mail: john.lindstrom@ltu.se

L. Redding and R. Roy (eds.), *Through-life Engineering Services*,
Decision Engineering, DOI 10.1007/978-3-319-12111-6_12

up to 20–30 years, which forces providers to consider through-lifecycle aspects, or long-term challenges, at an early stage, so as to avoid excessive cost levels and diminished incomes later on. FP has been defined by Lindström et al. [9–11] as integrated hardware, software, a service-support system and management of operation. The software component grows as the requirements for monitoring, remote management and maintenance, as well as software upgradability become further sophisticated [12, 13]. The software is often integrated with the hardware and service-support system and, depending on type of functional product, can also be seen as a stand-alone entity providing its own value to the delivery of the function. The service-support system is needed to keep the hardware and software operable, and the triad is combined to provide a complete function to customers [3, 14]. Alonso-Rasgado et al. [3] add that the service-support system is much more than maintenance, often including decision-making, operations planning, remanufacture and education. Throughout the FP lifecycle, operation of the FP must be managed, modelled, simulated, optimized and further developed.

Among many corporations, there is an increased interest in use of the FP business model as a means to improved competiveness and sustainable profitability. Recent research [15, 16] indicates that a provider can, if having an adequate set-up in terms of network relations, skills and competences, increase the profitability significantly when offering integrated products and services or functions—compared to add-on services for products which might not be profitable at all. Moving into FP provision with a specified availability level requires identification of a future win–win situation with the customer, since the ownership of the FP is foreseen to remain with the provider or consortium providing the FP. Thus, the provider of an FP needs to be able to simulate at least the most important parts of the system (or if possible the whole). The simulation is necessary in order to assess the cost for different availability levels in order to find the cost drivers early on in the FP process and be able to manage them [17, 18].

The most current research on FP through-lifecycle aspects, e.g. Lindström et al. [11], using a structural perspective, addresses aspects including: business cases and business modelling, lifecycle management, asset management, availability, risks, monitoring, maintenance, etc. Further, similar such aspects for FP have been covered comprehensively by, among others, Alonso-Rasgado et al. [3], Lindström et al. [9] and Karlsson et al. [19]. Karlsson et al. [19] address the aspects from a visionary perspective, whereas Alonso-Rasgado et al. [3] and Lindström et al. [9] look at them in development and partially operational contexts. Some of the aforementioned aspects such as business cases, lifecycle management and asset management can be considered very inclusive and comprise a lot of aspects. However, this paper uses a detail-oriented view and brings up all relevant aspects found. Some aspects can be foreseen and planned for during the development. However, there must necessarily be an attentiveness to look for additional new aspects from the start of operation of the FP—and manage the discovered ones adequately and in a timely manner so as to prevent them from becoming future major (unsolvable) problems.

There is a lot of related research within the closely related concepts TES, PSS and IPS^2 addressing one or a few through-lifecycle aspects relevant for FP. However, looking for further comprehensive studies covering multiple such aspects, little work has been published, with the exception of Meier et al.'s [4] overview of the IPS^2 concept. In addition, TES and its through-lifecycle aspects have been addressed by for instance Toossi et al. [20] using a causal loop diagram to map through-life challenges and opportunities. Further, Redding et al. [21] identify a number of maintenance-repair-overhaul (MRO) challenges in TES. Within the PSS concept, Baines et al. [22, 23] and Isaksson et al. [24] provide, from a general level, a number of relevant through-lifecycle aspects and challenges ranging from business cases, business modelling, lifecycle management (environment and costs) and risks to maintenance and monitoring, etc. As previously mentioned, Meier et al. [4] look at, from a high-level perspective, the general IPS^2 through-lifecycle aspects of business modelling, lifecycle management, costs, design and development, operational/organizational structures, maintenance, knowledge, risks and legalities. Additionally, Meier [25] adds planning and development in terms of business modelling, strategy, lifecycle impact and development processes and methods, while Rese et al. [26] bring up business models, lifecycle management and risk while discussing an ontology of business models for IPS^2.

Issues may turn into small or large challenges as the contracted timeframe exceeds the expected life time, one or several times, for key constituents/portions/components part of the FP offering. Further, as time passes the same goes for general technology and its related human skills and knowledge as their lifecycle/availability/actuality ends on the market for other contexts. Other matters, which are hard to control for individual providers, but may have large impact may include changes in requirements, usage/application, legal/regulatory frameworks, etc. Further, often the expected lifecycle (both economic and technical) is extended much longer than originally foreseen due to economic reasons. Examples of such are public transportation systems such as railways and undergrounds [27]. Baker [28] posits a different view, and argues that as lifecycles are shortened for many offers, their through-life plans no longer assume mid-life upgrades or technology insertions. However, if such offerings are sold as FP instead of products and services, it is up to the provider to keep them in shape to honour the availability level agreed upon in the contract. Thus, the provider of FP needs to be compensated for the foreseen costs and risks in order to create a sustainable long-term win-win situation between the provider and customer. There are some special situations where the provider is locked and the possibilities for re-design and re-manufacturing are small or at least very costly, and that is when any change may require a re-certification or very thorough re-testing of the whole function. Placed upon a provider, such limitations may give rise to very high costs and limit improvements during the contract.

Currently, there is a dearth of a literature providing a comprehensive overview on through-lifecycle aspects regarding development and operation of FP. However, several relevant through-lifecycle aspects have already been covered in related literature, and these are compiled into an comprehensive overview in this paper.

Thus, the research question addressed in this paper can be formulated as: which potential through-lifecycle aspects need to be considered during development and operation of FP?

The purpose of the paper is to highlight which through-lifecycle aspects are, or may be, relevant to consider for FP customers, providers and researchers. The paper begins with an introduction (this section), followed by sections with research approach, literature review, and finally discussion and conclusions.

12.2 Research Approach

The research approach was based on a literature review. The literature review was initially limited to literature since 2011 for the following concepts: FP, TES, PSS and IPS^2, in order to capture the latest developments regarding through-lifecycle aspects (and there is plenty of recent research). For a start, proceedings from some of the most relevant conferences[2] related to the concepts and problem addressed were manually screened, in an open search, to be able to find as many relevant through-lifecycle aspects as possible (although they sometimes were addressed quite differently). In total, more than 330 papers were manually screened, after which a complementary search using search engines and databases was conducted. The searches used the aspects found, "long-term challenges or issues or aspects", and "future challenges or issues or aspects" as search parameters. The main purpose of these on-line searches was to find complementary literature for those concepts (i.e. FP, TES, PSS and IPS^2) where few or no papers had already been found. If no relevant literature was found for a concept and aspect within that timeframe, the search was extended another 2 years at a time back until 2005. In total, approximately 600 papers, book chapters and books were screened. The main objective of the literature has been *to find and compile a comprehensive set of potential FP through-lifecycle aspects* and not to list all existing/relevant literature. Thus, some relevant literature may be missing in Table 12.1, while most potential FP through-lifecycle aspects have hopefully been found.

12.3 Literature Review—Existing and Potential New Through-Lifecycle Aspects for Functional Products

The result of the literature review is a set of potential through-lifecycle aspects to consider during development and operation for FP providers and customers. Some of the aspects are of a general nature, and can be seen as comprising, or partly

[2] TES—1st and 2nd International Through-life Engineering Services Conferences.
IPS^2—3rd–5th CIRP International Conferences on Industrial Product-Service Systems.

Table 12.1 Set of potential FP through lifecycle aspects found in literature

Through-lifecycle aspect	References	Conc
(PN) Business model and business case – to model the FP business as well as create and maintain an FP business case. The FP business case should outline for instance the business logic, rationale, strategy and sustainability of an FP offer. Further, the business case should describe the value for customers, how to keep or improve the level of value perceived by the customers, and how to reach a necessary win-win situation. The business case should be active from initial business modelling to later refinements and potential extensions using real data while the FP is in operation at customers until end of the lifecycle.	[9, 10, 19]	FP
	[20]	TES
	[22–24, 29–33]	PSS
	[4, 25, 26, 34–44]	IPS^2
(PN) Cost drivers – it is necessary to investigate and simulate the cost drivers in an FP scenario due to the often long contracted duration in order to find economically sustainable options. The cost drivers need to be identified as early as possible, be quantified, and used as input to simulations. Of particular interest is availability (i.e. the ratio between level of availability and cost). Other major potential cost-drivers are, for instance, the service-support system and its set-up, asset and obsolescence management, power/fuel (e.g. electricity), insurance, and ICT including communication networks. Further, costs to develop, build and maintain: knowledge, information, trust and relations should not be underestimated from start.	[3, 9, 19]	FP
	[20]	TES
	[33, 45–49]	PSS
	[4, 37, 47, 50–54]	IPS^2
(MOD) Lifecycle engineering and management – due to the often long durations of FP lifecycles, it is advisable to use a model to evaluate, for instance, the economic- (LCC), environmental (LCA) and where adequate also societal impact and sustainability of an FP at early as well as later stages of the lifecycle. Further, lifecycle management systems, such as PLM/PDM, integrated with additional information systems/sources, can assist decision-making during the lifecycle of the FP.	[3, 11, 55]	FP
	[56]	TES
	[22, 24, 32, 45, 48, 49, 57–70]	PSS
	[4, 25, 26, 36, 37, 43, 52–54, 61, 71–73]	IPS^2
(E) Availability management – in FP scenarios it is necessary to honour the availability level in contracts while keeping a sustainable ratio between the level of availability and costs. Availability can be seen as a function of reliability and maintainability. Thus, if the reliability is poor, the FP should be designed to be maintainable to lower the costs as more maintenance efforts are likely required during the operational part of the lifecycle. Consequently, if the reliability is high, less maintenance will be required and maintenance costs will be lower. Commonly, availability is the premier performance indicator for FP in operation. However, there can of course be additional ones related to productivity or other forms of customer value.	[3, 9–14, 17–19, 55, 74]	FP
	[7, 20, 75]	TES
	[32, 76, 77]	PSS
	[7, 50, 78]	IPS^2

(continued)

Table 12.1 (continued)

Through-lifecycle aspect	References	Conc
(E) Risk management – to manage and mitigate risks is crucial in most FP scenarios and these activities need to be conducted continuously until the end of the lifecycle. The risk management may span, for example: development related risks, residual risks post initial development, operational risks, obsolescence risks, market/business risks, financial risks, environment risks, supplier coordination risks, communication risks, service-support system risks, and innovation risks. Risks can on a high level be divided into three segments according to when they occur: during creation, delivery or capture of value [79]. Another aspect of risk in an FP scenario is how the risks should be shared between the provider (or provider consortium) and the customer. This needs to be agreed upon in the contract and potentially be updated as more is learnt on the risks and their management during the operational part of the FP lifecycle.	[3, 9–11, 19, 55, 79]	FP
	[7, 80]	TES
	[22, 29, 45, 63, 77, 81–84]	PSS
	[4, 26, 34, 36, 50, 51, 53, 81, 85, 86]	IPS^2
(PN) Asset management – asset management involves, among other things, keeping track of what assets an organization has and where, as well as planning for stocking/rolling stock/procurement of components/sub-systems, obsolescence and replacement schedules, recycling issues, etc. In a FP scenario, the three main constituents: hardware, software and service-support system (comprises many sub-constituents of a tangible nature), are the main concern for asset management – which can be seen as an umbrella activity. Hardware, the tangible asset/physical artefact/product/equipment, needs to be kept operational to meet the contracted level of availability. Maintenance/service, replacement parts, re-manufacturing and re-development, etc. must be considered from the view point of economy and practical sustainability. Software may in an FP scenario span operating systems, communication software, tools, portals and development/run-time environments, etc. The software part may involve software that is owned, licensed or bought as a service (i.e. software- or platform-as-a-service). In addition, a variety of actors such as the FP providers' own IT-personnel, consultants, software vendors, and service providers can be expected to participate in software- (and hardware-) related activities. Further, how to run, maintain, update and patch software over time needs to be considered as well. The service-support system can in an FP context span a number of assets such as on-line and off-line monitoring systems, analytic systems, warning/notification systems, maintenance and support instructions/manuals, information and information systems, knowledge and knowledge management systems, education and training courses/materials, etc. Further, the service/support organization (which comprises for instance: structures, processes, procedures, humans as well as vehicles, tools, spare parts and repair kits) can be considered to contain a number of assets that need to be cared for.	[11]	FP
	[27, 87–89]	TES
	[61, 83, 84, 90, 91]	PSS
	[61]	IPS^2

(continued)

Table 12.1 (continued)

Through-lifecycle aspect	**References**	**Conc**
A global ICT infrastructure can in an FP scenario be seen as an asset and integrator of hardware, software and the service-support system as it connects them into the overall operational framework for FP. It may comprise networks and infrastructure, which is owned, leased, bought as a service (for instance platform-as-a-service), or hybrids of the aforementioned. It is necessary for FP providers (and customers) to manage the ICT infrastructure well and follow the development of technology and standards in order to ensure communication with the FP and all its necessary linked constituents. If such pro-activeness is missing, likely costly bridges to old technology stacks are necessary.		
(PN) Obsolescence management – as FP lifecycles can be long, e.g. up to 30 years, original portions such as components, sub-systems or technologies in the FP main constituents hardware, software and service-support system will need updates as well as replacements. If such portions become obsolete (i.e. unavailable or not possible to use due to other reasons) before the FP demand for them is exhausted, there will be an obsolescence management problem. Parts of this problem can be solved with stocking components, but it is expensive, binds capital, requires storage space and related efforts. For FP, a trade-off on stocking versus continuous re-development or re-manufacturing needs to be made. This may not be easy if the FP comprises a lot of new or recently developed systems/sub-systems, which do not yet have a service/support history. Further, the potential software parts will likely require a faster update scenario than the hardware-related, and the full software platform needs to be factored in as suppliers of operating systems, compilers, software frameworks, application servers or run-time environments, etc. used may get out-dated and vendor support stopped unless a certain update schedule is followed. The service-support system may also rely on certain hardware and software, for instance, sensors, data extraction components, network components/appliances, communication networks, highly scalable analytic tools, databases and various information systems, etc. Thus, in an FP context, in addition to long-term planning, there may be hardware, software and service-support system considerations that need to be made. In addition, soft items such as human skills and knowledge need to factored in as FP lifecycles exceed the lifecycles or hardware and software technologies used. Therefore, the foreseen cost related to obsolescence management needs to be factored in when pricing and negotiating the length of the FP contract as well as the level of availability that can be honoured over time.	NRRF	FP
	[7, 20, 28, 80]	TES
	[61, 83, 90–92]	PSS
	[61]	IPS^2

(continued)

Table 12.1 (continued)

Through-lifecycle aspect	References	Conc
(PN) Self-healing systems and autonomous systems – self-healing systems try to detect issues and then mitigate and recover from any failures. This makes it possible to reduce maintenance (by humans) through improved autonomy or autonomous maintenance. Autonomous systems need to be self-configurable to adapt to the context and application, etc. In FP scenarios, self-healing and autonomous systems may be used where critical to uphold availability for the function delivered and it is hard to maintain or change/reconfigure. However, as these are likely more expensive to develop and produce, it is not likely they will be used elsewhere until hard data from operations show it is necessary.	NRRF	FP
	[7, 93–95, 146]	TES
	NRRF	PSS
	[78]	IPS^2
(PN) Change management – covers change, variations, change orders, change management processes, and control of change. Change management is necessary to apply both during initial develop as well as later during the operational part of the FP lifecycle to manage changes in requirements and other necessary or mandatory changes. Change management can be seen as an umbrella for configuration management and vice versa [101].	[9]	FP
	[96]	TES
	[70, 73, 97]	PSS
	[25, 71, 98]	IPS^2
(PN) Configuration management – applying adequate processes, resources, tools, etc. in order to achieve consistency/integrity between requirements, constituents (which in an FP context means hardware, software and service-support system) and configuration information. Further, traceability of changes to configurations over time is also often needed or regulated.	[99]	FP
	[100, 101]	TES
	[102]	TES/ PSS
	[83, 97, 103]	PSS
	[104]	IPS^2
(E) Remote on-line and off-line monitoring – real-time analysis of pertinent data streams (originating from sensors and other data sources with extractors) combined with meta data and stored/historical data is necessary or preferred to monitor and diagnose the FP - to be able to react quickly if there are signs of breakdowns or need for reactive/corrective maintenance, as well as planning for proactive maintenance. Further, the analysis can provide input to re-designs/re-development and re-manufacturing if weak areas/patterns or components/sub-systems are found during operation.	[11, 13, 18, 19]	FP
	[21, 87, 89, 105–108]	TES
	[24, 76, 109, 110]	PSS
	[4, 78, 111, 112]	IPS^2

(continued)

Table 12.1 (continued)

Through-lifecycle aspect	References	Conc
If it is possible to achieve planned and proactive (predetermined and condition-based) maintenance, instead of reactive/corrective, maintenance cost-effectiveness is likely improved and operational losses minimized, since unplanned interruptions are minimized.		
(E) Maintenance, self-service and support – to carry out general service and support, which may include maintenance/repair/overhaul (MRO), self-service, support, etc., requires that a number of matters are planned and coordinated, e.g. service-support organization structure, levels of service, general support, supply chain integration and potential partnering/outsourcing, etc. Asset management and availability management have a large impact on maintenance and its set-up as well as planning. In complex scenarios modelling and simulation of the FP service-support system may be necessary to optimize cost versus availability, and thus achieve an overall sustainable FP service-support system. Design for maintenance or maintainability is of great interest for the FP context (as availability can be seen as a function of reliability and maintainability). Another perspective on maintenance is to use risk based maintenance. Further, the data, collected manually or automatically, on maintenance and other related activities should preferably be stored digitally to be easily searchable and analyzable (see information and knowledge management in Table 14.1, below), used proactively, and when appropriate turned into information or knowledge.	[3, 11, 13, 18, 55, 74]	FP
	[7, 20, 21, 75, 89, 106, 113–115]	TES
	[116]	FP/ PSS
	[23, 24, 29, 47, 59, 76, 81, 83, 90, 91, 97, 117, 118]	PSS
	[4, 7, 34, 47, 81, 111]	IPS^2
(E) Information and knowledge management – may in an FP scenario comprise information systems, secure knowledge management systems, knowledge and know-how sharing, education and training. As time passes and information systems, database software, etc. grow old, it will be hard to keep information and knowledge stored digitally unless preserved and made available (in potentially new data formats) for new technologies. Regarding the human factor, if there is a high level of employee turnover – it will be tough to keep an adequate level of knowledge and share it unless it is written down in a pedagogic manner combined with pictures or videos, etc. and continuously studied/taught/practiced with mentors. Further, service engineers, customers, operators/users often manually gather information about the installed base of machines or equipment, etc. which, if stored digitally and managed adequately, can provide intelligence and insight for potential improvements or benefits. Thus, manual notes and other observations should when appropriate be stored in an information system. In particular, deviations, malfunctions and continuous issues should be stored to be searchable for statistical and pattern recognition purposes – to enable improvements and optimizations.	[3, 11, 55, 119]	FP
	[120–123]	TES
	[124–126]	PSS
	[42, 43, 127–130]	IPS^2

(continued)

Table 12.1 (continued)

Through-lifecycle aspect	References	Conc
(E) Financial issues and processes – the potentially long FP lifecycles expected make financial planning and forecasting difficult. To be able to plan for and solve financial issues during the whole lifecycle, financial plans and processes need to be developed and adapted to support the FP and its long-term management of operation. FP ownership issues, which can involve many partners, can further be considered as part of management of financial and contract issues.	[10, 11]	FP
	[27, 131]	TES
	[29, 47, 63, 84, 90, 97]	PSS
	[47, 127]	IPS^2
(E) Contract and contract management – the contract between the FP provider and customer regulates the business relationship over time. It is important to keep the contract up to date, as the surrounding environment will likely change over time and it is necessary to maintain a win-win situation. Thus, the business relation and contract terms may require changes or additions to remain sustainable. Contracts are necessary to formalize matters such as: terms, obligations, liabilities/penalties, extension/termination of business relationship, revenue/cost/risk sharing, management of intellectual property, renegotiation of contract, etc. Further, FP ownership issues, management of brand/status/image, as well as provider consortium and partner contracts, can be seen as part of this.	[3, 10, 11]	FP
	[7]	TES
	[29, 47, 63, 81, 97, 132]	PSS
	[4, 47, 81]	IPS^2
(E) Management and transfer of intellectual property – during the FP lifecycle the FP will likely be further developed and additional or new intellectual property created. The intellectual property needs to be managed and some may be transferred to consortium partners or customers. In addition, some intellectual property may be transferred from the customer to the consortium to improve the FP. Intellectual property management is necessary to keep and sustain the trust among the partners and customers involved, as well as to protect the competiveness of the FP and its future revenue. Contract management should cover mutual intellectual property management to avoid problems and unnecessary disputes. Another aspect of management of intellectual property, which also affects asset/obsolescence management and maintenance, is counterfeit parts or sub-systems. If FP can detect "not recommended" or counterfeit parts, based on a built-in verification/validation of the parts, sub-systems and function offered, that enables use of maintenance/service partners in areas with weak legal support against counterfeiting or illegal copying. The detection is well suited for critical parts or sub-systems, which may have serious impact on the availability (if breaking, failing or malfunctioning).	[3, 11]	FP
	NRRF	TES
	[33, 81, 133–135]	PSS
	[42, 43, 81, 144]	IPS^2
(E) Building up trust and relations – in a long-term contract the trust and relations must be developed and maintained, between the provider side and the customer, in order to reach a sustainable win-win scenario. If the balance is skewed, and one part is on the losing side, a long-term contract will be hard to execute and bad-will may arise. In addition, continuous network learning and change of mind-set are necessary among the parties involved. Further, on the provider side, network or consortium management is necessary to keep the development and operational activities/resources aligned with the contract's obligations over time.	[3, 10, 11, 136]	FP
	NRRF	TES
	[29, 81, 137–139]	PSS
	[34, 47, 81, 140, 141, 145]	IPS^2

overlapping, other aspects in an FP scenario. Examples of such "inclusive" aspects are: business case, cost drivers, lifecycle management, availability management, risk management, asset management, etc. In Table 12.1, such inclusive aspects have been split apart into relevant high-level aspects. In Table 12.1, the aspects found have been roughly grouped together based on commonalities and the groups are separated by grey horizontal bars. Further, aspects that can be regarded as holistic or affecting whole organizations, e.g. quality, information security, safety, digital preservation, etc., are not brought up as stand-alone and inclusive aspects but highlighted where appropriate in Table 12.1.

All relevant aspects found are briefly described, and references from the FP, TES, PSS and IPS^2 concepts are provided where available. The through-lifecycle aspects which are already part of the current FP definition are marked as existing (E) and the potential new ones as (PN). One existing aspect which has been modified is marked as (MOD). If no references were found for a concept, that is marked with no relevant-reference-found (NRRF). The word "concept" has been abbreviated "Conc" in the header of Table 12.1.

The management and coordination of the FP through-lifecycle aspects are, of course, crucial for achieving a sustainable result. Lindström et al. [9, 11] propose the use of a specific main constituent, i.e. management of operation, to manage, coordinate and control what is necessary for a successful operation of the FP over time. Meier et al. [142] propose an Execution System for IPS^2 (IPS^2-ES), to assist in the coordination of tasks, etc. among a potential set of partners. The IPS^2-ES can further be integrated with an IPS^2 Control System (IPS^2-CS), as proposed by Uhlmann and Raue [143], to enhance coordination of support and service processes. Thus, having a through-lifecycle perspective already from the very start of the FP planning and development is necessary to avoid costly and potentially fatal issues and problems later on during the lifecycle.

The seven potential new aspects found during the literature review, which are not already part of the proposed current FP definition [11], are: business model and business case, cost drivers, asset management, obsolescence management, self-healing systems and autonomous systems, change management, and configuration management. These aspects need further verification prior to being proposed to extend the current FP definition. One existing aspect, lifecycle management, was modified to also include engineering, i.e. resulting in lifecycle engineering and management.

The literature review indicates that some of the aspects in Table 12.1 may need additional research within the FP context and, in some cases, for the TES, PSS and IPS^2 contexts as well. In particular, the following aspects seem to need further context-specific research within FP: asset management, obsolescence management, self-healing and autonomous systems, change management, configuration management, and financial issues and processes.

12.4 Discussion and Conclusions

The paper makes a contribution to theory by corroborating the current definition of FP [11] as well as highlighting a set of seven potential new through-lifecycle aspects. These seven ones need further verification prior to being proposed to extend the current FP definition. Further, one existing aspect was modified.

The current FP definition and set of potential new FP through-lifecycle aspects can be used by FP customers, providers and researchers. The customers can use them in their reasoning with FP providers, while considering requirement specifications during procurement or negotiations, or when analyzing how to optimize any current FP operations and planning. The providers can use them when considering the development and operation of their current FP offers, or in analyses and considerations of new FP offers. Further, researchers can use them when addressing FP in general as well as the related through-lifecycle aspects. Parts of the results may, depending on context and application, also be applicable for offers based on the TES, PSS, and IPS^2 concepts.

The service-support system and management of operation are essential main constituents for FP and their lifecycles, and a means to help find a sustainable win–win situation between providers and customers. In particular, it is necessary to be able to manage the through-lifecycle aspects where large costs and risks are involved, e.g. asset management, and do it in an elaborately considered manner. Further, it seems necessary for FP providers to interact and collaborate with researchers on development, operation and innovation regarding FP to inspire, challenge the existing perceptions, and trigger new ideas and perspectives in order to improve key parts or the wholeness of the FP offered or planned for. Through-lifecycle innovation is, of course, something that is of interest for FP as well as most other areas. Innovation can be seen as an integrated topic which is addressed by FP providers, customers and researchers in order to continuously improve the wanted win–win situation and further increase sustainability. If or when the innovation slows down or stops, the lifecycle will likely be shortened and the possibilities for charging a premium diminished.

Additional aspects, such as quality, information security, safety, digital preservation, etc., which can be seen as holistic aspects affecting large parts of FP, may be considered as either requirements or activities within the initial or post development efforts or as potential standalone FP constituents. The paper has highlighted these additional aspects as part of the through-lifecycle aspects listed in Table 12.1. A further aspect of interest may be big data, which depend on how much data is generated in terms of development/modelling/simulation/test data, meta data, monitoring data, stored statistical data from the operation, etc., during the lifecycle of the FP. Today, practitioners are only beginning to see what big data, generated from FP and their constituents, combined with global ICT-infrastructures and advanced analytic tools can be used for within FP.

A reflection on the literature covered is that a large part of the literature is of a conceptual nature, and there is a need, where appropriate, to verify and validate results in order to make the result additionally usable for practice in industry. However, this requires that industry is willing to participate in such activities as well. A further reflection regarding the literature review is that there are a number of aspects (see Table 12.1) which may need additional research within the FP context, as well as in some cases for the TES, PSS and IPS^2. In particular, for FP, the following aspects seem to need further context-specific research: asset management, obsolescence management, self-healing and autonomous systems, change management, configuration management, and financial issues and processes. Some of these aspects can be considered as driving costs, while others are seen to control/minimize costs, and are therefore significant for a sustainable FP business.

The type of FP, the application, the duration of the planned lifecycle, as well as the context and surrounding environment determine which through-lifecycle aspects are relevant for FP customers and providers. Table 12.1 highlights a number of through-lifecycle aspects, existing as well as potential new ones, to consider and investigate if they are relevant for a specific context as well as to what degree. Additional new aspects found by providers and customers during development and operation of FP should be added to the set of aspects that must be continuously considered.

Acknowledgments This work was conducted at the VINNOVA Excellence Centre the Faste Laboratory at Luleå University of Technology, Sweden.

References

1. Stahel W (1997) The functional economy: cultural and organizational change. In: Richards DJ (ed) The industrial green game: implications for environmental design and management. National Academy Press, Washington DC
2. Thoben K-D, Eschenbächer J, Jagdev H (2001) Extended products: evolving traditional product concepts. In: Proceedings of 7th international conference on concurrent enterprising, Bremen, 27–29 June 2001
3. Alonso-Rasgado T, Thompson G, Elfstrom B-O (2004) The design of functional (total care) products. J Eng Des 15(6):515–540
4. Meier H, Roy R, Seliger G (2010) Industrial product-service systems—IPS2. CIRP Annals Manufact Technol 59(2):607–627
5. White AL, Sloughton M, Feng L (1999) Servicizing: the quiet transition to extended product responsibility. Tellus Institute, Boston
6. Tomiyama T (2001) Service engineering to intensify service contents in product life cycles. In: Proceedings of EcoDesign 2001, 2nd International symposium on environmental conscious design and inverse manufacturing, Tokyo, Japan
7. Roy R, Shaw A, Erkoyuncu JA, Redding L (2013) Through-life engineering services. Measur Control 46(6):172–175
8. Tukker A, Tischner U (2005) New business for old Europe—Product-service development, competitiveness and sustainability. Greenleaf Publishing, Sheffield
9. Lindström J, Karlsson L, Löfstrand M, Karlberg M (2012) A development process for functional products: hardware, software, service support system and management of operation. Int J Prod Dev 16(3/4):284–303

10. Lindström J, Plankina D, Nilsson K, Parida V, Ylinenpää H, Karlsson L (2013) Functional products: business model elements. In: Proceedings of 5th CIRP international conference on industrial product-service systems, Bochum, 14–15 Mar 2013
11. Lindström J, Sas D, Lideskog H, Löfstrand M, Karlsson L (in press) Defining 'Functional Products' through their constituents. Int J Prod Dev 20(1)
12. Alzghoul A, Löfstrand M (2011) Increasing availability of industrial systems through data stream mining. Comput Ind Eng 60:195–205
13. Alzghoul A, Löfstrand M, Karlsson L, Karlberg M (2011) Data stream mining for increased functional product availability awareness. In: Proceedings of 3rd international conference on industrial product service systems, Braunschweig, pp 237–241, 5–6 May 2011
14. Li J-P, Thompson G, Alonso-Rasgado T (2009) Simulation based Reliability assessment of services in the context of functional products. J Saf Reliab Soc 29(4):47–78
15. Kothamäki M, Partanen J, Parida V, Wincent J (2013) Non-linear relationship between industrial service offering and sales growth: the moderating role of network capabilities. Ind Mark Manage 42(8):1374–1385
16. Parida V, Sjodin DR, Wincent J, Kohtamäki M (2014) Mastering the transition to product-service provision: insights into buisness models, learning activities, and capabilities. Res technol manage 57(3), 44–52
17. Löfstrand M, Andrews J, Karlberg M, Karlsson L (2011) Functional product system availability: simulation driven design and operation through coupled multi-objective optimization. Int J Prod Dev 13(2):119–131
18. Löfstrand M, Backe B, Kyösti P, Lindström J, Reed S (2012) A model for predicting and monitoring industrial system availability. Int J Prod Dev 16(2):140–157
19. Karlsson L, Lindström J, Löfstrand M (2012) Functional products—Goodbye to the industrial age. Ericsson Bus Rev 81(2):21–24
20. Toossi A, John P, Shehab E, Mackey T (2012) Through-life challenges and opportunities of designing high-value complex systems. In: Proceedings of 1st international through-life engineering services conference, Cranfield, 5–6 Nov 2012
21. Redding LE, Tiwari A, Roy R (2012) Identifying the challenges in through-life engineering services. In: Proceedings of 1st international through-life engineering services conference, Cranfield, 5–6 Nov 2012
22. Baines TS, Lightfoot HW, Evans S, Neely A, Greenough R, Peppard J, Roy R, Shehab E, Braganza A, Tiwari A, Alcock JR, Angus JP, Bastl M, Cousens A, Irving P, Johnson M, Kingston J, Lockett H, Martinez V, Michele P, Tranfield D, Walton M, Wilson H (2007) State-of-the-art in product-service systems. Proc Inst Mech Eng, Part B: J Eng Manuf 221:1543–1552
23. Baines TS, Lightfoot HW, Benedettini O, Kay JM (2009) The servitization of manufacturing: a review of literature and reflection on future challenges. J Manuf Technol Manag 20 (5):547–567
24. Isaksson O, Larsson TC, Öhrwall Rönnbäck A (2009) Development of product-service systems: challenges and opportunities for the manufacturing firm. J Eng Des 20(4):329–348
25. Meier H (2012) Planning and development of industrial product-service systems. In: Proceedings of 4th CIRP international conference on industrial product-service systems, Tokyo, 8–9 Nov 2012
26. Rese M, Meier H, Gesing J, Bosslau M (2012) An ontology of business models for industrial product-service systems. In: Proceedings of 4th CIRP international conference on industrial product-service systems, Tokyo, 8–9 Nov 2012
27. Norris M (2013) Planning to extend the life of major assets. In: Proceedings of 2nd international through-life engineering services conference. Procedia CIRP, Cranfield, 5–6 Nov 2013
28. Baker A (2013) Configurable obsolescence mitigation methodologies. In: Proceedings of 2nd international through-life engineering services conference. Procedia CIRP, Cranfield, 5–6 Nov 2013

29. Vladimirova D, Evans S, Martinez V, Kingston J (2011) Elements of change in the transformation towards product service systems. In: Proceedings of 3rd CIRP international conference on industrial product-service systems, Braunschweig, 5–6 May 2011
30. van Ostaeyen J, Neels B, Duflou JR (2011) Design of a product-service systems business model: strategic analysis and option generation. In: Proceedings of 3rd CIRP international conference on industrial product-service systems, Braunschweig, 5–6 May 2011
31. van Ostaeyen J, Kellens K, Van Horeneek A, Duflou JR (2012) Quantifying the economic potential of a PSS: methodology and case study. In: Proceedings of 4th CIRP international conference on industrial product-service systems, Tokyo, 8–9 Nov 2012
32. Kurak CDF, Barquet APB, Rozenfeld H (2013) Challenges for PSS implementation: identification and classification. In: Proceedings of 5th CIRP international conference on industrial product-service systems, Bochum, 14–15 Mar 2013
33. van Ostaeyen J, Kerremans Y, Van Goethem G, Duflou JR (2013) A PSS model for diamond gemstone processing: economic feasibility analysis. In: Proceedings of 46th CIRP conference on manufacturing systems. Procedia CIRP 7, Setubal, 29–31 May 2013
34. Meier H, Völker O, Funke B (2011) Industrial product-service systems (IPS2). Int J Adv Manuf Technol 52:1175–1191
35. Meier H, Funke B, Bosslau M (2011) Flexible resource planning in the context of dynamic IPS2 business models. In: Proceedings of 3rd CIRP international conference on industrial product-service systems, Braunschweig, 5–6 May 2011
36. Meier H, Steven M, Funke B, Bosslau M, gen. Schulte JK (2011) Complexity and flexibility of IPS2 across various planning levels. In: Proceedings of 3rd CIRP international conference on industrial product-service systems, Braunschweig, 5–6 May 2011
37. Steven M, Alevifard S, genannt Schulte JK (2011) Economic relevance of IPS2 flexibility. In: Proceedings of 3rd CIRP international conference on industrial product-service systems, Braunschweig, 5–6 May 2011
38. Barquet APB, Cunha VP, Oliveira MG, Rozenfeld H (2011) Business model elements for product-service system. In: Proceedings of 3rd CIRP international conference on industrial product-service systems, Braunschweig, 5–6 May 2011
39. Uhlmann E, Gabriel C, Raue N, Stelzer C (2011) Influences of the IPS2 business model on the development of a micro milling spindle. In: Proceedings of 3rd CIRP international conference on industrial product-service systems, Braunschweig, 5–6 May 2011
40. Kondoh S, Mishima N (2011) Case based idea generation method for sustainable business. In: Proceedings of 3rd CIRP international conference on industrial product-service systems, Braunschweig, 5–6 May 2011
41. Meier H, Bosslau M (2012) Design and engineering of dynamic business models for industrial product-service systems. In: Proceedings of 4th CIRP international conference on industrial product-service systems, Tokyo, 8–9 Nov 2012
42. Cedergren SI, Elfving SW, Eriksson J, Parida V (2012) A road map for future research on industrial product-service systems (IPS2): a systematic review. In: Proceedings of 4th CIRP international conference on industrial product-service systems, Tokyo, 8–9 Nov 2012
43. Cedergren SI, Elfving SW, Eriksson J, Parida V (2012) Analysis of the industrial product-service systems (IPS2) literature: a systematic review. In: Proceedings of 2012 IEEE international conference on management of innovation and technology, Sanur Bali, 11–13 June 2012
44. Kishita Y, Hirosaki M, Mizuno Y, Wada H, Fukushige S, Umeda Y (2012) Formalizing scenario design processes to plan long-term business strategies for sustainability. In: Proceedings of 4th CIRP international conference on industrial product-service systems, Tokyo, 8–9 Nov 2012
45. Erkoyuncu JA, Roy R, Shehab E, Wardle P (2009) Uncertainty challenges in service cost estimation for product-service systems in the aerospace and defence industries. In: Proceedings of the 1st CIRP industrial product-service systems (IPS2) conference, Cranfield, 1–2 Apr 2009

46. Erkoyuncy JA, Durugbo C, Roy R (2012) Service-oriented cost uncertainties and contracting for availability. In: Proceedings of 4th CIRP international conference on industrial product-service systems, Tokyo, 8–9 Nov 2012
47. Sundin E, Öhrwall Rönnbäck A, Sakao T (2010) From component to system supplier: strategic warranty management as a key to efficient integrated product/service offering. CIRP J Manufact Sci Technol 2:183–191
48. Chun Y-Y, Kondoh S, Mishima N, Lee K-M (2011) A study on comparison of different PSS concepts based on value creation efficiency. In: Proceedings of 3rd CIRP international conference on industrial product-service systems, Braunschweig, 5–6 May 2011
49. Rapaccini M, Porcelli I, Saccani N, Cinquini L, Lugarà A (2012) LCCA and TCO: a how-to approach to assess the costs in the customer's eye. In: Proceedings of 4th CIRP international conference on industrial product-service systems, Tokyo, 8–9 Nov 2012
50. Datta PP, Roy R (2010) Cost modelling techniques for availability type service support contracts: a literature review and empirical study. CIRP J Manufact Sci Technol 3:142–157
51. Erkoyuncy JA, Roy R, Shehab E, Cheruvu K (2011) Understanding service uncertainties in industrial product-service systems cost estimation. Int J Adv Manuf Technol 52:1223–1238
52. Lanza G, Behmann B, Werner P, Vöhringer S (2011) Simulation of life cycle costs of a product service system. In: Proceedings of 3rd CIRP international conference on industrial product-service systems, Braunschweig, 5–6 May 2011
53. Roy R, Erkoyuncu JA (2011) Service cost estimation challenges in industrial product-service systems. In: Proceedings of 3rd CIRP international conference on industrial product-service systems, Braunschweig, 5–6 May 2011
54. Kondoh S, Komoto H (2012) Reuse/sharing business design method considering the difficulties for product transfer. In: Proceedings of 4th CIRP international conference on industrial product-service systems, Tokyo, 8–9 Nov 2012
55. Alonso-Rasgado T, Thompson G (2006) Rapid design process for total care product creation. J Eng Des 17(6):509–531
56. Lee J, Lapira ER, Yang S, Kao A, Yen N (2012) New PHM techniques and trends for asset lifecycle management. In: Proceedings of 1st international through-life engineering services conference, Cranfield, 5–6 Nov 2012
57. Bey N, McAloone T (2006) From LCA to PSS—Making leaps towards sustainability by applying product/service-system thinking in product development. In: Proceedings of 13th CIRP international conference on life cycle engineering: LCE2006-towards a closed loop economy, Leuven, 31 May–2 June 2006
58. Hepperele C, Orakowski R, Nolte BD, Mörtl M, Lindemann U (2010) An integrated lifecycle model of product-service-systems. In: Proceedings of 2nd CIRP IPS2 conference, Linköping, 14–15 Apr 2010
59. Annamalai G, Hussain R, Cakkol M, Roy R, Evans S, Tiwari A (2011) An ontology for product-service systems. In: Proceedings of 3rd CIRP international conference on industrial product-service systems, Braunschweig, 5–6 May 2011
60. Isaksson O, Larsson TC, Johansson P (2011) Towards a framework for developing product/service systems. In: Proceedings of 3rd CIRP international conference on industrial product-service systems, Braunschweig, 5–6 May 2011
61. Fernandes P, Roy R, Mehnen J, Harrision A (2011) An overview on degradation modelling for service cost estimation. In: Proceedings of 3rd CIRP international conference on industrial product-service systems, Braunschweig, 5–6 May 2011
62. Thompson AW, Larsson TC, Broman G (2011) Towards sustainability-driven innovation through product-service systems. In: Proceedings of 3rd CIRP international conference on industrial product-service systems, Braunschweig, 5–6 May 2011
63. Sakao T, Paulsson S, Müller P (2011) Integrated evaluation of a PSS business case and a PSS design method—application of the PSS layer method to an industrial drilling solution. In: Proceedings of 3rd CIRP international conference on industrial product-service systems, Braunschweig, 5–6 May 2011

64. Khumboon R, Kara S, Ibbotson S (2011) A simplified decision making model for employing product service system in industry at a preliminary planning stage. In: Proceedings of 3rd CIRP international conference on industrial product-service systems, Braunschweig, 5–6 May 2011
65. Hiraoka H, Ueno T, Arita M, Shigeji Y, Horii K, Kawaharada H (2012) Promoting reuse of mechanical parts using part agents. In: Proceedings of 4th CIRP international conference on industrial product-service systems, Tokyo, 8–9 Nov 2012
66. Matsumoto E, Ohtake J, Okada J, Takata S (2012) A method for selecting delivery modes in environmentally benign product service system design. In: Proceedings of 4th CIRP international conference on industrial product-service systems, Tokyo, 8–9 Nov 2012
67. Amaya J, Lelah A, Zwolinski P (2012) Environmental benefits of PSS strategies: a bicycle sharing system case study. In: Proceedings of 4th CIRP international conference on industrial product-service systems, Tokyo, 8–9 Nov 2012
68. Kwak M, Kim H (2013) Economic and environmental impacts of product service lifetime: a life-cycle perspective. In: Proceedings of 5th CIRP international conference on industrial product-service systems, Bochum, 14–15 Mar 2013
69. Schulte JK, Waltemode S, Steven M (2013) Lifecycle oriented flexibility assessment of customized solutions in capital goods industry: a state of art. In: Proceedings of 5th CIRP international conference on industrial product-service systems, Bochum, 14–15 Mar 2013
70. Marquès G, Chalal M, Boucher X (2013) PSS production systems: a simulation approach for change management. In: Emmanouilidis C, Taisch M, Kiritsis D (eds) Advances in production management systems. Competitive manufacturing for innovative products and services IFIP advances in information and communication technology, vol 398. Springer, Berlin, pp 377–384
71. Abramovici M, Neubach M, Schulze M, Spura C (2009) Metadata reference model for IPS2 lifecycle management. In: Proceedings of 1st CIRP industrial product-service systems (IPS2) conference, Cranfield, 1–2 Apr 2009
72. Abramovici M, Dang HB, Wolf M (2013) IPS2-KOP:IPS2 knowledge-based service-oriented lifecycle management platform. In: Proceedings of 5th CIRP international conference on industrial product-service systems, Bochum, 14–15 Mar 2013
73. Abramovici M, Aidi Y, Dang HB (2013) Knowledge-based lifecycle management approach for product service systems (PSS). In: Bernard A, Rivest L, Dutta D (eds) Product lifecycle management for society IFIP advances in information and communication technology, vol 409, pp 239–248
74. Reed S, Löfstrand M, Karlsson L, Andrews J (2013) Service support system modelling language for simulation driven design of functional products. In: Proceedings of 2nd international through-life engineering services conference. Procedia CIRP, Cranfield, 5–6 Nov 2013
75. Chandler SR, Matthews PC (2013) Through-life systems engineering design and support with SysML. In: Proceedings of 2nd international through-life engineering services conference. Procedia CIRP, Cranfield, 5–6 Nov 2013
76. Hussain R, Locket H, Vasantha GA (2011) Industry practices and challenges in using product in use data to inform PSS conceptual design. In: Proceedings of 3rd CIRP international conference on industrial product-service systems, Braunschweig, 5–6 May 2011
77. Kumar AA, Trinh GC, Sakao T (2012) How is uncertainty provided and managed in design by PSS providers?—Relation to PSS types provided. In: Proceedings of 4th CIRP international conference on industrial product-service systems, Tokyo, 8–9 Nov 2012
78. Roy R, Erkoyuncu, JA, Shaw A (2013) The future of maintenance for industrial product-service systems. In: Proceedings of 5th CIRP international conference on industrial product-service systems, Bochum, Germany
79. Reim W, Parida V, Lindström J (2013) Risks for functional products—Empirical insights from two Swedish manufacturing companies. In: Proceedings of 2nd international through-life engineering services conference. Procedia CIRP, Cranfield, 5–6 Nov 2013

80. Sandborn P (2013) Design for obsolescence risk management. In: Proceedings of 2nd international through-life engineering services conference. Procedia CIRP, Cranfield, 5–6 Nov 2013
81. Lindahl M, Sakao T, Öhrwall Rönnbäck A (2009) Business implications of integrated product and service offerings. In: Proceedings of 1st CIRP industrial product-service systems (IPS2) conference, Cranfield, 1–2 Apr 2009
82. Schmitt R, Schumacher S, Zentis T (2013) Quality-oriented risk-management-approach for service innovations. In: Proceedings of 5th CIRP international conference on industrial product-service systems, Bochum, 14–15 Mar 2013
83. Smith L, Maull R, Ng ICL (2014) Servitization and operations management: a service dominant-logic approach. Int J Oper Prod Manag 34(2):242–269
84. Byers SS, Groth JC, Sakao T (in press) Using portfolio theory to improve resource efficiency of invested capital. J Cleaner Prod
85. Steven M, Alevifard S (2012) Management of flexibility in IPS2-business-relationships. In: Proceedings of 4th CIRP international conference on industrial product-service systems, Tokyo, 8–9 Nov 2012
86. Rese M, Pick D, Maiwlad K (2012) Reference marketing for industrial product service systems. In: Proceedings of 4th CIRP international conference on industrial product-service systems, Tokyo, 8–9 Nov 2012
87. Mooren Ceng PJH, van Dongen LAM (2013) Application of remote condition monitoring in different rolling stock life cycle phases. In: Proceedings of 2nd international through-life engineering services conference. Procedia CIRP, Cranfield, 5–6 Nov 2013
88. Parada Puig JE, Basten RJI, van Dongen LAM (2013) Investigating maintenance decisions during initial fielding of rolling stock. In: Proceedings of 2nd international through-life engineering services conference. Procedia CIRP, Cranfield, 5–6 Nov 2013
89. Norden C, Hribernik K, Ghrari Z, Thoben K-D, Fuggini C (2013) New approaches to through-life asset management in maritime industry. In: Proceedings of 2nd international through-life engineering services conference. Procedia CIRP, Cranfield, 5–6 Nov 2013
90. Tan AR, Matzen D, McAloone TC, Evans S (2010) Strategies for designing and developing services for manufacturing firms. CIRP J Manufact Sci Technol 3:90–97
91. Yang L, Xing K, Lee S-H (2010) Framework for PSS from service' perspective. In: Proceedings of international multi conference of engineers and computer scientists 2010 (IMECS 2010), vol III, Hong-Kong, 17–19 Mar 2010
92. Romero Rojo FJ, Roy R, Shehab E, Wardle PJ (2009) Obsolescence challenges for product-service systems in aerospace and defence industry. In: Proceedings of the 1st CIRP industrial product-service systems (IPS2) conference, Cranfield, 1–2 Apr 2009
93. Pierce SG, Dobie G, Summan R, MacLeod CN (2012) Reconfigurable robotic platforms for structural non-destructive evaluation. In: Proceedings of 1st international through-life engineering services conference, Cranfield, 5–6 Nov 2012
94. Puvani M, Cabri G, Frei R (2102) Self-healing in ensambles' adaptive collaboration patterns. In: Proceedings of 1st international through-life engineering services conference, Cranfield, 5–6 Nov 2012
95. Schiefer P, McWilliam R, Purvis A (2013) Creating a self-configuring finite state machine out of memory look-up tables. In: Proceedings of 2nd international through-life engineering services conference. Procedia CIRP, Cranfield, 5–6 Nov 2013
96. Stasis A, Whyte J, Dentten R (2013) A critical examination of change control processes. In: Proceedings of 2nd international through-life engineering services conference. Procedia CIRP, Cranfield, 5–6 Nov 2013
97. De Coster R (2010) A collaborative approach to forecasting product-service systems (PSS). Int J Adv Manufact Syst 52:1251–1260
98. Abramovici M, Bellalouna F, Goedel JC (2010) Towards adaptable industrial product-service systems (IPS2) with an adaptive change management. In: Proceedings of 2nd CIRP international conference on industrial product-service systems, Linköping, 14–15 Apr 2010

99. Lindström J, Karlsson L, Löfstrand M, Karlberg M (2012) Functional product development: what information should be shared during the development process? Int J Prod Dev 16 (2):95–111
100. Ali U, Kidd C (2012) Configuration management process capabilities. In: Proceedings of 2nd through-life engineering services conference. Procedia CIRP, Cranfield, 5–6 Nov 2012
101. Xu Y, Malisetty MK, Round M (2013) Configuration management in aerospace industry. In: Proceedings of 2nd international through-life engineering services conference. Procedia CIRP, Cranfield, 5–6 Nov 2013
102. Müller P (2013) Configuration management—A core competence for successful through-life systems engineering and engineering services. In: Proceedings of 2nd international through-life engineering services conference. Procedia CIRP, Cranfield, 5–6 Nov 2013
103. Burger A, Bittel V, Awad R, Ovtcharova J (2011) Design for customer—Sustainable customer integrations into the development processes of product-service system providers. In: Proceedings of MECIT international conference on applied information and communications technology, Muscat, 22–23 Mar 2011
104. Rojo FJR, Shehab E, Cheruvu E, Blackman I, Rumney GA (2010) Key challenges in managing software obsolescence for industrial product-service systems (IPS2). In: Proceedings of 2nd CIRP international conference on industrial product-service systems, Linköping, 14–15 Apr 2010
105. Mori M, Fujishima M, Riedemann R (2012) Remote monitoring and maintenance system. In: Proceedings of 1st international through-life engineering services conference, Cranfield, 5–6 Nov 2012
106. Widjanarko T, Tinsley L, Roy R, Mehnen J (2012) Characterisation and performance assessment of a pulsed-thermography camera system for component degradation inspection. In: Proceedings of 1st international through-life engineering services conference, Cranfield, 5–6 Nov 2012
107. Arnaiz A, Konde E, Alarcon J (2013) Continuous improvement on information and on-line maintenance technologies for increased cost-effectiveness. In: Proceedings of 2nd international through-life engineering services conference. Procedia CIRP, Cranfield, 5–6 Nov 2013
108. Uhlmann E, Geisert C, Hohwieler E, Altmann I (2013) Data mining and visualization of diagnostic messages for condition monitoring. In: Proceedings of 2nd international through-life engineering services conference. Procedia CIRP, Cranfield, 5–6 Nov 2013
109. Hameed B, Minguez J, Wörner M, Hollstein P, Zor S, Silcher S, Dürr F, Rothermel K (2011) The smart real-time factory as a product service system. In: Proceedings of 3rd CIRP international conference on industrial product-service systems, Braunschweig, 5–6 May 2011
110. Lelah A, Brissaud D (2012) Clarifying frontiers of PSS and information and communication technologies. In: Proceedings of 4th CIRP international conference on industrial product-service systems, Tokyo, 8–9 Nov 2012
111. Biege S, Schröter M, Weissenberger-Eibl M (2011) Product adaptation for industrial product-service systems—characteristics, motives and challenges. In: Proceedings of 3rd CIRP international conference on industrial product-service systems, Braunschweig, 5–6 May 2011
112. Rese M, Everhartz J (2012) Condition monitoring of industrial product service systems—helpful selling argument or potential marketing pitfall? In: Proceedings of 4th CIRP international conference on industrial product-service systems, Tokyo, 8–9 Nov 2012
113. Uhlmann E, Bilz M, Baumgarten J (2013) MRO—Challenge and chance for sustainable enterprises. In: Proceedings of 2nd international through-life engineering services conference. Procedia CIRP, Cranfield, 5–6 Nov 2013
114. Brennan F (2013) Risk based maintenance for offshore wind structures. In: Proceedings of 2nd international through-life engineering services conference. Procedia CIRP, Cranfield, 5–6 Nov 2013
115. Mori M, Fujishima M (2013) Sustainable service system for machine tools. In: Proceedings of 2nd international through-life engineering services conference. Procedia CIRP, Cranfield, 5–6 Nov 2013

116. Reed S, Andrews J, Löfstrand M, Karlsson L (2012) The design and modelling of maintenance during the development of functional products or product service systems. In: Proceedings of 1st international through-life engineering services conference, Cranfield, 5–6 Nov 2012
117. Uhlmann E, Otto F, Geisert C (2012) Improving maintenance services for machine tools by integrating specific software functions. In: Proceedings of 4th CIRP international conference on industrial product-service systems, Tokyo, 8–9 Nov 2012
118. Akasaka F, Nguyen HN, Nemoto Y, Lenz B, Shimomura Y, Stark R (2012) Adaptation of the service self checklist for industrial product-service systems. In: Proceedings of 4th CIRP international conference on industrial product-service systems, Tokyo, 8–9 Nov 2012
119. Lindström J, Plankina D, Lideskog H, Löfstrand M, Karlsson L (2012) Functional product development: criteria for selection of design methods on strategic and operational levels. In: Proceedings of 4th CIRP international conference on industrial product-service systems, Tokyo, 8–9 Nov 2012
120. McMahon CA, Ford G, Thangarajah U, Rowley C (2012) From information repositories to business knowledge through the exploitation of unstructured data. In: Proceedings of 1st international through-life engineering services conference, Cranfield, 5–6 Nov 2012
121. McMahon C, Ball A (2013) Information systems challenges for through-life engineering. In: Proceedings of 2nd international through-life engineering services conference. Procedia CIRP, Cranfield, 5–6 Nov 2013
122. Rämänen J, Mahlamäki K, Borgman J, Nieminen M (2013) Human role in industrial installed base information gathering. In: Proceedings of 2nd international through-life engineering services conference. Procedia CIRP, Cranfield, 5–6 Nov 2013
123. Manonen P, Hölttä V (2013) Where the knowledge goes? information gathering and managing practices in a global technical support center. In: Proceedings of 2nd international through-life engineering services conference. Procedia CIRP, Cranfield, 5–6 Nov 2013
124. Holmqvist J, Wenngren J, Ericsson Å, Johansson C, Thor P (2011) Sharing expertise: easier said than done. In: Proceedings of 3rd CIRP international conference on industrial product-service systems, Braunschweig, 5–6 May 2011
125. Chirumalla K, Bertoni A, Ericsson Å, Isaksson O (2012) Knowledge-sharing network for product-service system development: is it atypical? In: Proceedings of 4th CIRP international conference on industrial product-service systems, Tokyo, 8–9 Nov 2012
126. Durugbo C, Erkoyuncu JA (2012) Managing service requirements engineering: the role of information systems. In: Proceedings of 4th CIRP international conference on industrial product-service systems, Tokyo, 8–9 Nov 2012
127. Sundin E, Ölundh Sandström G, Lindahl M, Öhrwall Rönnbäck A, Sakao T, Larsson TC (2009) Challenges for industrial product/service systems: experiences from a learning network of large companies. In: Proceedings of the 1st CIRP industrial product-service systems (IPS2) conference, Cranfield, 1–2 Apr 2009
128. Rese M, Strotmann W-C, Karger M, Gesing J (2011) Learning about product service systems —conveying information to improve the benefit evaluation and stabilize preferences. In: Proceedings of 3rd CIRP international conference on industrial product-service systems, Braunschweig, 5–6 May 2011
129. Bochnig H, Uhlmann E, Gegusch R, Seliger G (2011) Knowledge feedback to the IPS2 development. In: Proceedings of 3rd CIRP international conference on industrial product-service systems, Braunschweig, 5–6 May 2011
130. Beverungen D, Uhlmann E, Stelzer C, Raue N, Gabriel C, Knackstedt R (2012) A workshop method for identifying information flows in IPS2. In: Proceedings of 4th CIRP international conference on industrial product-service systems, Tokyo, 8–9 Nov 2012
131. Datta PP, Srivastava A, Roy R (2013) A simulation study on maintainer resource utilization of a fast jet aircraft maintenance line under availability contract. Comput Ind 64(5):543–555
132. Lingegård S, Lindahl M, Svensson N (2011) PSS contracts for rail and road infrastructure. In: Proceedings of 3rd CIRP international conference on industrial product-service systems, Braunschweig, 5–6 May 2011

133. Bohr C, Kranz J-N, Aurich JC (2012) Life cycle oriented prevention of product piracy in technical product-service systems. In: Proceedings of 4th CIRP international conference on industrial product-service systems, Tokyo, 8–9 Nov 2012
134. Schenkl SA, Schneider D, Mörtl M, Lindemann U (2012) PSS for preventing product imitation. In: Proceedings of 4th CIRP international conference on industrial product-service systems, Tokyo, 8–9 Nov 2012
135. Sakao T, Öhrwall Rönnbäck A, Ölundh Sandström G (2013) Uncovering benetifts and risks of integrated product service offerings—Using a case of technology encapsulation. J Syst Sci Syst Eng 22(4):421–439
136. Parida V, Rönnberg-Sjödin D, Wincent J, Ylinenpää H (2013) Win–Win collaboration, functional product challenges and value-chain delivery: a case study approach. In: Proceedings of 2nd international through-life engineering services conference. Procedia CIRP, Cranfield, 5–6 Nov 2013
137. Durugbo C, Bankole O, Erkoyuncu JA, Tiwari A, Alcock JR, Roy R, Shehab E (2010) Product-service systems across industry sectors: future research needs and challenges. In: Proceedings of 2nd CIRP IPS2 conference, Linköping, 14–15 Apr 2010
138. Lelah A, Mattieux F, Brissaud D (2011) Building a network of SME for a global PSS infrastructure in complex high-tech systems: example of urban applications. In: Proceedings of 3rd CIRP international conference on industrial product-service systems, Braunschweig, 5–6 May 2011
139. Mougaard K, Neugebauer L, McAloone TC, Bey N, Andersson JB (2012) Collaborative product/service-systems—On conceptualization of PSS offerings and business nets. In: Proceedings of 4th CIRP international conference on industrial product-service systems, Tokyo, 8–9 Nov 2012
140. Schweitzer E, Aurich JC (2010) Continuous improvement of industrial product-service systems. CIRP J Manufact Sci Technol 3:158–164
141. Rese M, Maiwald K (2011) Organizational buying behaviour in case of IPS2. In: Proceedings of 3rd CIRP international conference on industrial product-service systems, Braunschweig, 5–6 May 2011
142. Meier H, Morlock F, Dorka T (2013) Functional specification for IPS2-execution systems. In: Proceedings of 5th CIRP international conference on industrial product-service systems, Bochum, 14–15 Mar 2013
143. Uhlmann E, Raue N (2012) IPS2 control system for the integrated support and service processes. In: Proceedings of 4th CIRP international conference on industrial product-service systems, Tokyo, 8–9 Nov 2012
144. Meier H, Uhlmann E, Völker O, Geisert C, Stelzer C (2010) Reference architecture for dynamical organization of IPS2 service supply chains in the delivery phase. In: Proceedings of 2nd CIRP IPS2 conference, Linköping, 14–15 Apr 2010
145. Meier H, Lagemann H, Dorka T (2012) Requirements for transfer and application of IPS2 resource planning—case-study of a global machine tool manufacturer. In: Proceedings of 4th CIRP international conference on industrial product-service systems, Tokyo, 8–9 Nov 2012
146. Frei R, Macbeth S, Pitt J, di Marzo Serugendo G (2012) Simulation of complex emergent structures in presage. In: Proceedings of 1st international through-life engineering services conference, Cranfield, 5–6 Nov 2012

Chapter 13
Understanding Maintenance Decisions: How to Support Acquisition of Capital Assets

Jorge E. Parada Puig, Rob J.I. Basten and Leo A.M. van Dongen

Abstract This chapter contributes with theoretical and practical insights on maintenance decision making during acquisition of capital assets. We give theoretical insights about maintenance decision making by reviewing the literature, while our practical insights come from examples of the decisions made at a maintenance organization for rolling stock: NedTrain. We find that strategic maintenance decisions are more relevant before contracting than tactical or operational decisions, and they have the largest potential to impact Life Cycle Costs. The research on strategic maintenance decisions is too broad to review individual decisions, and therefore we review papers that structure decisions in the form of frameworks. We find that according to the literature, assets and their maintenance services should be developed concurrently. However, we find in practice that NedTrain's approach is to fit new assets into the existing maintenance services, while there are parallel continuous improvement processes for those services. From practice, we conclude that strategic maintenance decisions are not concurrent to rolling stock design decisions. We also conclude that there is a need for methods and tools to support strategic maintenance decision making during early stages of acquisition, especially before contracting.

13.1 Introduction

Capital assets are trains, airplanes, power plants or MRI scanners. They are ubiquitous in industrial societies, who rely heavily on their use to sustain all kinds of public services such as transportation, energy supply or health care. Organizations

---

J.E. Parada Puig (✉) · R.J.I. Basten · L.A.M. van Dongen
Faculty of Engineering Technology, University of Twente, Enschede, Netherlands
e-mail: j.e.paradapuig@utwente.nl

R.J.I. Basten
e-mail: r.j.i.basten@utwente.nl

L.A.M. van Dongen
e-mail: leo.vandongen@nedtrain.nl; l.a.m.vandongen@utwente.nl

L. Redding and R. Roy (eds.), *Through-life Engineering Services*,
Decision Engineering, DOI 10.1007/978-3-319-12111-6_13

use an acquisition process to specify, select, contract, develop and field these assets. Also, the services required to provide support for the asset throughout its life cycle are determined during this process. Therefore, acquisition decisions should include making decisions about the asset and about the maintenance that it requires.

Research has pointed to the importance of considering maintenance in planning the product life cycle, (see, e.g., [47] but this emphasis has received limited attention [42, 43]. Acquisition costs only represent around 40 % of the Life Cycle Cost (LCC) of an asset. However, upon contract award, 85 % of the cost of ownership decisions are made [21]. During asset acquisition, maintenance decision making helps to ensure that the asset can be properly supported, e.g., deciding on facilities, equipment or training of the workforce [4, 16, 19, 21]. However, maintenance decision making during acquisition has many challenges.

Acquisitions involve many criteria, and also require a lot of effort, time and knowledge from the organizations involved. Decisions made during acquisition have a strong strategic impact on operational performance of capital assets. Therefore, performance outcomes are only visible in the long term, making it difficult to assess good decisions. Nevertheless, there are opportunities to exploit. During acquisition, there is much more flexibility to make maintenance decisions than in the operating stage of the life cycle. The impact of decision support can be much larger.

Decision support can help improve the decision making process to be more effective (do the right things) and efficient (do things right), helping to make better decisions. Better decisions during acquisition result in assets that will be supportable through-life. This will help organizations to improve the services and therefore obtain better performance at lower cost. Good support can help service organizations give better advise to their clients. Achieving these potential benefits in the context of the acquisition of rolling stock is the main motivation for our research.

Rolling stock are vehicles that move on a railway, i.e., trains. In 2010, rolling stock represented a worldwide market volume of more than EUR 70 billion, and was set to grow by more than 4 % by 2015 [23]. Rolling stock is considered both a focal point of customer experience and of the operational performance of the railways. Rolling stock is also an important contributor to the cost of travel. In the UK, for example, the annual cost of rolling stock represented approximately 15 % of the total railway operating costs, amounting to 1.9 billion (at 2009/10 prices) [2].

Maintenance is a fundamental part of supporting rolling stock through-life. It is important for safety, quality and convenience of rail transportation, and it is required to set the conditions for high operational performance. Maintenance costs are a significant proportion of the life cycle costs of rolling stock. For the Netherlands Railways, maintenance and cleaning of rolling stock represent approximately 25 % of the direct operating costs (personal communication, NS Reizigers). Railway companies regularly purchase rolling stock in order to match demand for passenger service, withdrawal for maintenance and obsolescence. It is understandable that rolling stock acquisition is a fundamental activity for railway companies.

This chapter focuses on maintenance decision making in the context of acquisition of rolling stock. The maintenance function can play a key role by supporting decision making during rolling stock acquisition [32]. We contribute by improving the understanding about the theory and practice of maintenance decision making during acquisition of rolling stock. We also contribute by identifying opportunities to provide support for maintenance decision making during acquisitions. We review the literature looking for insights about the theory of maintenance decision making. We refer to frameworks on maintenance decision making as a source for structure about choices, decision areas, methods or the tools employed. We give empirical insights from one company that maintains passenger service rolling stock (trains and locomotives): NedTrain, the largest maintenance service provider for rolling stock in the Netherlands, and subsidiary of the Netherlands Railways (NS).

The structure of this chapter is as follows. Section 13.2 discusses the research methodology. Next, the results of the literature review are presented in Sect. 13.3. Section 13.4 presents results of a case of maintenance decisions during acquisition at NS/NedTrain. Section 13.5 discusses our results. Finally, in Sect. 13.6 we give our conclusions.

13.2 Methodology

This research aims at improving the current understanding of maintenance decisions made during the acquisition of rolling stock, and identifying opportunities to support decision making in companies like NedTrain. Our main question is therefore *what maintenance decisions are made during acquisition according to the literature, according to practice, and what decisions need support*? Our research combines a literature review and a case study. With this approach, we examine both academic research and practice to identify opportunities for decision support during acquisition.

The literature review includes research from several research disciplines in the fields of decision sciences and engineering. We review journal papers from the disciplines of operations research (OR), production and operations management (POM), maintenance engineering (ME), reliability engineering (RE) and systems engineering (SE). Relevant sources are queried using electronic database services such as Scopus, Science Direct, IEEE Xplorer and Web of Science. We use keywords such as *management*, *decision*, and *decision making*. Furthermore, we include *maintenance*, and several synonyms for maintenance—*repair*, *replace/replacement*, *upkeep*, *mro*, *spare part*, *service*, for example. Finally, we include in our search additional categories such as *strategic*, *tactical* and *operational* maintenance decisions. We explicitly focus on research on *maintenance decision frameworks* as additional set of search terms. However, we exclude most *conceptual frameworks*, because they are used in empirical papers for measuring operationalized constructs and give little guidance about decision making (e.g., [37]).

We include books known to discuss the topic of *maintenance decision making*, *maintenance management* or *logistics*.

For the case study we combine interviews, company documents and plant visits as primary sources of data. We analyze the case data and go back to the literature to clarify the decision making process, and to compare practice and theory. We focus on the maintainer. The case uses individual semi-structured interviews. The sample includes management—e.g., senior, acquisition, support, maintenance or fleet managers, mostly with an engineering background—and technical staff that has been involved or consulted for improvement projects, acquisition projects or maintenance assessment. In total, 18 experts have been interviewed. Ease of access to the organization also allowed for many informal conversations with key informants. Some experts in key positions have been interviewed several times. Interviews have been recorded, transcribed and analyzed to extract decisions mentioned by experts. We have carefully documented these decisions, giving special attention to when these decisions were made, who is either responsible or has a stake in the decision and what criteria are important to make the decision.

13.3 Results of Literature Review

This section reviews the literature on maintenance decision making. Our goal is to describe what maintenance decisions are made during acquisition of capital assets according to the literature. We begin in Sect. 13.3.1 describing the typical hierarchical decomposition of decisions found in the literature. There, we find that strategic maintenance decisions are most relevant during early stages of acquisition. Next, Sect. 13.3.2 reviews the strategic decisions. We find that according to the literature, assets and their maintenance services should be developed concurrently. We also find that existing research on strategic maintenance decision making is dedicated to later phases of acquisition, especially after contracting.

Maintenance decision making is an active research area. For generally useful books and reviews see, e.g., Ben-Daya et al. [3], Garg and Deshmukh [11], Murthy and Kobbacy [28], Pintelon and Gelders [35], Pintelon and van Puyvelde [36]. Operations research offers several reviews on maintenance decision making from the perspective of maintenance optimization (see, e.g., [18, 31]). The research on strategic maintenance decisions is too broad to review individual decisions, and therefore we review papers that structure decisions in the form of frameworks. We find this to be a good focus for our review of strategic decision areas. Frameworks are meta-models that structure the theory with its models, and typically outline the (i) connection between decisions, (ii) choices, (iii) decision areas, (iv) methods for making decisions and/or (v) tools for making decisions. We use these attributes to examine the research. As such, frameworks are commonly used in research to structure decision processes, and to position methods used to support decision making.

Before continuing with our review in the next section, we first give some definitions. We adopt the definitions of maintenance concept, policy and action from Pintelon and van Puyvelde [36]. A maintenance concept is the *set of maintenance policies and actions of various types and the general decision structure in which these are planned and supported*. A maintenance policy is a *rule* or a *set of rules describing the triggering mechanism for the different maintenance actions*. Maintenance actions are *basic maintenance interventions, elementary tasks carried out by a technician*. The maintenance program documents the planned maintenance actions, and shows the schedule needed for the implementation of maintenance interventions.

13.3.1 Strategic, Tactical and Operational Decisions

Decomposing decisions into hierarchies of decisions has been widespread in the operations management, supply chain management and operations research literature [9]. It appears that strategic decisions have the largest impact during early stages of acquisitions. Table 13.1 shows the typical decision areas in maintenance decision making. The strategic decision areas of the table are drawn from operations management research [17, 34, 46], while tactical and operational decisions are inspired by Pintelon and Gelders[35] and Pyke and Cohen [38].

Strategic decisions give shape to the *maintenance strategy*. The planning horizon for strategic decisions is typically two to five years (some spanning 10+ years). The maintenance strategy refers to the *structured pattern of decisions* made in specific *decision areas* to develop its maintenance *capabilities* [36]. Strategic decisions typically have the highest level of associated uncertainty, and they involve the conceptual design of maintenance services.

Tactical decisions are policy decisions that are revised annually or bi-annually. These decisions are typical of the maintenance supply chain organization. Tactical decisions tend to prescribe information and material flows, throughput or inventory levels, and this includes repair (job) shop planning. Material flow decisions are typically decisions on (i) batch size, (ii) timing of a production or shipment request, (iii) setting dispatch or allocation rules, and (iv) the presence of interference mechanisms for expediting or handling of emergency orders [38]. Operational

Table 13.1 Main decision areas involved in strategic, tactical and operational decisions

Planning level	Decision areas
Strategic	(1) Capacity, (2) Facilities, (3) Technology, (4) Integration, (5) Organization, (6) Policy, (7) Human resources, (8) Design, (9) Production control, (10) Performance measurement and reward
Tactical	(1) Information flows, (2) Material flows, (3) Throughput, (4) Buffers
Operational	(1) Schedules, (2) Coordination (3) Task grouping

decisions are weekly or daily decisions. They involve choices about task schedules and handling of work orders, e.g. arranging the sequence in which work orders will be executed and by whom [35].

13.3.2 Strategic Maintenance Decisions

The literature discusses two phases of acquisition during which maintenance decisions for new capital assets are made: one phase during the functional phase of system development (first phase), and one phase focusing on developing support for the design solution (second phase) [21]. The first phase entitles the buyer with the role of developing requirements. Once requirements are issued in a Request For Proposals (RFP), it is expected that the original equipment manufacturer is responsible for developing support options that will be submitted with the proposals. Design decisions for the maintenance and the asset are made concurrently. The support options will be decided upon when the buyer selects the best proposal [4, 21]. Second phase decisions involve the detail design of the asset and the required services. To the best of our knowledge, existing research on strategic maintenance decision making is dedicated to this phase.

Few papers give an overview with several decision areas in strategic maintenance decisions. Notable exceptions are Pinjala et al. [34], Tsang [46]. Horenbeek et al. [18], Jardine and Tsang [20] review maintenance optimization models and give an overview framework that illustrates the typical decision areas (output) of maintenance optimization models. These frameworks tend to explain the areas of strategic decision making, but they either do not show connections between decisions, or do not show the methods and tools required to make them. In contrast, Al-Turki [1] use their framework to describe the alignment to higher level corporate strategy for the selection of the appropriate strategies regarding service delivery mode, type of contracts for outsourcing, organization, work structure or maintenance management methodology.

Table 13.2 organizes the papers found in the literature according to the strategic decision areas of Table 13.1. The table shows that most papers provide frameworks for General/Asset Management, Maintenance Concept and Maintenance Policy decisions. We include the category of General/Asset Management in Table 13.1 for frameworks that give insight into the management system. The decision areas of Capacity, Facilities and Integration can be found in several reviews that include maintenance manpower (see, e.g., [52]), network design (see, e.g., [26, 41] or outsourcing decisions(see, e.g., [36]). We found only one framework in the area of Technology [15]. Although it is not specific to maintenance, it gives insights that can be easily extended to the selection of Computerized Maintenance Management Systems.

Maintenance Policy selection frameworks tend to use quantitative approaches. For example, Faccio et al. [10] give a quantitative framework to develop maintenance policies. The authors use several cost models that include spare parts, labor,

Table 13.2 Research papers proposing frameworks for strategic maintenance decisions

Decision areas	Literature sources
General/asset management	Strategic Maintenance Management [29], Maintenance Management Framework [6, 7, 24, 35], Distribution Network Service Providers Framework [13], Life Cycle Maintenance Framework [42, 43], Asset Management Framework [44, 45]
(1) Capacity	Maintenance Resource Requirements [20]; Maintenance manpower supply (see, e.g., [52])
(2) Facilities	Maintenance Resource Requirements [20], Network Design (see, e.g., [26, 41])
(3) Technology	IT/IS Evaluation Framework [15]
(4) Integration	Maintenance Resource Requirements [20], Outsourcing (see, e.g., [36])
(5) Organization	Network Design (see, e.g., [26, 41])
(6a) Policy	Risk Based Maintenance Policy Framework [8], Quantitative Framework (Faccio et al. [10], Multi Criteria Decision Making Framework [12])
(6b) Concept development	Reliability Centered Maintenance Framework [39], Total Productive Maintenance Framework [25], CIBOCOF [49–51], Value Driven Maintenance Planning [40], Individualising Maintenance Concept [30])
(7) Human resources	Training Program [14]
(8) Design	Integrated Logistics Support [4, 16, 21]
(9) Production control	Spare Parts Planning [5, 9]
(10) Performance measurement	Maintenance Performance Measurement Framework [22, 27, 33]

missing production costs, and other indirect costs. The proposed framework has three phases, (i) equipment analysis, (ii) survival data collection and analysis and (iii) decision making process. The general decision scope includes a mix of strategic and tactical decisions.

Frameworks for Concept Development give a general decision structure for planning/supporting the set of policy decisions for an asset or a group of assets. Maintenance concept development is the focal point of maintenance decision making, and we briefly discuss two classic approaches: Reliability Centered Maintenance (RCM) and Total Productive Maintenance (TPM). RCM provides a very structured approach to maintenance concept development and it is prescriptive in the methods and tools used. A good example is given by Rausand's twelve steps to concept development with RCM [39]. A typical TPM program has four implementation phases. The first and second phases of decisions in TPM focus on design modifications and training [25]. The policy decisions are made in the third and fourth phase of a TPM program. Naughton and Tiernan [30] challenge the notion of using resource intensive, non-generic and pre-existing frameworks for concept development. Based on their experience, these are often complex and specific to one particular industry. Some research has therefore focused on tailoring maintenance concepts for individual organizations (e.g., [30, 49]).

We categorize the Integrated Logistics Support (ILS) framework in the area of Design. ILS is used in the defense sector for structuring decisions on the logistic support concept. The emphasis is that maintenance service design and asset design should go together, starting from the initial stages of asset design [21]. Production Control frameworks are mainly focused on tactical decisions. For example, in Cavalieri et al. [5] authors propose five decision making steps for spare parts control of first line maintenance. Authors in Driessen et al. [9] extend that framework by including the repair shop—second line maintenance—and its control.

We found only one framework in the area of Human Resources. Gramopadhye et al. [14] give an overview framework about designing visual inspection training programs in aviation. Performance Measurement and Reward frameworks give insight into alignment of the maintenance function with operational and corporate goals, and they help position the performance management of maintenance services (e.g., [27]).

13.4 Results of Case Research

This section presents the results of the empirical case. The goal of the case is to describe practice, and identify the possible directions for developing support. The context of the case is a company that maintains passenger service rolling stock: NedTrain. van Dongen [48] gives insights about the maintenance strategy of NedTrain; Parada Puig et al. [32] give more detail about its maintenance operations. This section is structured as follows. Section 13.4.1 describes the current way in which NedTrain plans the maintenance for new rolling stock. This section shows how NedTrain fits new rolling stock to their existing maintenance system. The following two sections organize the results in a similar way to the literature review. Firstly, Sect. 13.4.2 discusses strategic, tactical and operational maintenance decisions at NedTrain. We find there that many maintenance decisions result in continuous improvement projects to manage the existing fleet. Secondly, Sect. 13.4.3 presents the case results about strategic maintenance decisions. There we show that the strategic process improvements have developed independently of rolling stock acquisitions.

13.4.1 New Approach to Planning Maintenance During Acquisition

The most recent approach to acquisition at the Netherlands Railways (NS) involves NedTrain as collaborator during acquisition. With knowledge collected from past projects, the maintainer is now seen as an important stakeholder in the life cycle of rolling stock. NedTrain now supports decision making during the entire acquisition process, starting with maintenance requirements definition. To fit the new trains to the maintenance services of NedTrain, the decision process follows a so-called *maintenance assessment*. We find that these decisions intend to fit new rolling stock to an existing maintenance infrastructure.

There are multiple stakeholders involved in active collaboration during the life cycle of rolling stock. Within NedTrain, three departments have a leading role in decision making. Fleet Services is accountable for (i) equipment performance monitoring, (ii) management functions—maintenance, configuration, RAMSHE/LCC management—, (iii) maintenance development and (iv) engineering—maintenance, reliability and systems engineering—. Operations Management is in charge of the production workshops. Supply Chain Management handles acquisition and logistics of the maintenance supply chain, i.e., supply chain operations.

NedTrain has knowledge that can help NS to select the best candidate from a maintenance point of view. For this reason, NedTrain collaborates with suppliers and the NS during acquisition by performing maintenance assessment. Figure 13.1 shows the place of maintenance assessment in acquisition projects of NS. We focus on the decisions made during maintenance assessment, as they are most influential in making a good fit of new assets to NedTrain.

Suppliers receive a detailed description of NedTrain's maintenance strategy. This is a part of the maintenance requirements definitions. These requirements are made during the tender, and they are included in a Request For Proposals/Request For Quotations (RFP/RFQ). RFP/RFQ is an opportunity for the supplier to provide an offering that fits the maintenance concept of NedTrain. NedTrain then gives a preliminary design assessment focused on asset and subsystem level. For NedTrain, this would allow decision making for the design of the maintenance services before contract. For NS this would give a more realistic RAMS/LCC commitment, both from the suppliers and the maintainer of the asset.

The decisions made during maintenance assessment can be described by: *given the (forecasted) demand associated with the new train series, allocate resources—of the existing maintenance infrastructure—to support the new trains*. The method of analysis consists of decisions about maintenance manpower—the *who*—, content —the *what*—, interval—the *when*—, maintenance level—the *where*—, RAMS/LCC and quality—the *why*—and the procedures—the *how*. It is desirable for NedTrain to adapt the original manufacturer's maintenance program with the goal of obtaining a smooth demand pattern for the resources of its workshops.

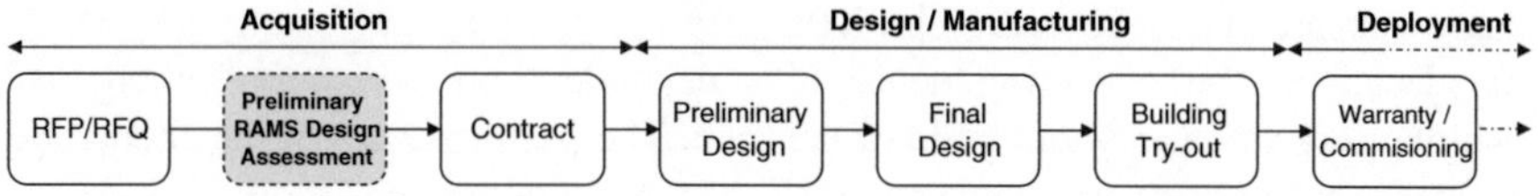

Fig. 13.1 Maintenance assessment in acquisition projects. *RFP/RFQ* request for proposals/request for quotations. *RAMS* reliability, availability maintainability and supportability

The initial *asset life cycle plan* is the key result from the decision making process during acquisitions. It contains, for example, information on asset configuration, the maintenance program, technical KPIs, obsolescence management and technology refreshment plans. van Dongen [48] offers more detail. After (commissioning) fielding the new train series, the individual life cycle plan is coupled to the life cycle plan of the entire fleet. RAMS/LCC are the most important performance criteria for maintenance assessment. However, RAMS/LCC performance data has a level of aggregation that makes operational costs difficult to estimate before contracting. Therefore, many of the strategic decisions of designing the maintenance services are being shifted to the initial fielding of the train, when the contract is already signed and there is less room for change.

13.4.2 Strategic, Tactical and Operational Decisions

At NedTrain, many decisions are made to optimize the maintenance services. The sum of these decisions results in the configuration of a bundle of services that we call the *maintenance service package*. We find that these decisions are not made during acquisitions, but to manage the existing fleet. A fleet life cycle plan bundles the life cycle plans of individual train series and gives the insights that NedTrain managers use to choose improvement targets.

An important question for NedTrain is *how do we improve the maintenance concept*? Figure 13.2 shows NedTrain's improvement cycle centered in the plan-do-check-act cycle. Planning is the responsibility of maintenance engineers (ME). Maintenance engineering is a function of asset management responsible for drawing up the maintenance tasks, the maintenance program and the maintenance concept. Implementation and execution of the maintenance concept is the "do", in the lower part of Fig. 13.2; this becomes the responsibility of Operations Management. Reliability engineers (RE) have a leading role in the "check" part of the cycle. This role includes measuring performance against contract parameters, analyzing train behavior and drawing-up improvement proposals. Finally, in the "act" part of the cycle, asset management together with operations is responsible for determining improvement actions.

Figure 13.3 shows in more detail the process of developing the maintenance concept. There are five phases within the decision process, namely the (i) inventorying technical maintenance tasks, (ii) task clustering, (iii) forecasting or estimating, (iv) concept evaluation and (v) detail planning. The first three phases are the responsibility of Fleet Management. Task inventorying produces the maintenance management inventory. The KPIs safety, reliability and quality are used in this decision. The clustering decision is dependent on skills, tools/equipment, resulting task frequency and build of the train. The performances of the alternatives are measured in costs per hour, material costs and downtime.

A rolling stock series "*materiel team*", is responsible for the fourth phase: concept evaluation. This requires group decisions and agreements between

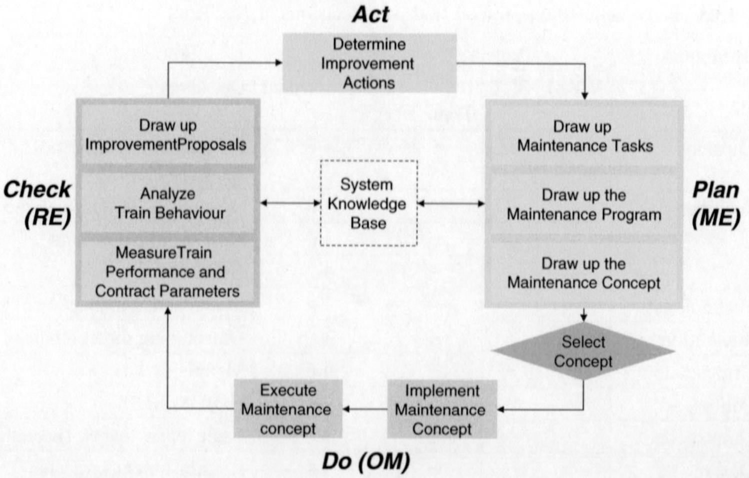

Fig. 13.2 Plan-do-check-act cycle to design/approve a maintenance concept. Reliability Engineering (*RE*), Maintenance Engineering (*ME*) and Operations Management (*OM*) roles

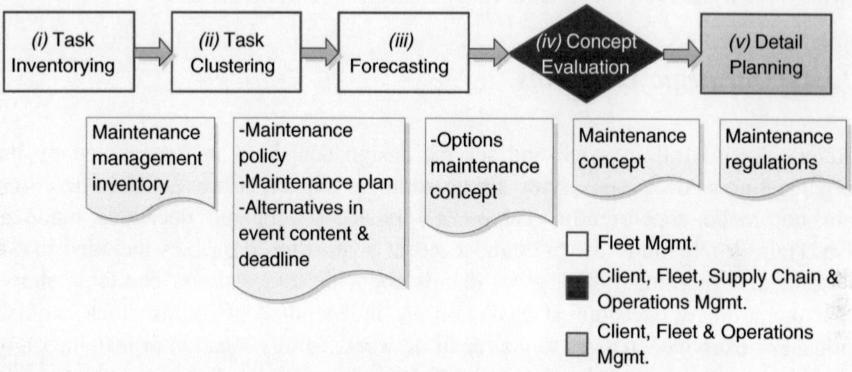

Fig. 13.3 NedTrain's maintenance concept development

Operations Management, Supply Chain Management and the client, NS. The team is normally composed of at least six experts: a rolling stock (equipment) manager, a maintenance engineer, a reliability engineer, a configuration manager, and account manager (NS) and a business controller. Therefore, this decision takes into account cross-functional and cross-hierarchical aspects. The decision of the fourth stage has a focus on costs and availability. Upon approval of the improved maintenance concept, detailed operational planning will follow. Maintenance operations has to determine the task allocation approach, the best routing policy (maintenance routing of trains) and the required documentation—work orders and work descriptions, for example.

Table 13.3 Three acquisition projects and past decisions at NedTrain

Decision areas	Project			Process	
	VIRM	SLT	HS-Train	Improvement	Decisions
(1) Capacity	–	++	–	++	Functional expansion of Leidschendam
(2) Facilities	–	–	++	++	Opening of Watergraafsmeer
	–	++	–	++	New technical centers
	++	++	–	++	GOIDS
(3) Technology	–	–	++	++	Maintenance management system
(4) Integration	–	–	–	++	Outsourcing direct repair of OBIS
(5) Organization	–	–	–	++	Material teams
(6) Policy	–	–	–	++	GOIDS, RBM
(7) Human Res	–	++	–	++	Fault repair teams, specialization
(8) Design	++	–	–	++	Refurbishment projects
(9) Prod. Ctrl	–	–	–	++	GOIDS, item-level RFID
(10) Performance	–	–	–	++	RBM

– 'no relation between decision and acquisition project'
++ 'acquisition project influenced the decision'
GOIDS 'No Withdrawals During Rush Hour'. *RBM* Risk Based Maintenance

13.4.3 Strategic Decisions

Acquisition of rolling stock and service design decisions are recognized by the organization as decisions having strong strategic impact. However, these decisions are not made concurrently. Table 13.3 positions strategic decisions made at NedTrain in relation to the three most recent acquisition processes included in our case data. Van Dongen [48] gives details about these decisions. The table shows that the strategic decisions at NedTrain are independent of rolling stock acquisitions. For example, changes to the repair network, facility location or installation of resources are not generally driven by acquisition projects. Strategic maintenance decisions are focused on process improvements to develop the maintenance services.

13.5 Discussion

The review in Sect. 13.3 results in the organization of the research as shown in Table 13.2. In Sect. 13.3.1 we find the focus on strategic decisions. The research is too broad to review individual decisions. Therefore, in Sect. 13.3.2 we review papers that structure decisions in the form of frameworks. These papers refer to managing built and in-use assets, and are not specific to decisions made during the acquisition. Few papers give an overview of all relevant decisions. Most papers discuss strategic maintenance decisions of a single decision area from Table 13.1, e.g., maintenance *policy* selection.

Our interviews and case data, described in Sect. 13.4, show little evidence of strategic maintenance decision making during acquisition. The case results show two distinct types of decision processes: one type was shown in Sect. 13.4.1 as a process that intends to fit new assets to existing services, while another type was shown in Sects. 13.4.2 and 13.4.3 showing decisions about continuous improvement of the maintenance services. Decisions in Sect. 13.4.1 focus mainly on optimizing maintenance before fielding, not before contracting. Table 13.3 positioned several strategic decisions of NedTrain, showing that these decisions are mostly independent of acquisition projects.

The approach of NedTrain to maintenance assessment has elements of the framework of Logistics Support Analysis (see, e.g., [4, 16, 21]). Such analysis may influence the choice of equipment as well as the nature and structure of supporting services [16]. Jones [21] provides rules for the selection of the comparison system. The new system and the comparison system must have similar performance functions, operational environment and support environment. This can hardly be met if any changes are to be implemented, either in the new asset itself or on the way it will be maintained compared to older systems. Because of the importance of the early design decisions in this functional phase, the maintenance organization has an important role in decision making. However, most of the effort of decision making is currently invested in the second stage of decisions involving detailed design. During the phase of functional design, decision makers are steered into a premature reduction of the tradespace.

13.6 Conclusions

This chapter has presented research to improve our understanding of the theory and practice of maintenance decisions made during acquisition of rolling stock, a special type of capital asset. Our aim was to identify which decisions required support. Firstly, we have reviewed the literature on maintenance decision making. Secondly, we have presented a case of NedTrain, the largest maintenance service provider for rolling stock in the Netherlands. The review resulted in an overview of strategic, tactical and operational levels, and then has focused on strategic maintenance decisions. The papers providing frameworks to structure maintenance decisions were organized into decision areas. The case has described decisions involved in the design of the maintenance services for passenger service rolling stock.

According to the literature, maintenance should be an integral part of systems engineering. The maintenance concept should be developed with the asset, and most maintenance decisions could be made at an early stage of development. In practice, however, strategic maintenance decisions are not made during acquisition of rolling stock by the maintenance organization. Logistic support design decisions are delegated to the original equipment manufacturer. The maintenance system of NedTrain is not designed concurrently with new rolling stock being acquired: the supplier optimizes rolling stock locally; the rolling stock maintenance company optimizes services locally. New rolling stock is fitted to the existing maintenance system, and the maintenance system is gradually improved.

There is a need to support strategic maintenance decisions made at NedTrain. These decisions involve large risks, a full causal model is not understood, and the range of possible outcomes for operational performance cannot be predicted. A specific issue that has been identified is the need to broaden the decision tradespace: exploring more maintenance concept options during acquisition. However, complexity of rolling stock, and the limited detail information about final design that exists during acquisition calls for alternative implementation of the traditional maintenance decision models found in the literature. Case-based decision analysis using rigorous analogies or quantitative multiple scenario tools could be very useful to support decision makers in this context.

Acknowledgments This research is part of the "Rolling Stock Life Cycle Logistics" applied research and development program, funded by NS/NedTrain. We thank the experts at NedTrain for their invaluable help and insights. The second and third authors gratefully acknowledge the support of the Lloyd's Register Foundation (LRF). LRF helps to protect life and property by supporting engineering-related education, public engagement and the application of research. We also extend our words of gratitude to our colleagues from the chair of Maintenance Engineering and to the reviewers for their invaluable help in improving this chapter.

References

1. Al-Turki U (2011) A framework for strategic planning in maintenance. J Qual Maintenance Eng 17(2):150–162. doi:10.1108/13552511111134583
2. Atkins (2011) Rail value for money study: asset management and supply chain management assessment of GB rail, issue 1.1 edn. Atkins, London
3. Ben-Daya M, Duffuaa SO, Raouf A, Knezevic J, Ait-Kadi D (2009) Handbook of maintenance management and engineering. Springer, Berlin
4. Blanchard BS (2004) Logistics engineering and management., Prentice-Hall international series in industrial and systems engineeringPearson Prentice Hall, NJ
5. Cavalieri S, Garetti M, Macchi M, Pinto R (2008) A decision-making framework for managing maintenance spare parts. Prod Plann Control 19(4):379–396. doi:10.1080/09537280802034471
6. Crespo Márquez A, Gupta JND (2006) Contemporary maintenance management: process, framework and supporting pillars. Omega 34(3):313–326. doi:10.1016/j.omega.2004.11.003
7. Crespo Márquez A, Moreu De León P, Gómez Fernández J, Parra Márquez C, López Campos MA (2009) The maintenance management framework: a practical view to maintenance management. J Qual Maintenance Eng 15(2):167–178. doi:10.1108/13552510910961110
8. Dawotola AW, Trafalis TB, Mustaffa Z, van Gelder PHAJM, Vrijling JK (2013) Risk-based maintenance of a cross-country petroleum pipeline system. J Pipeline Syst Eng Pract 4 (3):141–148. doi:10.1061/(ASCE)PS.1949-1204.0000121

9. Driessen M, Arts J, van Houtum GJ, Rustenburg JW, Huisman B (2014) Maintenance spare parts planning and control: a framework for control and agenda for future research. Prod Plann Control, 1–20. doi:10.1080/09537287.2014.907586
10. Faccio M, Persona A, Sgarbossa F, Zanin G (2014) Industrial maintenance policy development: a quantitative framework. Int J Prod Econ 147:85–93. doi:10.1016/j.ijpe.2012.08.018
11. Garg A, Deshmukh SG (2006) Maintenance management: literature review and directions. J Qual Maintenance Eng 12(3):205–238. doi:10.1108/13552510610685075
12. Ghosh D, Roy S (2010) A decision-making framework for process plant maintenance. Eur J Ind Eng 4(1):78. doi:10.1504/EJIE.2010.029571
13. Gómez Fernández JF, Crespo Márquez A (2009) Framework for implementation of maintenance management in distribution network service providers. Reliab Eng Syst Saf 94 (10):1639–1649. doi:10.1016/j.ress.2009.04.003
14. Gramopadhye AK, Drury CG, Prabhu PV (1997) Training strategies for visual inspection. Hum Factors Ergon Manuf 7(3):171–196. doi:10.1002/(SICI)1520-6564(199722)7:3<171:AID-HFM2>3.0.CO;2-0
15. Gunasekaran A, Ngai E, McGaughey R (2006) Information technology and systems justification: a review for research and applications. Eur J Oper Res 173(3):957–983. doi:10.1016/j.ejor.2005.06.002
16. Hastings NAJ (2010) Logistic support. In: Physical asset management. Chap 4 Springer, London, pp 179–195. doi:10.1007/978-1-84882-751-6
17. Hayes RH, Wheelwright SC, Clark KB (1988) Dynamic manufacturing: creating the learning organization. Free Press, NY
18. Horenbeek A, Pintelon L, Muchiri P (2011) Maintenance optimization models and criteria. Int J Syst Assur Eng Manag 1(3):189–200. doi:10.1007/s13198-011-0045-x
19. INCOSE (2010) Systems engineering handbook. A guide for system life cycle processes and activities. INCOSE, Seatle
20. Jardine A, Tsang A (2013) Maintenance, replacement, and reliability: theory and applications, mechanical engineering, 2nd edn. Taylor & Francis, London
21. Jones JV (2006) Supportability engineering handbook: implementation, measurement, and management. SOLE Logistics Press/McGraw-Hill, New York
22. Kutucuoglu KY, Hamali J, Irani Z, Sharp JM (2001) A framework for managing maintenance using performance measurement systems. Int J Oper Prod Manag 21(1/2):173–195. doi:10.1108/01443570110358521
23. Leenen M, Wolf A (2010) World market to reach 160 billion by 2015. Int Railway J (Sep 2010) pp 2–3
24. López Campos MA, Crespo Márquez A (2011) Modelling a maintenance management framework based on PAS 55 standard. Qual Reliab Eng Int 27(6):805–820. doi:10.1002/qre.1168
25. McKone K, Weiss E (1998) TPM: planned and autonomous maintenance: bridging the gap between practice and research. Prod Oper Manag 7(4):335–350
26. Melo M, Nickel S, Saldanha-da Gama F (2009) Facility location and supply chain management a review. Eur J Oper Res 196(2):401–412. doi:10.1016/j.ejor.2008.05.007
27. Muchiri P, Pintelon L, Gelders LF, Martin H (2011) Development of maintenance function performance measurement framework and indicators. Int J Prod Econ 131(1):295–302. doi:10.1016/j.ijpe.2010.04.039
28. Murthy DNP, Kobbacy KAH (eds) (2008) Complex system maintenance handbook. Springer series in reliability engineeringSpringer, London. doi:10.1007/978-1-84800-011-7
29. Murthy DNP, Atrens A, Eccleston JA (2002) Strategic maintenance management. J Qual Maintenance Eng 8(4):287–305. doi:10.1108/13552510210448504
30. Naughton MD, Tiernan P (2012) Individualising maintenance management: a proposed framework and case study. J Qual Maintenance Eng 18(3):267–281

31. Nicolai RP, Dekker R (2008) Optimal maintenance of multi-component systems: a review. In: Murthy D, Kobbacy K (eds) Complex system maintenance handbook. Chap 11 Springer, London, pp 263–286. doi:10.1007/978-1-84800-011-7_11
32. Parada Puig JE, Hoekstra S, Huisman B, van Dongen LAM (2011) Supportability and purchasing decisions for capital assets: positioning paper. In: IET and IAM asset management conference 2011. IET, London, pp P07–P07. doi:10.1049/cp.2011.0566
33. Parida A, Kumar U (2006) Maintenance performance measurement (MPM): Issues and challenges. J Qual Maintenance Eng 12(3):239–251
34. Pinjala SK, Pintelon L, Vereecke A (2006) An empirical investigation on the relationship between business and maintenance strategies. Int J Prod Econ 104(1):214–229. doi:10.1016/j.ijpe.2004.12.024
35. Pintelon L, Gelders LF (1992) Maintenance management decision making. Eur J Oper Res 58 (3):301–317. doi:10.1016/0377-2217(92)90062-E
36. Pintelon L, van Puyvelde F (2006) Maintenance decision making, 1st edn. Acco, Leuven
37. Prasad Mishra R, Anand G, Kodali R (2006) Development of a framework for world-class maintenance systems. J Adv Manuf Syst 5(2):141–165. doi:10.1142/S0219686706000807
38. Pyke DF, Cohen MA (1990) Push and pull in manufacturing and distribution systems. J Oper Manag 9(1):24–43. doi:10.1016/0272-6963(90)90144-3
39. Rausand M (1998) Reliability centered maintenance. Reliab Eng Syst Saf 60(2):121–132. doi:10.1016/S0951-8320(98)83005-6
40. Rosqvist T, Laakso K, Reunanen M (2009) Value-driven maintenance planning for a production plant. Reliab Eng Syst Saf 94(1):97–110. doi:10.1016/j.ress.2007.03.018
41. Srivastava S (2008) Network design for reverse logistics. Omega 36(4):535–548. doi:10.1016/j.omega.2006.11.012
42. Takata S (2013) Maintenance-centered circular manufacturing. Procedia CIRP 11:23–31. doi:10.1016/j.procir.2013.07.066
43. Takata S, Kimura F, van Houten FJAM, Westkamper E, Shpitalni M, Ceglarek D, Lee J (2004) Maintenance: changing role in life cycle management. CIRP Ann Manuf Technol 53 (2):643–655. doi:10.1016/S0007-8506(07)60033-X
44. Tam AS, Price JW (2008) A generic asset management framework for optimising maintenance investment decision. Prod Plann Control 19(4):287–300. doi:10.1080/09537280802034042
45. Tam AS, Price JW, Beveridge A (2007) A maintenance optimisation framework in application to optimise power station boiler pressure parts maintenance. J Qual Maintenance Eng 13 (4):364–384. doi:10.1108/13552510710829461
46. Tsang AH (2002) Strategic dimensions of maintenance management. J Qual Maintenance Eng 8(1):7–39. doi:10.1108/13552510210420577
47. Umeda Y, Takata S, Kimura F, Tomiyama T, Sutherland JW, Kara S, Herrmann C, Duflou JR (2012) Toward integrated product and process life cycle planning an environmental perspective. CIRP Ann Manuf Technol 61(2):681–702. doi:10.1016/j.cirp.2012.05.004
48. van Dongen LAM (2014) Through-life engineering services: the NedTrain case. In: Redding L, Roy R (eds) Through-life engineering services: motivation, theory and practice chap 3, Springer, Cranfield
49. Waeyenbergh G, Pintelon L (2002) A framework for maintenance concept development. Int J Prod Econ 77(3):299–313. doi:10.1016/S0925-5273(01)00156-6
50. Waeyenbergh G, Pintelon L (2004) Maintenance concept development: a case study. Int J Prod Econ 89(3):395–405. doi:10.1016/j.ijpe.2003.09.008
51. Waeyenbergh G, Pintelon L (2009) CIBOCOF: a framework for industrial maintenance concept development. Int J Prod Econ 121(2):633–640. doi:10.1016/j.ijpe.2006.10.012
52. Yan S, Yang TH, Chen HH (2004) Airline short-term maintenance manpower supply planning. Transp Res Part A Policy Pract 38(9–10):615–642. doi:10.1016/j.tra.2004.03.005

Chapter 14
Integration of Operational Data into Maintenance Planning

Peter Schuh, Christian Perl and Kirsten Tracht

Abstract In machines a broad range of operational and failure information, like hours of operation, temperatures of components or information about surrounding conditions are available. However, this information is barely used for failure prediction or maintenance planning. At the same time, product life cycles shorten and machine variants increase, making estimation of replacement instances challenging. Stochastic models offer the opportunity of integrating operational and failure information and thereby utilize them for more accurate planning. Within this chapter, a literature overview about existing stochastic prognosis methods and an approach for cost minimal replacement are presented. Within that method data preprocessing, interpretation and utilizing takes place. It can be applied to any system exposed to mechanical wear. The novel planning approach is applied to wind energy turbine data and verified by comparison to established methods.

14.1 Introduction

Ensuring availability and minimising unit time costs are the major aims of maintenance. Unit time costs are accumulated acquisition and maintenance costs during the time of operation of a unit [1]. The longer a unit operates, the smaller are depreciation costs, but the higher are maintenance costs. This relationship has been utilized in classical planning approaches that estimate the most economic instant of renewal by comparing long term depreciation costs and costs of maintenance. In a dynamic field of operation facing wavering feasibility or wavering utilization and output quantity, these approaches do not succeed in estimating cost minimal replacement instances.

P. Schuh (✉) · C. Perl · K. Tracht
Bremen Institute for Mechanical Engineering (BIME),
University of Bremen, Badgasteiner Straße 1, 28359 Bremen, Germany
e-mail: schuh@bime.de

L. Redding and R. Roy (eds.), *Through-life Engineering Services*,
Decision Engineering, DOI 10.1007/978-3-319-12111-6_14

If component states could be predicted more accurate by considering available information about operation, condition and failures, cost minimal replacement instances could be achieved. Some researchers already highlighted the advantages of integrating surrounding conditions, operator behaviour or internal machine states into lifetime estimation for improving prognosis accuracy [2–4]. However, restrictions regarding time dependency and amount of data have been imposed.

During operation of machines a variety of operational and condition data (OCD) is generated. These depend on the sensors mounted in the monitored machine. For example there are sensors for recording component temperatures, hours of operation, vibration signals or number of oil particles. Furthermore, log files are automatically generated by the machine control and maintenance records are kept by maintenance technicians. This broad range of operation and condition monitoring information has mostly been neglected in prognosis.

Within this chapter a state of the art presentation of stochastic and probabilistic methods for integrating condition monitoring information into prognosis takes place. Based on that, an approach will be presented, which utilizes the Proportional Hazards Model (PHM), and allows for maintenance planning under consideration of OCD in an environment of wavering costs and returns. For the application of the PHM, covariates are extracted systematically from condition monitoring systems (CMS). This data pre-processing method is also presented.

Herewith, external as well as internal influences and operational information are integrated into maintenance planning. The proposed method is validated by means of an exemplary application to data of wind turbines.

14.2 State of the Art

14.2.1 Life Expectancy Models

Life expectancy models describe lifetime of technical units with the help of failure distributions. These distributions are derived from historical failure data [5, 6]. Usually, the failure process of mechanical components follows the Weibull distribution [7], whereas the failure mechanism of electronic devices is based on the Exponential distribution [8].

Every failure distribution features a density $f(t)$, a cumulative density function $F(t)$ and a hazard rate $h(t)$. Due to its importance and flexibility, the density function of the Weibull distribution is presented in Eq. (14.1).

$$f(t) = \frac{\beta}{T}\left(\frac{t}{T}\right)^{\beta-1} e^{-\left(\frac{t}{T}\right)^{\beta}} \tag{14.1}$$

It represents the probability of failure at the instant t. The shape parameter β and the scale parameter T fully parameterize the two-parametric Weibull distribution. If the shape parameter equals one, Eq. (14.1) also represents the density function of the

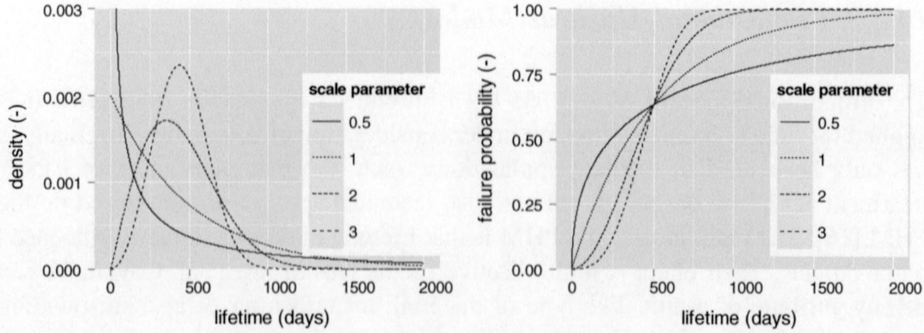

Fig. 14.1 Density (*left*) and cumulative density function (*right*) of the Weibull distribution

Exponential distribution [7]. The influence of the shape parameter is highlighted in Fig. 14.1 (T = 500). The higher the shape parameter is, the more probable a failure due to wear is. If the shape parameter is smaller than one, it indicates early failures that can be caused by wrong material or mistakes during assembly.

The integrated density function equals the cumulative density function *F(t)*. It describes the probability of failure during the interval from zero to *t*. Equation (14.2) shows the cumulative density function of the Weibull model, which is also visualized on the right hand side of Fig. 14.1 [7].

$$F(t) = 1 - e^{-\left(\frac{t}{T}\right)^{\beta}} \tag{14.2}$$

The hazard function is calculated by means of the density and the cumulative density function Eq. (14.3). It represents the probability of failure at the instant of t, conditionally the unit survived up to date.

$$h(t) = \frac{\beta}{T}\left(\frac{t}{T}\right)^{\beta-1} \tag{14.3}$$

In terms of forecast accuracy it is crucial to select the correct failure distribution for units under investigation. To ensure the correct choice of failure distribution, the sample size should be analyzed by means of the Chi^2-Test or the Kolmogorov-Smirnov test [9].

The Weibull distribution has been introduced by Rosin and Rammler [10] in 1933 for calculating the size of coal particles. In the field of maintenance planning it is applied for failure probability estimation today. Große [11] utilized the distribution for maintenance planning of steam machines. The author parameterized the failure distribution with the help of expert judgements. Mazhar et al. [12] combined the probabilistic method with a neuronal network approach, which allowed for integrating condition monitoring data. Combining lifetime prediction based on failure distributions with spare parts supply realized Lanza and Niggeschmidt. They varied shape parameters according to machine load that allowed for consideration of dynamic load effects in the failure distribution [13].

14.2.2 Proportional Hazards Model

The Proportional Hazard Model has been introduced by Cox in 1972 [14]. It is applied for time to failure estimation under consideration of covariates. Originally is has only been used in medical applications, such like risk assessment of cancer treatment [15]. In more recent studies, also technical applications are based on the PHM [16]. The main idea of the PHM is that lifetime of an individual is influenced by covariates. Their effect is multiplicative on the hazard rate [17]. Covariates can be any attribute of a unit, like type of material, manufacturer or also surrounding temperature in operation. These are influencing covariates, which are independent of the condition of the unit. There are also covariates that are influenced by the condition, like number of oil particles, temperature of components or vibration signals. In Eq. (14.4) the semi-parametric model is presented. It consists of an unspecified baseline hazard function $h_0(t)$ and the covariate term ($\exp(\alpha z_m)$).

$$h(t, z) = h_0(t) \exp\left(\sum_{m=1}^{m} \alpha_m z_m\right) \tag{14.4}$$

Without assuming the baseline hazard function, the regression parameter α is calculated with the help of the Maximum Likelihood method. With it, the effect of the covariate z on the hazard rate $h(t, z)$ is assessed. It is utilized for calculation of the survival probability $S(t; z; z(t))$ at the instant t Eq. (14.5).

$$S(t; z; z(t)) = \exp\left(-\int_0^t h(t; z; z(t))\right) \tag{14.5}$$

For maintenance planning the remaining useful lifetime (RUL) is often the criterion for renewal Eq. (14.6) [4, 18]. It is derived from the survival probability of Eq. (14.5) or any other continuous failure distribution.

$$RUL(t) = \frac{\int_t^\infty S(t)dt}{S(t)} \tag{14.6}$$

Banjevic and Jardine implemented the time varying covariate "number of metal particles" to perform survival analysis of a mining truck gear box. Louit et al. also used this information and initiate spare part planning [3, 19]. However; economic effects caused by lowered operational availability and maintenance planning are not integrated in the model. Maintenance planning has been realized by Banjevic and Jardine. They enhanced the PHM with the help of a specified failure distribution and implemented information gathered at inspections [19].

The expected spare parts demand has also been the result of Ghordrati's thesis. He combined a renewal process with the survival probability extracted from the PHM.

The findings highlight the necessity of investigating into surrounding and operational conditions when assessing survival probabilities [2]. In addition to Louit et al., Ghordrati, and Zhao et al. developed a method for cost minimal renewal. The authors assume that wear is directly integrated into the model, but don't describe how it is measured and how condition information of sensors are considered. This enhanced model is used for maintenance planning and is validated with the help of a Monte Carlo Simulation [20]. Vlok also performed maintenance planning in a periodic inspection scenario. The author considers for repairable and non-repairable items and derives an universally valid proportional intensity model [4, 18]. This model is utilized for assessing vibrational data that are measured at inspections. The resulting RUL is then applied in maintenance planning. If the RUL is smaller than the period of Inspection, the renewal is performed at the instant of investigation.

14.2.3 Kalman Filter

The Kalman Filter represents a discrete state space model and works similar to a feedback control system, in which the initial state is continually updated, if new information is available. This filtering technique makes use of the current state, the expected or estimated state as well as information derived from sensors. The recursive algorithm integrates the last known state and the latest sensor value into prognosis, even if input data are stochastically disturbed [21].

Originally, the Kalman Filter has been applied in signal processing, but there are also a number of applications in the technical domain of RUL prognosis.

Wang and Christer estimated the RUL of an inductor furnace. The authors assess the system state at the instant of inspection for calculating the RUL with the Kalman Filter. Above that they define costs for preventive and corrective maintenance that allow for an assessment of the instant of cost minimal replacement [22].

Gomes et al. predicted the RUL of hydraulic pumps by considering the current pressure. By investigating into level of significance and observed end of life they proved that condition signals combined with the Kalman Filter are applicable to failure prediction [23].

Wang combined stochastic filtering and a Markov Chain for wear process prediction. For that the amount of metal particles in oil indicates the condition, which is simulated with the help of the Markov Chain. Thereby, the hidden wear state of the application has been predicted [24].

14.2.4 Hidden Markov Models

Hidden Markov Models (HMM) have been invented by Baum in 1966. They belong to the group of Markov Models, whereas the current state is hidden. There are several applications known in medicine, speech processing and lifetime prediction [25, 26]. Every state of the system is linked to each other and will be entered by a defined probability. By means of emissions or condition indicators, the probability of being in a predefined state is estimated [27].

14.3 Applications

Buruah applied the HMM for prognosis of wear on a metal cutting tool, by evaluating drilling force and torque. Based on four predefined states (good, median, bad and worst) diagnostics and prognostics is performed [25]. The results of the model heavily depends on the training of the HMM with historical demand data.

Bunks et al. [28] modelled states for a helicopter gear box, which are considered in a HMM. Information about the current state is utilized in RUL estimation by evaluating transition state probabilities. The authors neglected a quantification of transitions between states based on data, due to missing historical information. Instead, expert knowledge has been utilized for evaluation of transitions.

Zhou et al. combined an expert system with the HMM, allowing for integration of surrounding conditions of a continuous stirred tank. The surrounding conditions are constituted by the initial temperature and the flow rate of the fluid and the current stirred tank temperature represents the condition indicator. The results are promising in terms of prognosis accuracy, whereas application to more complex technical systems needs to be investigated [29].

Neves et al. combined prognosis and maintenance decision making within the HHM. With the help of the Baum Welch algorithm an iterative update of model parameters takes place. The approach of maintenance decision making integrates costs for corrective and preventive activities. However, model accuracy could not be verified [30].

14.4 State of the Art Assessment

All stochastic methods base on historical data and all of them are suitable for basic lifetime analysis. However, they differ in flexibility and modelling effort.

A high amount of historical failure events is necessary for applying HMMs, because of the state definition and the quantification of transitions between them. This amount of data is usually only available in laboratories. Furthermore, goodness of model fit cannot be assessed due to hidden states.

The application of the Kalman Filter is often prohibited because of the necessary simplifications of the model. For utilizing the Kalman Filter, linear behaviour of the technical application is required. Similar to the HMM, deriving model dependencies of real data is a challenge.

Modelling effort for the PHM is low and it is easy and flexible to apply, but there are a lot of constraints regarding OCD that are integrated into the model.

All stochastic approaches have the common risk that technically non-existing coherences can be modelled. This risk needs to be considered when interpreting prognosis results. The advantages of the PHM in modelling effort over the other methods will be used for prognosis in the following. The weakness of the PHM that results from high requirements on OCD will be faced with a detailed presentation of a data processing approach.

Decision making on replacement in known approaches is either performed by utilizing long term costs or by calculating the RUL [1]. Assessing long term unit time costs neglects the current condition and machine load. Vlok, Banjevic and Jardine [4, 19] introduced a RUL-based approach that utilizes the PHM for integration of condition data, derived from vibration sensors in a scenario that allowed only monitoring during periodic inspections. RUL based approaches estimate the most probable lifetime, but do not give information about the cost effectiveness of replacement. If online CMS steadily record information, potential in prognosis is revealed that remains unused by RUL-based approaches in periodic inspection environments. Known methods in literature do not provide the required functionality of integrating online CMS information and integrating them into replacement decision making. Therefore, costs as well as probability of occurrence are utilized in a framework basing on the PHM.

14.5 Approach

The approach for cost minimal replacement in an environment of wavering machine utilization or wavering feasibility of maintenance processes combines CMS and failure information as well as an extension of the PHM, the Weibull-Proportional Hazards Model (WPHM). It is an enhancement of the forecast model proposed by Tracht et al. [31]. Historical data are used to parameterize the WPHM once. If further components fail, computational steps for parameterization have to be repeated, to keep regression parameters of the model up to date (Fig. 14.2).

With the help of the parameterized WPHM, the individual hazard rate of monitored components is calculated. For that, current operational and condition monitoring data need to be processed and utilized in the model. During time window of non-feasibility or high machining utilization the probability of failure, conditionally the unit survived up to date, is computed. When preventive maintenance costs are compared to probably occurring corrective maintenance costs, the most economic instant of replacement is estimated. All computational steps are described in more detail in the following.

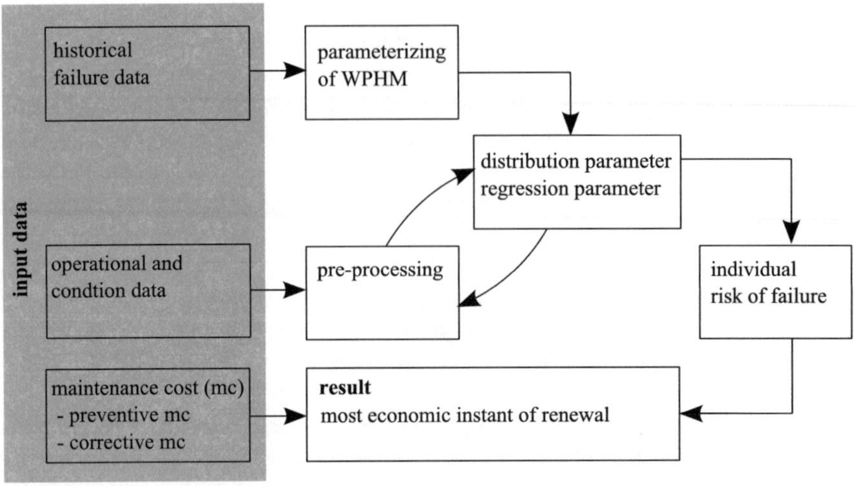

Fig. 14.2 Approach for condition based replacement

14.6 Data Processing

Raw data of CMS need to be checked if they are complete as well as formally and logically correct [32]. For that reason the data sets are filtered regarding outliers and wrong entries. Identification of outliers is performed with the help of threshold definitions or by application of the David, Hartley and Pearson test [5]. Furthermore, the examination of typical statistical values, like mean, standard deviation, minimal and maximal measured values are first indicators for correct and complete data sets. For example rapidly increasing component temperatures indicate a malfunction of a sensor, of an analogue input or a defective control system. It follows the deletion of the pertaining data base entries. Thereby, prognosis quality can be increased.

OCD are processed before utilized in the WPHM. The data needs to fulfil mathematical and technical boundary conditions. Mathematic requirements have been investigated by Kumar [33]:

- Monotone covariates must be avoided
- Dependent covariates must be described and modelled
- Relevant covariates must be included in the model

Monotone covariates steadily rise or fall during units' lifetime. Monotonic behaviour is detected by plotting covariates in chronological sequence. An example is presented in Fig. 14.3. The cumulated power output during operation is monotonic increasing; hence, it is not integrated into the PHM in its original form. Instead, the covariate needs to be split into strata, thus only hours of operation

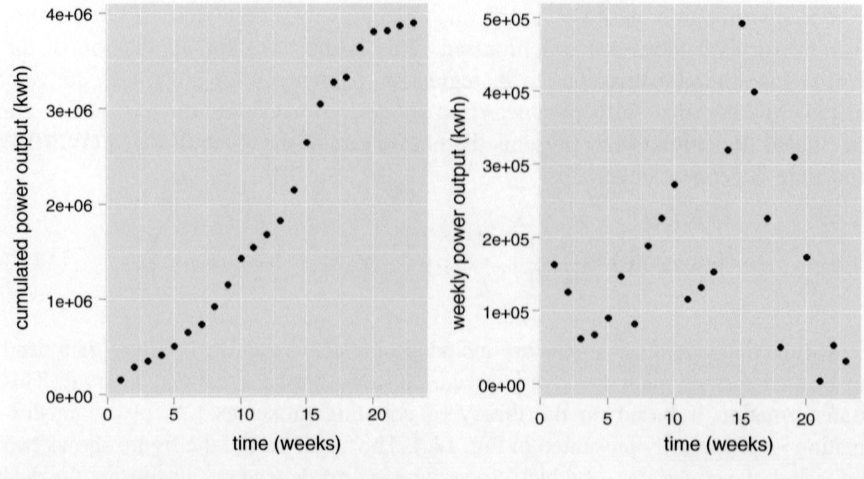

Fig. 14.3 Stratification of monotone covariates

during a predefined interval are cumulated. On the right hand side of Fig. 14.3 the result of stratification is presented. These covariates can be handled by the PHM.

Dependent covariates lead to biased hazard rates and prognosis results. Therefore, every covariate included needs to be tested for independency against each other. The Chi^2 test and the product momentum test are established methods for testing independencies [34]. If covariates depend on each other, they either cannot be integrated together or should be combined. The combination of two covariates for example comprises multiplication or division of each other. The resulting new covariate represents the information required that is integrated into the model.

The last mathematical requirement for relevant covariates is ensured by technical coherences and the application of test and validation data sets. For that, covariates, which are pre-processed in terms of independency and monotony, are integrated into the test and the validation model. If covariates are relevant, they will affect the resulting hazard rate of both models in a similar and significant way. Hence the covariate stays in the model [35].

14.7 Weibull PHM

Processed OCD are assigned to failure and demand information. However, demand and lifecycle data might be incomplete in terms of beginning or end of life. In case of an unknown instant of entry into operation the lifetime is referred to as left censored data. If the event of interest, a failure event for example, did not occur during investigation the lifetime is denoted as right censored. These units are still under the risk of failure [29, 36]. Right censored data are used in the PHM with the Partial Likelihood method, left censored data are neglected [37].

For investigation purpose, entry into operation of every unit is set to zero, thereby survival times are synchronised. This is the basis for application of the Partial Likelihood estimation. In it, regression parameters for covariates are estimated. In case of a fully parameterized model, also distribution parameters are calculated. Equation (14.7) presents the hazard rate of the Weibull PHM (WPHM) and time dependent covariates.

$$h(t,\ z;\ z(t)) = \frac{\beta}{T}\left(\frac{t}{T}\right)^{\beta-1} \exp\left(\sum_{m=1}^{m_1} \alpha_m z_m + \sum_{m=1}^{m_2} \delta_m z_m(t)\right) \tag{14.7}$$

If time dependent covariates are included in analysis data sets, failure data need to be transformed such as covariates are constant during a defined interval. This transformation is based on the theory of counting processes [38, 39]. Transformation is exemplarily presented in Fig. 14.4. The upper part of the figure shows two units and two covariates, of which covariate 2 is time dependent. Therefore, the data is transformed such as every week has only two constant covariates. The beginning and the end of the week are indicated by start and stop. If the unit fails at an instant, it is denoted by the numeric status. Status 1 marks a failure and status 0 indicates an operating unit at the end of the survival analysis period.

When data are available according to form, parameter outputs of the WPHM are applied to Eq. (14.5). With the help of the survival probability, the probability of failure during the prognostic interval $Fx1(x2)$ is performed with Eq. (14.8). The prognostic interval is either defined by the duration between two inspections, by the lead time of spare parts or by the expected duration of non feasibility.

$$Fx1(x2) = \frac{F(x2) - F(x1)}{S(x1)} = \frac{(1 - S(x2)) - (1 - S(x1))}{S(x1)} \tag{14.8}$$

Fig. 14.4 Data transformation for WPHM application

unit	age	status	cov 1	cov 2_1	cov 2_2	cov 2_3	cov2_4	cov 2_5	...
A	5	1	30	2	3	0	1	2	...
B	9	0	20	3	1	5	4	7	...

data transformation

unit	age	status	start	stop	cov 1	cov 2
A	34	0	0	1	30	2
A	34	0	1	2	30	3
A	34	0	2	3	30	0
A	34	0	3	4	30	...
A	34	1	4	5	30	2
B	9	0	0	1	20	3
B	9	0	1	2	20	1
B	9	0	2	3	20	5
B	9	0	3	4	20	4
B	9	0	4	5	20	7
...	...	...	...	...	...	...

14.7.1 Cost Minimal Replacement

The overall aim of maintenance, minimizing unit time cost, is mainly influenced by the instant of replacement. Units featuring exponential failure behaviour should never be renewed preventively if minimal unit time costs are required and the unit is not crucial in terms of machine safety [1]. Hence maintenance planning is not necessary.

In case of increasing hazard rates, for example in case of units exposed to wear, preventive maintenance tasks can be economically justified [40]. In this case, the cost minimal replacement of units depends on the instant of renewal. Usually planned maintenance is much cheaper than corrective maintenance because time for diagnosis and repair is shorter, which leads to reduced costs for machine downtime. In case of corrective maintenance, the unit operates longer compared to preventive maintenance strategy, but uncertainty about spare parts demand or amount and qualification of maintenance technicians will lead to prolonged down times. On the other hand, if the preventive renewal takes place too early, unit time costs rise because of high depreciation costs c_d (Eq. 14.9). They are influenced by the lifetime of the unit t_u, the interest rate z_s and by procurement costs c_s. The shorter the lifetime is, the higher are daily depreciation costs.

$$c_d = c_s(z_s + 1)^{t_u} \tag{14.9}$$

The criterion for replacement are the expected unit time costs that accumulate preventive renewal costs per day c_{pd} (Eq. 14.10) as well as corrective renewal costs per day c_{cd} (Eq. 14.11).

$$c_{pd}(x1) = \frac{c_p + c_d}{x1} \tag{14.10}$$

$$c_{cd}(x2) = \frac{c_c + c_d}{x2} \tag{14.11}$$

$$c_{pd}(x1) < c_{cd}(x2) \cdot Fx1(x2) \tag{14.12}$$

By comparing costs at the instant x1, which is the preventive maintenance instant, with the instant x2, the cost minimal maintenance instant is calculated. Equation (14.12) shows the probability of failure during a predefined time window, which is multiplied by costs for corrective maintenance. If these costs exceed preventive maintenance costs it is economically justified to perform a renewal at x1.

The method proposed in Fig. 14.2 optimizes expected costs during the next planning interval. Its length is determined by spare parts lead time for example, hence, purchase costs can be minimized with more accurate demand forecasts. The planning interval could also be determined by the time window weather conditions prohibit maintenance or by the interval of high machine utilization.

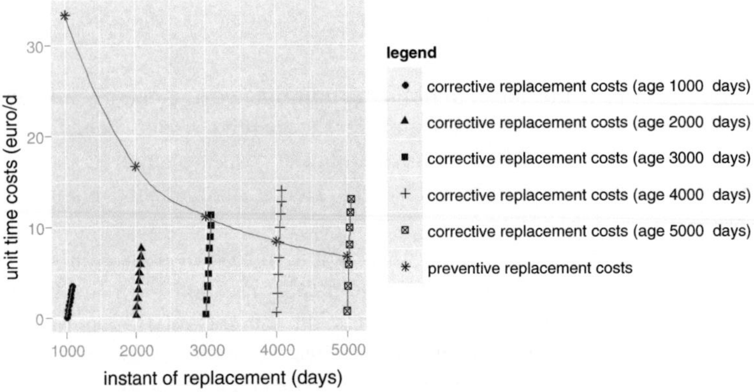

Fig. 14.5 Scenarios of cost minimal replacements (T = 2,000 days, $\beta = 3.2$)

Figure 14.5 presents the general coherences between instant of renewal, and unit time costs for renewal. It is depicted that increasing component age and prolonged forecast intervals lead to higher expected corrective maintenance costs. Within the parameter study components age of 1,000, 2,000, 3,000, 4,000 and 5,000 days have been chosen for planning dates. At these dates costs are calculated for different instances (10, 20, 30,…, and 100 days). If corrective maintenance costs exceed preventive maintenance costs, the replacement should be performed.

14.7.2 Application

The proposed method has been applied to generator bearings of wind energy turbines. CMS data have been pre-processed as described in the data processing section. All covariates are time dependent; hence, the analyzed data are split into intervals of 30 days. The following OCD have been investigated:

- Power output
- Temperature of bearing
- Temperature of stator 1
- Temperature of stator 2
- Wind speed

Data are processed by accumulating the amount of threshold exceeding during the interval of 30 days. The threshold is defined by the 0.99 level of significance (left side of Table 14.1). The tablet shows the model parameters for the most significant condition indicator "temperature stator 1". Bearing temperature and stator 2 temperatures showed no significant influence because available data have been incomplete. In total there have been twelve units, of which five units failed

Table 14.1 Output of the lifetime analysis

Weibull-PHM (including covariates)		Weibull model	
Parameter	Value	Parameter	Value
β	1.14	β	1.2
T	2,186 days	T	1,550 days
α	0.00439	α	–
p-value	0.0968	p-value	–

during investigation. The shape parameter of the model with (WPHM) and without (Weibull model) covariates indicate a wear process (Table 14.1).

The calculated hazard rate, costs and OCD are utilized for estimating the instant of replacement. First, costs for preventive and corrective maintenance are quantified. This is performed by estimating time to repair in case of preventive and in case of corrective failure, as well as adding costs for machine downtime. Second, covariate values within the planning horizon are predicted. For this scenario 100 temperature exceedances are expected in every wind turbine within the planning horizon of 100 days. The WPHM has been used for calculating failure probabilities, based on the predicted covariate value (Table 14.2). Based on the failure probability during planning horizon, costs for corrective maintenance per day c_{cd} and costs for preventive replacement per day c_{pd} are compared to each other.

Implementation of the cost parameters presented in Table 14.3 leads to the result that there is no replacement necessary during the investigated planning horizon. This is due to the short planning horizon and due to the age of the units, which both result in low conditional probabilities $Fx1x2$. Figure 14.5 highlights that the age of the units needs to be significantly higher than the characteristic lifetime T. Additionally, the low shape parameter postpones replacements.

For validation of the proposed method, parameters of Table 14.1 (Weibull model) and lifetime data presented in Table 14.2 are applied to the RUL-based approach Eq. (14.6). Comparing the estimated RUL (Table 14.4) with the prognosis

Table 14.2 Prognosis data and results

Input data			Output data				
Unit (wt)	X1 (days)	X2 (days)	Fx1 (–)	Fx2 (–)	Fx1x2 (–)	C_{cd} (€)	C_{pd} (€)
1	746	846	0.3646	0.4076	0.0678	1.02	12.96
2	1,318	1,418	0.5808	0.6115	0.0731	0.69	7.75
3	828	928	0.4000	0.4413	0.0687	0.95	11.77
4	985	1,085	0.4637	0.5014	0.0703	0.84	10.04
5	992	1,092	0.4665	0.5040	0.0704	0.84	9.98
6	1,318	1,418	0.5808	0.6115	0.0731	0.69	7.75
7	272	372	0.1332	0.1850	0.0597	1.98	33.96

Table 14.3 Cost parameters for scenario analysis

Parameter	Z_s	C_s	C_c	C_p
Value	0.04	8,000€	4,000€	1,000€

Table 14.4 Results of remaining useful lifetime analysis

Instant of prognosis (days)	Estimated RUL (days)
746	829
1,318	518
828	776
985	684
992	680
1,318	518
272	1,194

horizon, it becomes clear that replacement shouldn't be performed (RUL > 100 days). This indicates that the novel planning approach provides correct forecast results. With the help of the proposed method, weaknesses of existing methods for cost minimal replacement in dynamic surroundings are avoided. Furthermore, individual load situations are considered together with costs in replacement planning. Thereby, the result of the prognosis can be monetarily assessed and justified.

14.8 Summary

This chapter presented stochastic prediction methods allowing for integration of condition monitoring data. Due to its simplicity and flexibility, the semi parametric PHM has been combined with a planning framework. With it, cost minimal replacement instances of monitored components that are exposed to mechanical wear can be calculated. The application of a dataset of monitored wind turbines shows the applicability of the proposed method.

Acknowledgments The results presented were developed in the research project "Service logistics for wind turbines" funded by the German Federal Ministry of Economics and Technology (BMWI), Industrial Collective Research for SMEs (AiF), Bundesvereinigung Logistik e.V. (BVL).

References

1. Jorgenson DW, McCall JJ, Radner R (1967) Optimal replacement policy. North-Holland Publishing, Amsterdam
2. Ghordrati B (2005) Reliability and operational environment based spare parts planning. Lulea University of Technology, Lulea

3. Louit D, Rodrigo P, Banjevic D, Jardine AKS (2011) Condition-based spares ordering for critical components. Mech Syst Sig Process 25:1837–1848
4. Vlok P-J (2001) Dynamic residual life estimation of industrial equipment based on failure intensity proportions. University of Pretoria, Pretoria
5. Dietrich E (2009) Statistische Verfahren zur Maschinen—und Prozessqualifikation: mit 61 Tabellen, 6., vollst. überarb. Aufl. Hanser, München [u.a.]
6. Strunz M (2012) Instandhaltung, Grundlagen-Strategien-Werkstätten. doi:10.1007/978-3-642-27390-2
7. Abernethy RB (2006) The new Weibull handbook. Robert B. Abernethy Publishing, North Palm Beach
8. Epstein B, Sobel M (1954) Some theorems relevant to life testing from an exponential distribution. Ann Math Stat 25:373–381
9. Schenk M (2010) Instandhaltung technischer Systeme—Methoden und Werkzeuge zur Gewährleistung eines sicheren und wirtschaftlichen Anlagenbetriebs. Springer, Berlin
10. Rosin P, Rammler E (1933) The laws governing the fineness of powdered coal. Inst Fuel 7:29–36
11. Große G (1976) Instandhaltungsstrategien für einfache und komplexe Systeme mit Berücksichtigung periodischer Kontrollen. Ingenieurhochschule Zittau
12. Mazhar MI, Kara S, Kaebernick H (2007) Remaining life estimation of used components in consumer products: life cycle data analysis by Weibull and artificial neural networks. J Oper Manage 25:1184–1193. doi:10.1016/j.jom.2007.01.021
13. Lanza G, Niggeschmidt S, Werner P (2009) Optimization of preventive maintenance and spare part provision for machine tools based on variable operational conditions. CIRP Ann Manuf Technol 58:429–432. doi:10.1016/j.cirp.2009.03.030
14. Cox DR (1972) Regression models and life-tables. J Roy Stat Soc 34:187–220
15. Hastie T, Sleeper L, Tirshirani R (1992) Flexible covariate effects in the proportional hazards model. Breast Cancer Res Treat Issue 3(22):241–250
16. zu Eulenburg CHG (2009) Überleben, Progredienz und Tod: Multistate-Modelle zur Analyse konkurrierender und nachzeititger Ereignisse. Universität Hamburg, Hamburg
17. Klein JP, Moeschberger ML (2010) Survival analysis: techniques for censored and truncated data, 2nd edn. Springer, New York (corr. 3. print)
18. Vlok PJ, Coetzee JL, Banjevic D, Jardine AKS, Makis V (2002) Optimal component replacement decisions using vibration monitoring and the proportional-hazards model. J Oper Res Soc 53:193–202
19. Banjevic D, Jardine AKS (2006) Calculation of reliability function and remaining useful life for a Markov failure time process. J Manage Math 17:115–130
20. Zhao X, Bordes L, Fouladirad M, Bèrenguer C (2010) Condition-based inspection/replacement policies for non-monotone deteriorating systems with environmental covariates. Reliab Eng Sys Saf 95:921–934
21. Kalman RE (1960) A new approach to linear filtering and prediction problems. ASME-J Basic Eng 83:35–45
22. Wang W, Christer AH (2000) Towards a general condition based maintenance model for a stochastic dynamic system. J Oper Res Soc 51:145–155
23. Gomes JPP, Leao BP, Vianna WOL et al (2012) Failure prognostics of a hydraulic pump using kalman filter. Ann Conf Prognostics Health Manage Soc 3:1–5
24. Wang W (2007) A prognosis model for wear prediction based on oil-based monitoring. J Oper Res Soc 58:887–893
25. Baruah P, Chinnam RB (2005) HMMs for diagnostics and prognostics in machining processes. Int J Prod Res 43:1275–1293
26. Baum LE, Petrie T (1966) Statistical inference for probabilistic functions of finite state markov chains. Ann Math Stat 37:1554–1563
27. Rigoll G, Schenk J (2010) Mensch-Maschine-Kommunikation. Springer, Berlin
28. Bunks C, Mc Carhy D, Al-Ani T (2000) Condition-based maintenance of machines using hidden markov models. Mech Syst Sig Process 14:597–612. doi:10.1006/mssp.2000.1309

29. Zhou Z-J, Hu C-H, Xu D-L et al (2010) A model for real-time failure prognosis based on hidden Markov model and belief rule base. Eur J Oper Res 207:269–283. doi:10.1016/j.ejor.2010.03.032
30. Neves ML, Santiago LP, Maia CA (2011) A condition-based maintenance policy and input parameters estimation for deteriorating systems under periodic inspection. Comput Ind Eng 61:503–511. doi:10.1016/j.cie.2011.04.005
31. Tracht K, Goch G, Schuh P, Sorg M, Westerkamp JF (2013) Failure probability prediction based on condition monitoring data of wind energy systems for spare parts supply. CIRP Ann Manufact Technol 62:127–130. http://dx.doi.org/10.1016/j.cirp.2013.03.130
32. Sachs J, Hedderich L (2009) Angewandte Statistik: Methodensammlung mit R, 13., aktualisierte und erweiterte Auflage. Springer, Dordrecht
33. Kumar DBK (1994) Proportional hazards model: a review. Reliab Eng Sys Saf 44:177–188
34. Andy Field ZF, Miles J (2012) Discovering statistics using R. Sage, Thousand Oaks
35. Thall PF, Lachin JM (1986) Assessment of stratum-covariate interactions in cox's proportional hazards regression model. Stat Med 5:73–83
36. Kaplan EL, Meiers P (1958) Nonparametric estimation from incomplete observations. J Am Stat Assoc 53:457–481
37. Hosmer DW, Lemeshow S, May S (2008) Applied survival analysis: regression modeling of time-to-event data, 2nd edn. Wiley, Hoboken
38. Andersen PK, Gill RD (1982) Cox's regression model for counting processes: a large sample study. Ann Stat 10:1100–1120
39. Fleming TR, Harrington DP (1991) Counting processes and survival analysis. Wiley, New York
40. Beichelt F (1970) Zuverlässigkeit und Erneuerung: die optimale Instandhaltung. Verlag Technik, Berlin

Chapter 15
Integrated Maintenance System Trend and a Maintenance Scheduling System Application

Toshiharu Miwa, Toshiya Kaihara and Youichi Nonaka

Abstract Efficient operation and maintenance management for industrial facilities is one of the key issues of social infrastructure systems. This chapter describes some technology trends of integrated maintenance system, especially about monitoring, parameterizing, predicting, and control. Firstly, a trend from Time-Based Maintenance (TBM) to Condition-Based Maintenance (CBM) is discussed. TBM-based agreement between operation owner and maintenance provider is common one, but uncertainty of facility degradation requests CBM application as future service model. Secondly, this chapter describes an optimization method for facility maintenance scheduling, which focuses on the subject that the conflict between maintenance operations and net operations reduces system efficiency.

15.1 Introduction

As a result of the increased competition due to rapid globalization of the world economy, it is crucial for manufacturing firms to bring high quality and reliability products to the market. It is necessary not only to manage and control the quality and reliability in design and manufacturing processes but also to perform appropriate maintenance activities of products in operation while keeping high reliability throughout the product lifecycle. The product lifecycle management becomes increasingly important from the view point of reduction of environmental impact.

The advanced maintenance technologies utilizing information and communication technology are coming into practical use. For example, the remote and real-time detection of a symptom of product failure based on monitoring the operating

T. Miwa · Y. Nonaka (✉)
Yokohama Research Laboratory, Hitachi, Ltd, 292 Yoshida-Cho, Totsuka-Ku, Yokohama 2440817, Japan
e-mail: youichi.nonaka.ym@hitachi.com

T. Kaihara
Department of Computer Science and Systems Engineering, Graduate School of System informatics, Kobe University, 1-1 Rokkodai, Nada, Kobe 657-8501, Japan

L. Redding and R. Roy (eds.), *Through-life Engineering Services*,
Decision Engineering, DOI 10.1007/978-3-319-12111-6_15

condition enables to perform the maintenance before breakdown. Monitoring the operating condition of products distributed across the world enables to manage the supply and inventory of maintenance parts efficiently.

In particular, it is important to keep working at the high operating ratio in large-scale systems required vast investment such as the power plant and semiconductor production line from the view point of amortization reduction. The appropriate maintenance activities prevents from stoppage due to machine failure. On the other hand, the machine does not work generally during maintenance. Maintenance scheduling taking both the machine stoppage due to failure and maintenance into consideration leads to maximize the total operating ratio.

Section 15.2 describes the review of operation and maintenance and developed scheduling method optimizing operation and maintenance to maximize the effective productivity.

15.2 Reviews of Previous Studies

15.2.1 Changes of Maintenance Methods

In the past, when products (various machines and equipments) broke down, most of products were repaired as maintenance activity, otherwise products were discarded. The impact of product breakdown on society is getting bigger and bigger with increase in scale and complexity of product. The maintenance methods become advanced with the progress of condition monitoring and failure diagnosis technologies utilizing information and communication technology.

Figure 15.1 indicates the prime maintenance activities throughout product life-cycle [1]. The maintenance activity is expected to fix the gap between current function level and required level by means of treatment or upgrading. The product function is degrading gradually as cumulative operation time of product increases.

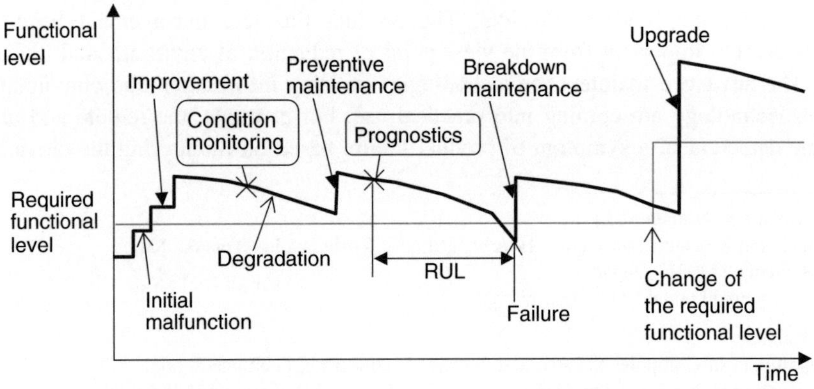

Fig. 15.1 Maintenance activities

Preventive maintenance is pre-formed before breakdown by monitoring the condition of functional degradation. On the other hand, breakdown maintenance is performed after failure with the progress of functional degradation.

Preventive maintenance is superior to breakdown maintenance in terms of avoiding the product stoppage. Therefore, preventive maintenance is applied in the product required high reliability, where the workflow and task in preventive maintenance are much complicated in comparison to breakdown maintenance. Firstly, Time-based Maintenance (TBM) was major method as preventive maintenance. TBM is executed based on cumulative operation time. There are some researches about TBM, such as DP based approach for flow-shop [2], branch and bound approach with column generation for parallel machines [3], approximation algorithm approach for single machine [4], and GA based approach for job-shop [5].

The more efficient Condition Based Maintenance (CBM) is coming into practical use with the progress of condition monitoring and failure diagnosis technologies. CBM is executed based on monitoring the operating condition of product. Several researches are reported about CBM, such as GA based approaches for single machine [6] and for parallel machine [7, 8], and exact algorithm for a single machine [9]. There are also some researches that try to combine TBM and CBM for flow-shop [10] and for job-shop [11].

The advanced ICT technologies enable to detect a symptom of product failure in real-time and remotely. The machine-to-machine (M2M) networks are commercialized as inexpensive wireless communication method among machines with significant growth of mobile network. Next-generation information platforms enable to process a large amount of data ("Big Data") being transmitted in real-time from machines via M2M network [12]. The "Big Data" processing technologies used in data centre provide the data analytics services [13].

There are many studies on Prognostics and Health Management (PHM), which enables to manage the future maintenance plan in advance based on predicting the remaining life time (RUL) using real-time monitoring data [14]. The physical and empirical approaches, which create the model of the physical phenomenon and behaviour in specified product using physical equation or empirical equation and predict the RUL accurately, are putting into practical application such as fault detection of aircraft engine [15] and degradation prediction of the lithium ion batteries [16]. There are also many studies on data driven approach as a general method without product knowledge. This approach focuses on the difference of behaviour in a large amount of sensing data obtained form product and detects the symptom of product failure using data mining and pattern recognition [17, 18]. The studies of PHM are mainly progressing in the area of aerospace industry required the high reliability.

On the other hand, preventive maintenance is not always best maintenance method in the industrial application from the view point of economical efficiency as described above. Takata and Kimura [19] advocated the design methodology of maintenance strategy based the product life cycle simulation. It is important to select the optimum maintenance strategy considering operation type, impact of product failure, maintenance cost and technological level of failure detection and diagnosis.

As described above, the state-of-the-art M2M technologies enable the real-time monitoring of the operating condition of products cost effectively. The optimum maintenance strategy considering the operation planning is becoming more important.

15.2.2 Optimization of Operation and Maintenance

This subsection describes the changes of business relating to operation and maintenance and technological trends of optimization for improving the efficiency of operation and maintenance activities.

Figure 15.2 shows the trend of business structure relating to operation and maintenance. As a first step, the user which operates the product owns its product. The user executes the operation and maintenance activities. In this business model, the product makers gain the profit by selling their products to the users. On the other hand, the users are required to educate the operators and keep their skill in order to maintain the complicated and high reliable product. Its activity to keep their maintenance capability costs the user. The product makers provide the maintenance services to users as the outsourcing business. As a result of the increased competition due to rapid globalization of the world economy, maintenance services throughput product lifecycle are attracting attention to secure the stable profitability. For example, it is reported that the service revenue such as maintenance parts replacement during product lifecycle exceeds the sales revenue in aircraft engine industry [20]. It is inevitable to create new business and generate new additional value taking operation and maintenance of products throughout product lifecycle into consideration from now.

User's mindset of products is changing from the ownership of products to the service to utilize the products. For example, many software application services using the cloud computing are commercialized without owing the software and related computer hardware. In this movement, the service providers enable the users to use the function of product when needed without ownership and maintenance of product. The business domain of outsourcing service is expanding to not only

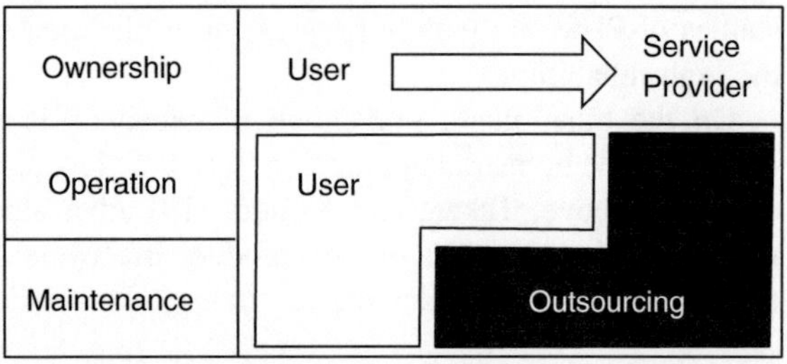

Fig. 15.2 Trend of operation and maintenance business

product maintenance but also total operation including the maintenance. The operation and maintenance services that are delivered through a cloud-based convergence of control and information technologies are developed. These O&M services offer the users to efficiently implement systems that collect and analyze the detailed information about the products, equipments, machines and the workers involved with operation and maintenance and use it to assist with optimization of operation and maintenance for improving the overall efficiency [21].

Next, the trend of research on optimization and scheduling technologies for improving the overall efficiency of operation and maintenance is explained. Figure 15.3 indicates the typical category of combination of product operation and maintenance. Here we assumed the product relating to large-scale and high reliable social infrastructure and this figure does not cover all kinds of products such as the consumer electronics and automobiles.

The first type is centralized management. Many products are working in one site and compose the large entire system. Various equipment in power plant and production line and large-scale construction machineries in mine development are typical examples. Many operators are necessary to operate and manage many equipment and machine. Operators in operation side also perform many maintenance activities such as inspection, check-up and simple repair and replacement. In some cases, operation and maintenance of entire system are outsourced to third party. The operation ratio and productivity are important performance indices since entire system has high total assets. For example, in the semiconductor production line, all of equipments are connected to Manufacturing Execution System (MES) and preventive maintenance based on on-line monitoring of equipment prevents the occurrence of product defects [22]. Operation and maintenance activities are

Fig. 15.3 Category of operation and maintenance

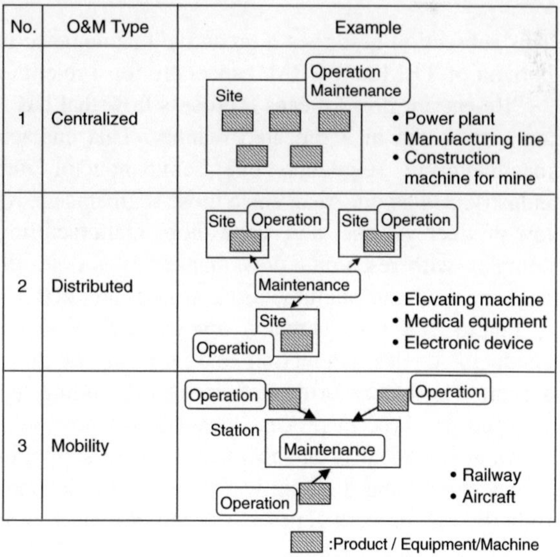

optimized to maximize the productivity and operating ratio complying with regulations as to operation and maintenance. In particular, there are many studies on scheduling methods to improve the operation ratio and productivity to since operations are complicated and timing of maintenance is flexible in the production line [23, 24]. Section 15.3 describes the developed maintenance scheduling for semiconductor production line.

The second type is distributed management. Products and machines are working at many sites. Elevating machines in building and medical equipment in hospital and electronic equipment in data centre are typical examples. In this type, users operate the product on each site. On the other hand, the maker dispatches maintainer and executes the maintenance on each site. The maker introduces the remote monitoring system of products distributed all over the world and manages the maintenance scheduling and parts inventory efficiently using remote and real-time monitoring data [25].

The third type is mobility management. Products are moving in operation and maintenance is performed at maintenance station. Railway and aircraft are typical examples. This type offers the customer the mobile service by operating the mobility product. The product moves to maintenance station for maintenance. In some cases, maintenance at station is outsourced to the maker. It is important to reduce the maintenance time at station working with operation management [26]. As explained above, different approaches to optimize the specific operation and maintenance are developed.

15.2.3 Proactive Maintenance for Re-entrant Flow Manufacturing

This subsection describes a proactive maintenance method research area as a new horizon of TBM and CBM, especially for a re-entrant flow manufacturing.

"Re-entrant flow" means a process flow that has some manufacturing processes are carried out in a repeated manner. This character can be found in so many manufacturing segments, like semiconductor industry, heavy manufacturing industry, and so on. Also, since those segments are requested high-product-mix and low-product volume, and since those manufacturing segments are extensive and complex with re-entrant flow manner, it is quite difficult to establish both short delivery time and manufacturing system productivity.

Figure 15.4 is a schematic view of the flow of production lots in a typical semiconductor device production line as a one of the re-entrant flow manufacturing system. The blocks in dashed lines in the figure indicate collections of machines engaged in a type of processing such as deposition, exposure, or etching.

More specifically, as shown in the case shown in Fig. 15.4, each set of workpieces comprising a single wafer production lot moves from the upper left to the right through process 1, process 2, and process 3. And the lot is repeatedly re-turned

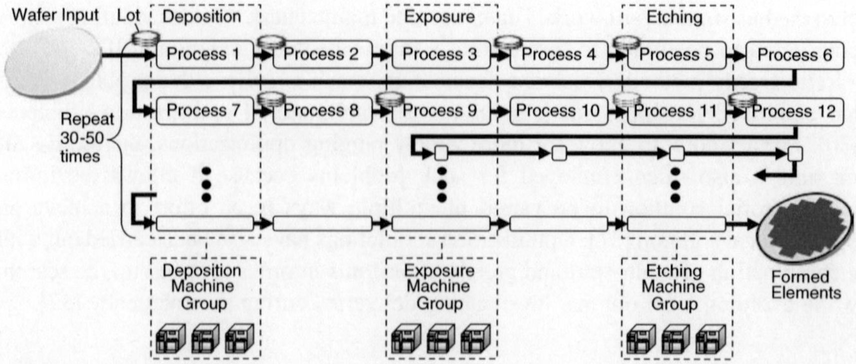

Fig. 15.4 A semiconductor manufacturing system

to the left end of the line to restart rightward progress after it reaches the right end. This production system is an example of a re-entrant flow, in which each set of machines carries out the same processes (i.e., deposition, exposure, and etching) for 30 to 50 cycles. The vital role that each of these machines plays in a process makes it clear that the availability factor of a single machine in a re-entrant flow has a significant effect on the production level of numerous processes.

One of the chief issues related to improving manufacturing system productivity is how to improve the availability factor. Especially, the nature of re-entrant flow process means that the availability of any single production machine has a significant influence on the productivity of multiple processes. Any effort to increase the availability factor of a machine involves several tasks, including reducing time spent on set-up and other auxiliary tasks, prevention of unforeseen stoppages such as process errors or machine malfunctions, and scheduling of regular maintenance work. The regular maintenance downtime for some machines, including start-up procedures, can extend to over a week.

Scheduling regular maintenance work for such machines is essential for improving manufacturing system productivity. Due to the extensive scale and complexity mentioned above, however, it is difficult to obtain effective solutions quickly with conventional scheduling tools. Instead, schedules are set by relying on the experience and instincts of veteran staff with long experience on site.

Conventionally, regular maintenance work on production systems is scheduled by an engineer using a standard manufacturing execution system (MES), on the basis of the available knowledge, and the results are recorded. The scheduling is generally calendar based, and there is no mutual coordination with the production lot [27]. Once the machine maintenance work has been placed on the schedule, an attempt is made to boost the productivity of the entire production system by coordinating the timing of the start of production of a lot with a dispatcher using the critical ratio algorithm [28].

If the production system in terms of a railway system is assumed, then the production lots are replaced with railroad cars and the machine maintenance is classified as station yard work. Thus, machine maintenance work scheduling can be considered to be similar to formulating a train scheduling diagram [29].

The greedy method, which is a separate approach to solving this problem, is often employed for approximate solutions to combinatorial optimization problems [30]. Furthermore, to allow for more widely ranging optimizations, the multi-start method is also often employed for such problems because it allows the initial combinatorial solution to be varied in multiple ways in an effort to achieve an optimal local solution [31]. Optimization simulations have also been carried out with a combination of multi-start and greedy algorithms in order to speed up the search, while avoiding local optima, for planning deliveries during an emergency [32].

15.3 A New Proactive Scheduling Method for Maintenance

15.3.1 Problem Definition

In this section, a new maintenance method is introduced for both short delivery time and manufacturing system productivity [24]. This maintenance method is supported by a proactive scheduling based on a multi-start/greedy method is proposed and compared to a conventional method based on Lagrangian decomposition coordination [23], in order to verify its effectiveness. Finally, the results obtained in a demonstration experiment performed to confirm the effectiveness of this procedure in an actual production system are described.

Now, the scheduling subjects of this production system are presented in Fig. 15.5. Three sequential processes are shown here in a vertical line and are numbered 1, 2, and 3. One of the machines from each process is assigned to each shop. Production time is expressed in the horizontal direction. Here, when a given production lot passes through processes 1, 2, and 3, requiring their respective lead times, the motion trace runs from the upper left corner of Fig. 15.5 to the lower right corner. When maintenance is performed in any of the three processes, it has the potential to interfere with the manufacturing schedule for the entire production lot,

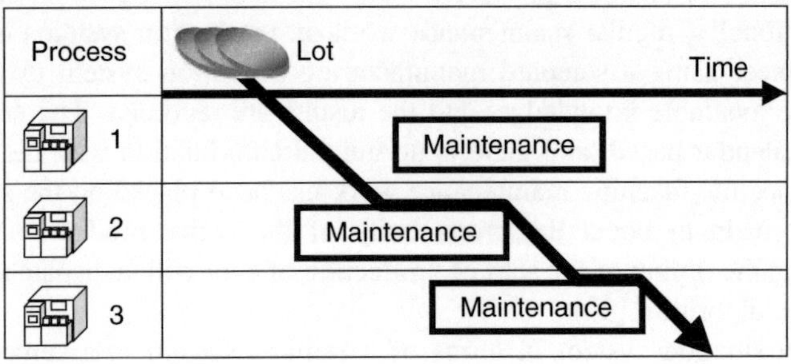

Fig. 15.5 A lot scheduling problem to machine maintenance

as shown in Fig. 15.5. This, in turn, will delay the manufacturing schedule by the length of the time in the maintenance schedule, increasing the manufacturing lead time indicated in the bent track of the arrow of the production lot in Fig. 15.5. This issue stems from the fact that the manufacturing schedule for a production lot is generally not synchronized with the machine maintenance schedules. It would be easy to synchronize the schedule with a system such as that in Fig. 15.5, with just three processes, but that synchronization is not so easy to accomplish with a large-scale re-entrant flow shop such as the semiconductor device production system shown in Fig. 15.4.

15.3.2 Solving Method

With regard to the above issue, observations of machine maintenance schedules in actual operations reveal that it is common to allow some flexibility in the time allowed for carrying out maintenance tasks within a scheduled period of time, based on the experience and instincts of veteran staff with regard to the actual tasks. In other words, the maintenance schedule is not locked to a certain day and time, but is allowed to slide forward or backward to some extent.

On the basis of these observations, in this study, it is focused on the fact that the maintenance schedule can be revised forward or backward somewhat, and hypothesized that if this could be done for each machine in a way that never interferes with the manufacturing schedule for a production lot, it would shorten production lead time. Figure 15.6 represents such an adjustment of Fig. 15.5; the maintenance schedule was moved back for the process 2 machine and forward for the process 3 machine, thus eliminating the interference with the manufacturing schedule. This proactive adjustment of the machine maintenance schedule allows it to mesh smoothly with the manufacturing schedule, and so it is called "proactive scheduling for maintenance" (PSM) in this study.

This study regards maintenance as one of the jobs that is assigned to all machines, and constructs a job allocation scheduling to machines in the manufacturing system. The objective function of the scheduling is to minimize total

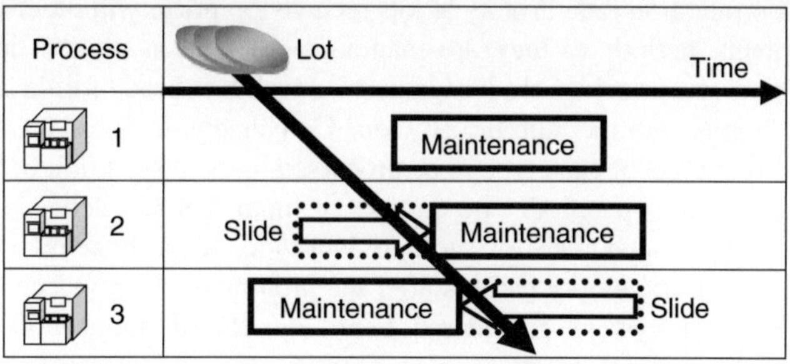

Fig. 15.6 A lot scheduling solution to machine maintenance

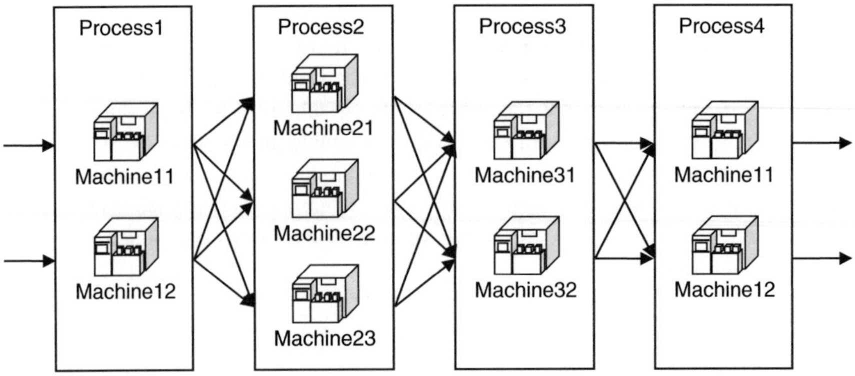

Fig. 15.7 PSM model

tardiness of machines, so that it is applicable to a semiconductor device production system that fits the scale and available calculation time.

The production system examined in this study is the re-entrant flow described in Fig. 15.4. It is defined as shown in Fig. 15.7 and the assignments of jobs to the machine are optimized. The goal of this process is to contribute to improving productivity of the production system by shortening the manufacturing lead time for a production lot.

Figure 15.7 shows the flow of production lots, i.e., the process route, in black arrows. As can be seen in the figure, the product moves from left to right beginning at process 1, proceeding to process 2, and then moving to process 3 and finally process 4. Processes 1 and 4 are performed in the same job shop, which consists of machines 11 and 12. A production lot is assigned for processing in process 1 to machine 11 or 12, to either machine 21, 22 or 23 for process 2, to either machine 31 or 32 for process 3, and to machine 11 or 12 for process 4. PSM scheduling is an approach in which production lots are assigned at each process step to whatever machine will accommodate them, as directed by these process routes. An available machine is selected with an algorithm that coordinates the production lot manufacturing schedule with the maintenance schedule for the production machines.

In this study, lots fed to the production system are classified as either prototype lots or mass production lots. Prototype lots receive top priority in order to minimize the development period, so they are manufactured with as short a lead time as possible. Therefore, priorities are designated for each lot ahead of time and machine assignment begins with the high-priority lots. Conversely, for mass production lots, the numbers of wafers of each type to be processed in the pre-set unit of time and by the deadline date are specified. The number of input lots and delivery date in the preset unit time are scheduled for each product type. Also, those mass production lots are scheduled into empty periods after the prototype lots have been scheduled.

The production system considered here consists of about 500 production machines and creates five products, each with its own process route. These routes consist of 100 to 1,000 processes. The production system operation calendar was established and machine maintenance task times and planned periods for maintenance are all scheduled for each machine. The time slots are defined in minutes and the production plan is created out to 3 months in the future for each product. Schedules are set for each machine at appropriate times without regard to statistical quantities such as the mean times to failure and/or repair. The limitations on productivity imposed by the skills and the number of available personnel were also disregarded.

15.3.3 PSM Computational Model

In this study, "job" will be used to denote not only any of the processing applied to a production lot, but also the tasks assigned to a machine, including the maintenance procedures addressed in this research and shutdown periods due to malfunctions or the facility calendar. "Slide" means that a job is moved forward or backward on the time axis. For example, sliding a job "forward" means that the job is moved in the negative direction on the time axis.

The symbols are defined in Table 15.1.

The following conditions are assumed in the model based on knowledge gained from actual manufacturing operations.

- The initial settings for maintenance are values selected on the basis of knowledge gained in previous manufacturing experience. The narrower the margin of allowed slide for maintenance, the more valid these values are.
- Plenty of time is left open between the maintenance period for any one machine and the maintenance period for any other machine, and no maintenance is ever scheduled in a way that forces maintenance rescheduling for another machine.
- Prototype lots are prioritized. Therefore, once the schedule is set for one of these lots, it will not be moved for the sake of any other job.

In the PSM calculation model, job r of prototype lot L_j in process s is modelled with other jobs in the same machine $A_{i,s}$. The modelling is carried out for two cases: when a job can be assigned for transfer from process $s-1$ to process s without wait time and when a job can be assigned for transfer after a designated wait time.

Table 15.1 Symbol definitions

Symbol	Definition
i	Machine number (1, 2, 3, …, n)
j	Prototype lot number (1, 2, 3, …, m)
k	Already-assigned job number (1, 2, 3, …, p)
r	Target job number for assignment (1, 2, 3, …, p)
s	Process number (1, 2, 3, …, q)
t	Scheduling time
t_j	Unique clock time reserved for production lot j

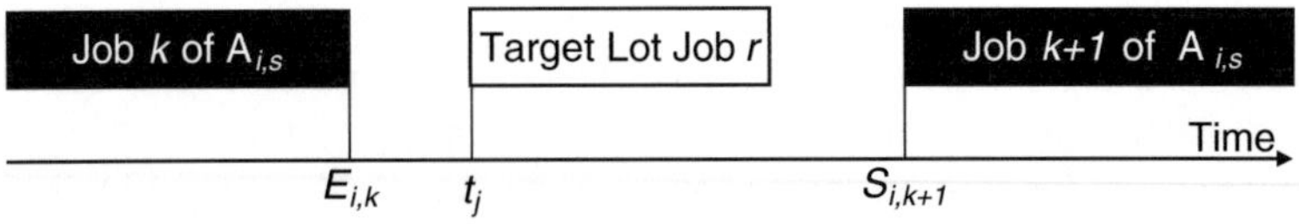

Fig. 15.8 PSM case 1

15.3.3.1 Transfer to Process *s* Without Wait Time

The case where job r of prototype lot L_j in process s assigned for transfer without wait time to process s can be further subdivided into two cases: the case where job r is assigned to machine $A_{i,s}$ without interfering with other jobs of machine $A_{i,s}$ and the case where maintenance job k of machine $A_{i,s}$ is moved forward or backward because job r of prototype lot L_j in process s interferes with maintenance job k.

First of all, the case where job r of prototype lot L_j in process s is examined to be assigned to machine $A_{i,s}$ without inference with other jobs.

Figure 15.8 shows the situation in which job r is assigned to machine $A_{i,s}$, whose scheduling time is represented by the horizontal axis, and does not interfere with any other jobs of machine $A_{i,s}$. In other words, the figure defines the case when Eqs. (17.1) and (17.2) below are satisfied simultaneously. Job r can then be assigned for transfer from process $s - 1$ to process s without wait time. When job r is assigned, the numbers for jobs $k + 1$ and above are each increased by 1, after which job r is registered as job $k + 1$.

$$E_{l,k} < t_j < S_{i,k+1} \tag{15.1}$$

$$t_j + WT_{j,r} \leq S_{j,k+1} \tag{15.2}$$

$S_{i,k}$ Start time for the k + 1-th job at machine i
$E_{i,k}$ End time for the k-th job at machine i
WTi, k processing time for the k-th job at machine i

Next, the case where maintenance job k of machine $A_{i,s}$ is moved forward or backward because job r of prototype lot L_j in process s is examined to be assigned to machine $A_{i,s}$, interferes with maintenance job k.

These conditions can be classified into the case when maintenance job k is moved forward for job r and when it is moved backward. Figure 15.9 presents the case when the job is moved forward. More precisely, Fig. 15.9 depicts the situation that results when maintenance job k is moved forward for job r on machine $A_{i,s}$ whose scheduling time is represented by the horizontal axis. This situation is defined when Eqs. (15.3) and (15.4) below are satisfied simultaneously. Maintenance job k is moved forward according to Eqs. (15.5) and (15.6), thus allowing job r to be assigned for transfer from process $s - 1$ without wait time.

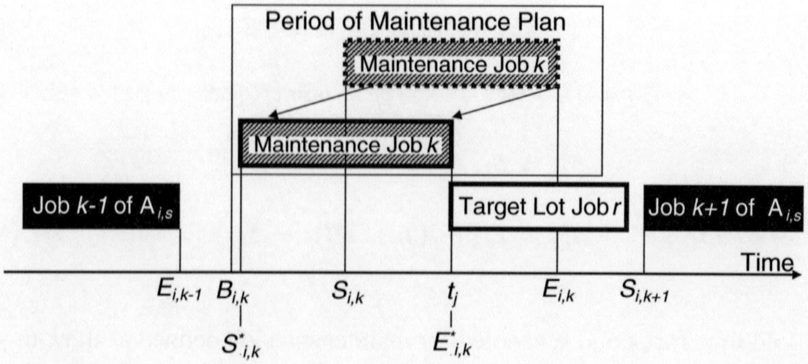

Fig. 15.9 PSM case 2

$$E_{i,k} - S_{i,k} + WT_{i,r} \leq S_{i,k+1} - E_{i,k-1} \tag{15.3}$$

$$max(B_{i,k}, E_{i,k-1}) \leq S_{i,k} - (E_{i,k} - t_j) \tag{15.4}$$

$$S^*_{i,k} = S_{i,k} - (E_{j,k} - t_j) \tag{15.5}$$

$$E^*_{i,k} = t_j \tag{15.6}$$

$B_{i,k}$ Scheduled start time for maintenance job defined as the k-th job of machine i
$S^*_{i,k}$ Start time for maintenance job k after rescheduling
$E^*_{i,k}$ End time for maintenance job k after rescheduling

Figure 15.10 presents the situation that exists when a maintenance job is moved backward for job r on machine $A_{i,s}$, whose scheduling time is represented by the horizontal axis. This situation is defined when Eqs. (15.7) and (15.8) below are satisfied simultaneously. Maintenance job k is moved backward according to Eqs. (15.9) and (15.10), thus allowing job r to be assigned for transfer from process $s - 1$ without wait time.

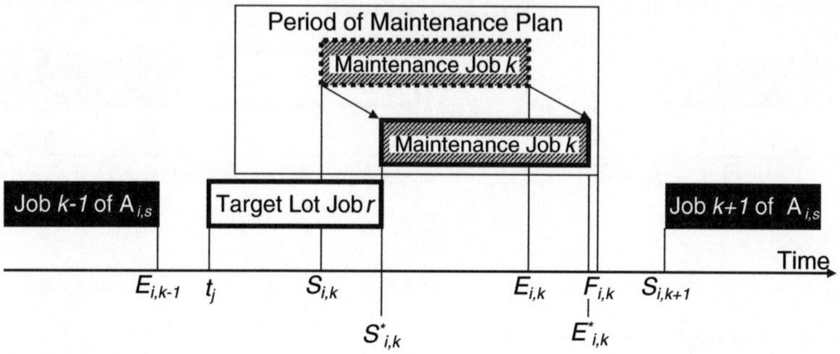

Fig. 15.10 PSM case 3

$$E_{i,k} - S_{i,k} + WT_{i,r} \leq S_{i,k+1} - E_{i,k-1} \tag{15.7}$$

$$E_{i,k} + \left(t_j + WT_{i,r} - S_{i,k}\right) \leq min\left(F_{i,k},\, S_{i,k+1}\right) \tag{15.8}$$

$$S^*_{i,k} = S_{i,k} + \left(t_j + WT_{i,r} - S_{i,k}\right) \tag{15.9}$$

$$E^*_{i,k} = E_{i,k} + \left(t_j + WT_{i,r} - S_{i,k}\right) \tag{15.10}$$

$F_{i,k}$ End time for period scheduled for maintenance job defined as the k-th job for machine i

15.3.3.2 Transfer to Process *s* After Some Wait Time

The situation that occurs when job r of prototype lot L_j in process s following some wait time can be further subdivided into two cases: moving maintenance job k of machine $A_{i,s}$ forward and job r backward, because of interference between the two jobs, and moving job r backward because it interferes with already-assigned prototype lot job k of machine $A_{i,s}$.

Firstly, the case is examined when maintenance job k is moved forward and job r is moved back because of the interference with maintenance job k of machine $A_{i,s}$.

Figure 15.11 shows the situation in which job r of machine $A_{i,s}$, whose scheduling time is represented by the horizontal axis, is moved backward and job k is moved forward. Here, the assumed conditions forbid moving job r forward and

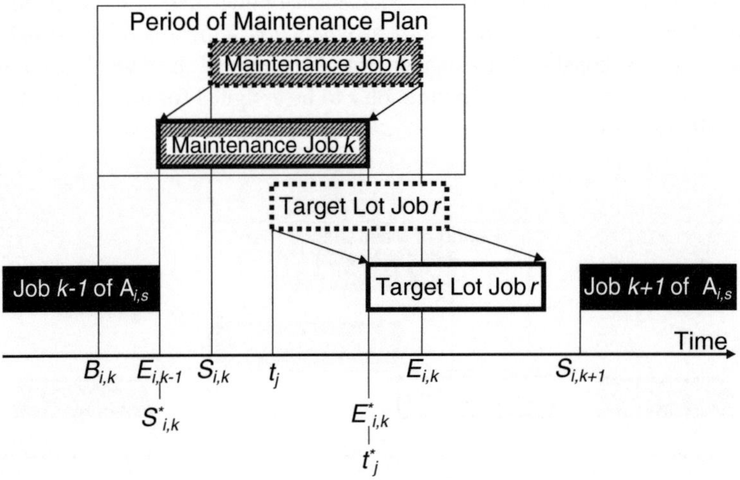

Fig. 15.11 PSM case 4

maintenance job k backward. The situation in Fig. 15.11 is defined by the simultaneous satisfaction of Eqs. (15.11) and (15.12) below. Job r and maintenance job k are moved according to Eqs. (15.13), (15.14), and (15.15) below, thus allowing us to minimize the wait time for transfer of job r from process $s - 1$ to process s after maintenance job k. The wait time due to this maintenance job k is thus reduced from $E_{i,k} - t_j$ to $E^*_{i,k} - t_j$.

$$E_{i,k} - S_{i,k} + WT_{i,r} \leq S_{i,k+1} - E_{i,k-1} \tag{15.11}$$

$$E_{i,k} - \{S_{i,k} - max(E_{i,k-1}, B_{i,k})\} + WT_{i,r} \leq S_{i,k+1} \tag{15.12}$$

$$S^*_{i,k} = max(E_{i,k-1}, B_{i,k}) \tag{15.13}$$

$$E^*_{i,k} = E_{i,k} - \{S_{i,k} - max(E_{i,k-1}, B_{i,k})\} \tag{15.14}$$

$$t^*_j = E^*_{i,k} \tag{15.15}$$

t^*_j Start time for job r after rescheduling

Next, the case is examined where job r is moved backward because it interferes with already-assigned prototype job k of machine $A_{i,s}$.

Figure 15.12 shows the situation where job r of machine $A_{i,s}$, whose scheduling time is represented by the horizontal axis, is moved backward in favor of already-assigned prototype job k. Here, the assumed conditions forbid sliding job r forward

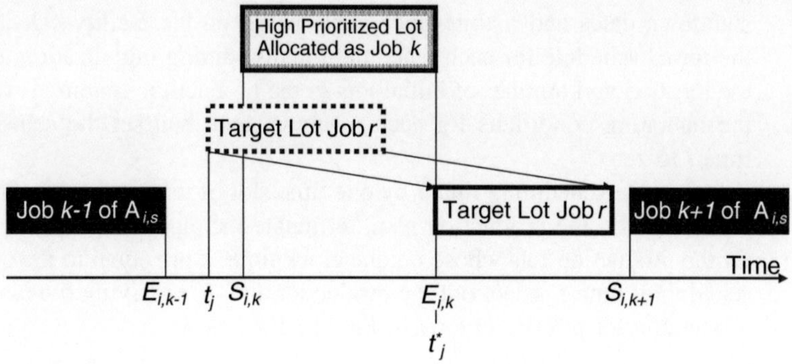

Fig. 15.12 PSM case 5

and already-assigned prototype job k backward. The situation in Fig. 15.12 can be defined by the simultaneous satisfaction of Eqs. (15.16) and (15.17) below. Job r is moved backward according to Eq. (15.18). Equation (15.18) indicates a wait time due to maintenance job k of $E_{i,k} - t_j$.

$$WT_{i,r} \le S_{i,k+1} - E_{i,k} \tag{15.16}$$

$$E_{i,k-1} < t_j < S_{i,k+1} \tag{15.17}$$

$$t_j^* = E_{i,k} \tag{15.18}$$

15.3.4 PSM Solution by Multi-Start/Greedy Method

15.3.4.1 Algorithm Flow in Proposed Method

The proposed method is described here using Fig. 15.13.

Step 1: Define the machines assigned to each job shop. Designate and assign the job shops and processes for each product to be manufactured to define the process routes. Next, define the production plans that set the order of priorities and the input and processing deadline dates for the prototype lots. For mass production lots, create the production plans by setting the numbers of input production lots into the system each day. Here, for the production system input date for a prototype lot, unique clock times t_j are set for each input date for each production lot. Next, define the scheduled periods and estimated time spans necessary for the tasks according to the shutdown dates and maintenance plans based on the facility calendar as the initial schedule for each machine. Finally, during initialization, define the location and number of initial lots in the production system, as well as the operating conditions for each machine, and then set the scheduling time t to zero.

Step 2: Advance the scheduling time t by one time slot. If this represents the final clock time on the production plan, terminate the algorithm.

Step 3: Of the production lots whose unique clock times t_j are equal to the current scheduling time t, select out the production lots L_j satisfying the assigned production lot priority criteria in Eq. (17.19) below.

$$j = \{L_x | min(RD_x/RW_x)\} \tag{15.19}$$

L_j Selected production lot
RD_x Time remaining until deadline date as of scheduling time t
RW_x Process time remaining as of scheduling time t

Fig. 15.13 PSM Multi-start/greedy algorithm

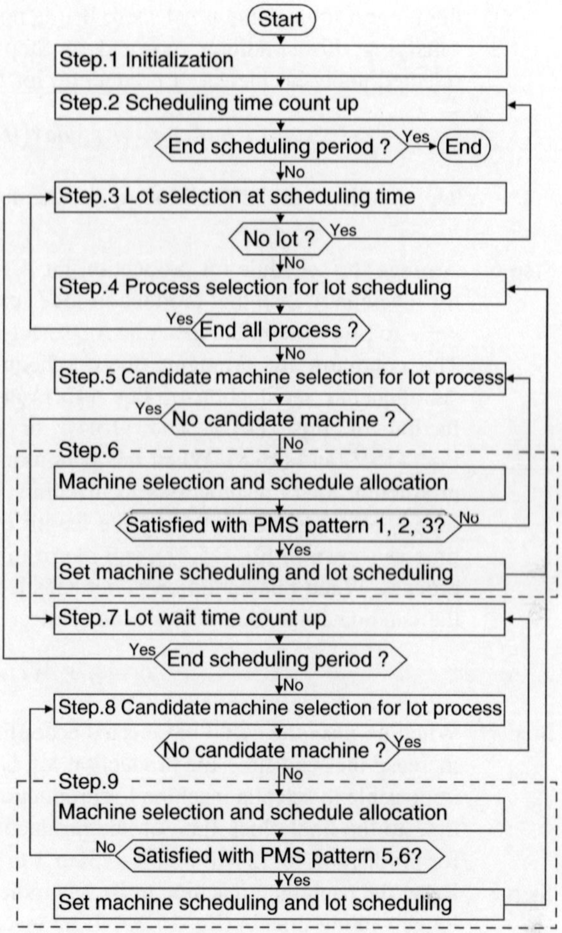

This means that the production lot with the minimum value for the critical ratio defined in Eq. (15.19), the time remaining as of the current scheduling time t until the deadline date divided by the remaining process time (i.e., the production lot with the least time margin), is designated with the highest priority. If no production lot satisfies this condition, return to Step 2, the step in which the scheduling time t is incremented.

Step 4: Transfer lot L_j to the next process s on the process route defined for the selected lots L_j. Then, once lot L_j has completed its entire process, return to Step 3, where a production lot assigned at the current scheduling time t is selected.

Step 5: Idling machines in the job shop responsible for process s, select machine $A_{i,s}$ satisfying the machine selection criteria given in Eq. (15.20) below. Equation (15.20) means that machine $A_{x,s}$ in the job shop that performs process s is idling at the current time t, and that machine $A_{i,s}$, which has the longest idling time is selected. If it is not possible to select a machine satisfying this condition, proceed to Step 7, where a machine $A_{i,s}$ is selected under an increased production lot L_j wait time.

$$A_{i,s} = \{A_{x,s} | max(ID_{x,s})\} \tag{15.20}$$

$ID_{x,s}$ Idling time of machine $A_{x,s}$ at the current time t

Step 6: Arrange the schedule for production lot L_j in process s with the schedule for machine $A_{i,s}$, so that production lot L_j can be transferred from process $s - 1$ to process s in the machine $A_{i,s}$ selected in Step 5 without wait time. The condition for arranging these schedules is one of the following: simultaneous satisfaction of Eqs. (15.1) and (15.2), simultaneous satisfaction of Eqs. (15.3) and (15.4), or simultaneous satisfaction of Eqs. (15.7) and (15.8). When this condition is satisfied, assign the job of production lot L_j in process s to machine $A_{i,s}$, advance the unique clock time t_j of production lot L_j by the length of the assigned job processing time as shown in Eq. (15.21), and return to Step 4 to continue to the next process. When this condition is not satisfied, return to Step 5 and selects the candidate machine with the next lower priority.

$$t_j = t_j + WT_{i,r} \tag{15.21}$$

Step 7: When no machine could be selected according to the conditions in Step 5, increase the wait time for production lot L_j by one time slot. If it is still impossible to select a machine for production lot L_j by increasing the wait time to the final clock time on the production plan, terminate scheduling for production lot L_j and return to Step 3 to select the next production lot.

Step 8: Reset the candidate machine to $A_{i,s}$, as indicated for the machine with the longest idling time in (Eq. 15.20). If no machine can be selected after this, return to Step 7.

Step 9: Arrange the schedule for production lot L_j in process s with the schedule for machine $A_{i,s}$ selected in Step 8 such that production lot L_j can be transferred from process $s - 1$ to process s with the minimum wait time. The condition for matching these schedules is either simultaneous satisfaction of Eqs. (15.11) and (15.12) or simultaneous satisfaction of Eqs. (15.16) and (15.17). When this condition is satisfied, assign the job of production lot L_j in process s to machine $A_{i,s}$, advance the unique clock time t_j of production lot L_j by the length of the assigned job processing time as shown in Eq. (15.21), and return to Step 4 to continue to the next process. When this condition is not satisfied, returns to Step 8 and selects the candidate machine with the next lower priority.

15.3.4.2 Optimization by Multi-Start/Greedy Method

This sub-subsection describes PSM optimization using the multi-start/greedy method. Under the PSM described in the previous paragraph, Eq. (15.19) defines the algorithm for scheduling that prioritizes the production lot with the least time margin before the deadline date as of scheduling time t. However, when the entire process from input to final is reviewed in a bird's-eye view, the algorithm in Eq. (15.19) cannot ensure the scheduling result put the highest production priority to the production lot with the least time margin before the deadline date, since the algorithm modifies the priority of production lot from input process to final process by process, with the condition of Eq. (15.19). That is, Eq. (15.19) for each process cannot ensure to pick up the highest priority production lot that has the least time margin before the deadline date, in front of one process of the entire process.

This study was conceived as an attempt to optimize the PSM using the multi-start method and the greedy method, two sub-optimization procedures, under the selection criteria for both production lots and machines.

Firstly, the method for deriving a suboptimal solution using the multi-start method with production lot selection criteria is described. In this method, Eq. (15.19) is rewritten as Eq. (15.22):

$$L_j = \{L_x | min(RD_x/RW_x)\, e^{-\alpha C} + RN_{0-1}\} \qquad (15.22)$$

$RN_{0\text{–}1}$ Random number between 0 and 1
α Weighting coefficient applied to increase randomness
C Number of starts

The objective function Z is defined by Eq. (15.23):

$$Z = min\left\{\sum_{j=1}^{j\le m}\sum_{i=1}^{i\le n} QT_{i,j} + \delta \sum_{j=1}^{j\le m} DT_j\right\} \qquad (15.23)$$

$QT_{i,j}$ Wait time for prototype lot j assigned to machine i
DT_j Delay time of prototype lot j to deadline date
δ Weighting coefficient defining how much delay time to apply with respect to deadline date

From Eqs. (15.22) and (15.23), the selection criteria for the production lot at each scheduling time t are intended to select out the production lot with the least time margin before the deadline date for highest priority. The weighting coefficients α and δ represent conferred randomness, according to criteria, and the number of attempts is given by the number of starts C. This prevents the selection criteria for a production lot from leading to a local optimum when searching for a schedule that provides a minimum delay beyond the deadline date for all prototype lots.

Secondly, the method for deriving a suboptimal solution when applying the greedy method for machine selection criteria is described. In this method, when production lot L_j is transferred to process s as of scheduling time t, the selection method of machine $A_{i,s}$ and job r allocation method of production lot L_j to machine $A_{i,s}$ are modified, in which machine $A_{i,s}$ has the longest idling time in Eq. (15.20). When Eq. (15.24) below is fulfilled, a degree of freedom in machine selection is assumed to become high, and job r of production lot L_j allocation to machine $A_{i,s}$ stops to apply, but the procedure returns to Step 3 to make the selection.

$$\beta < N_s \tag{15.24}$$

β Threshold value
N_s Number of machines in the job shop responsible for process s

15.3.5 Experimental Results and Application Trial

15.3.5.1 Effectiveness Verification by Numerical Experiment

An effectiveness of a PSM optimization using the multi-start/greedy method was validated by a numerical experiment based on data from an actual semiconductor production system. The sample data included a single type of prototype lot, subjected to 986 processes with 1.5 input lots per day on average. Also, thirty-two machines were scheduled for maintenance during this schedule period. Additionally, the mass production lot included two product types, with 3.0 input lots and 0.5 input lots per day on average for each.

Here, the ratio between the actually scheduled maintenance period $F_{i,k} - B_{i,k}$ and the maintenance period $E_{i,k} - S_{i,k}$, i.e., $\left(Fi_{,k} - Bi_{,k}\right)/\left(E_{i,k} - S_{i,k}\right)$ is named as the "slide margin ratio". The mean slide margin ratio in the data sample was 9.3, and the standard deviation was 7.2.

In 986 processes for the prototype lot, 65 % of the processes consist of only one machine for each process, 18 % of those consist of two machines, 8 % of those consist of three machines, 6 % of those consist of four machines. Thus, the degree of freedom in machine choice for each process expressed in Eq. (15.24) is ineffective in 65 % of processes. This indicates that the current data sample is more effectively treated with the multi-start method expressed with Eqs. (15.22) and (15.23) than with the greedy method expressed in Eq. (15.24).

Figure 15.14 shows the results of a further numerical experiment in which the threshold β was varied in value to see its effect on machine selection in the

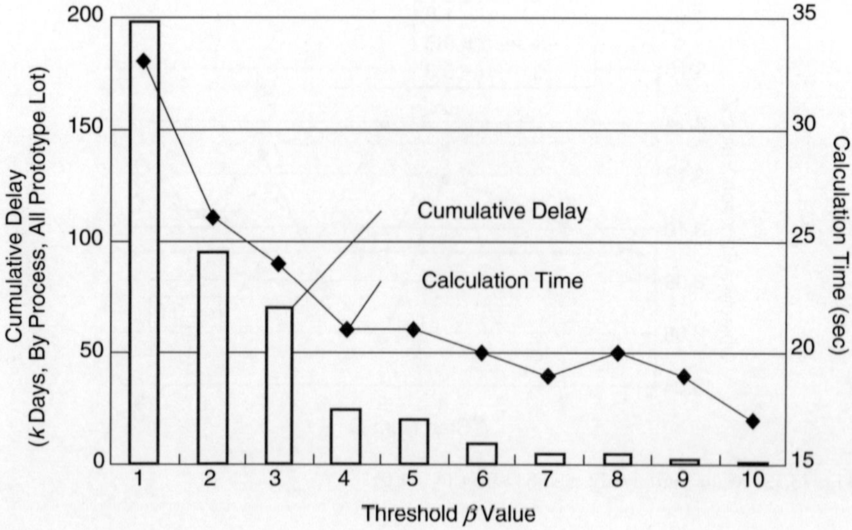

Fig. 15.14 Threshold β effect

multi-start/greedy method expressed in Eq. (15.24), using the same data sample. In this numerical experiment, the randomness weighting coefficient α in Eq. (15.22) concerning the multi-start method was set to zero and the number of starts C was set to 1.

The horizontal axis of Fig. 15.14 is the threshold β and the vertical axis on the left side is the total delay for the processes for prototype product $P1$ in number of days, found in multiple lots. The vertical axis on the right side is the time required for the calculation. From the figure, this data sample has the character such as the greater of the threshold β, the shorter the delay and calculation times.

From Fig. 15.14, job assignment by the prototype lot priorities, without the arrangement between the lot schedule and other lot schedules for each process, seems to reduce total delay time and calculation time. The priorities of the prototype lots, which were set according to the input values, were used to set the schedules for all the processes. Accordingly, it is not necessarily true that the objective function minimizing the overall sum of delays for all prototype lots, defined by Eq. (15.23), was the optimal one. Therefore, it was decided to fix the threshold β so that a more optimal solution could be found in each combination of coefficients α and C in the multi-start method. This was accomplished via Eq. (15.22), and also by considering the greedy method. In the following discussion, threshold β is set to the value of *6*.

The effects of the values for parameters α, β and C were examined in a numerical experiment using the multi-start/greedy method to select production lots with the same data sample as above. The value of C was fixed to *9*, β was set to 1.0 and 0.0. And α was set to 1.0, 0.5, and 0.3. Calculations were performed with all combinations of C, β, and α.

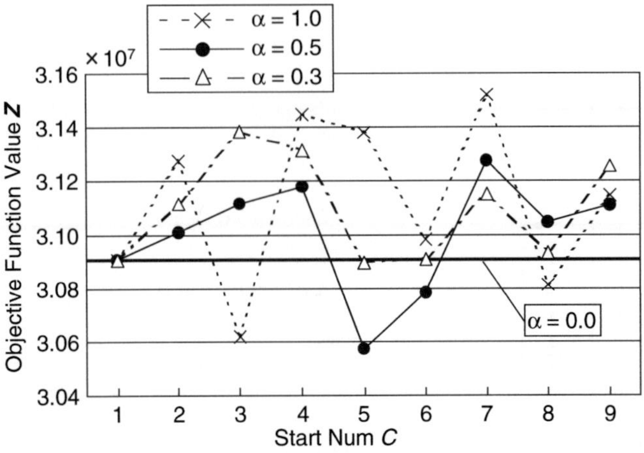

Fig. 15.15 Multi-start/greedy method effect (δ = 1.0)

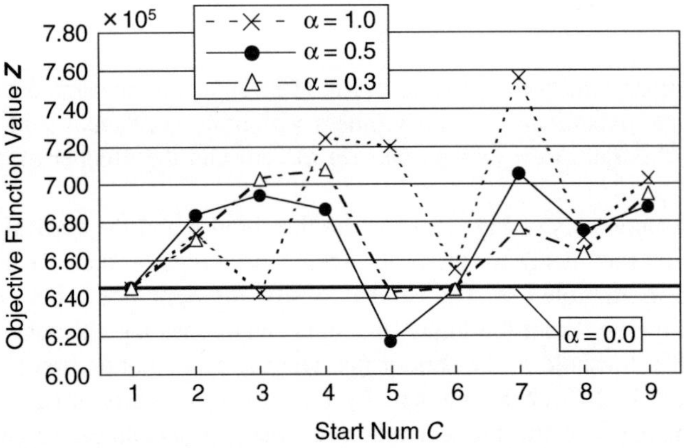

Fig. 15.16 Multi-start/greedy method effect (δ = 0.0)

Figure 15.15 and Fig. 5.16 show the results of the experiment, with the number of starts C on the horizontal axis and the objective function Z on the vertical axis. In Fig. 15.15, the weighting coefficient for deadline date δ has the value of 1.0, and in Fig. 15.16, δ is zero. The figures show the results while varying the weighting coefficient for randomness α at 1.0, 0.5, and 0.3. The horizontal line represents the value of the objective function when $\alpha = 0$, i.e., when no randomness is considered in the multi-start method. These results indicate a more optimal solution when the solution is beneath the horizontal line. From the figures, for this data sample, setting the randomness variable α between 0.5 and 1.0 provides more optimal solutions, whether or not the delay in the deadline date is accounted for.

15.3.5.2 Comparative Experiment Results with Lagrangian Decomposition Coordination

The multi-start/greedy method-based PSM developed here is compared with the Lagrangian decomposition coordination-based PSM [24], on the basis of a numerical experiment. The sample data included a single type of prototype lot, subjected to 108 processes. Also, five machines were scheduled for maintenance during this schedule period. Regards to the machines that have maintenance schedule, the mean slide margin ratio was 4.0, and the standard deviation was 1.9. In 108 processes for the prototype lot, 61 % of the processes consist of only one machine for each process, 17 % of those consist of two machines, 9 % of those consist of three machines, 3 % of those consist of four machines. The parameters for the multi-start/greedy method were $\alpha = 0.5$, $\beta = 6$, $\delta = 1.0$, and $C = 6$. In this experiment, just one prototype lot was cycled through the system; there was no mass production in terms of multiple lots. The experimental results are given in Tables 15.2 and 15.3.

Table 15.2 shows the calculation time, lead time for the production lot, total slide time on the maintenance schedule, and the total number of machines that were rescheduled. The rightmost column shows the ratios between the times required for the Lagrangian decomposition coordination algorithm to complete its calculations versus the times required by the multi-start/greedy method. Table 15.2 represents that the PSM's using multi-start/greedy method and the Lagrangian decomposition coordination algorithm obtained the same result for the manufacture lead time of the prototype lot, but the Lagrangian decomposition coordination approach took approximately 830 times as much calculation time as the multi-start/greedy approach. The Lagrangian decomposition coordination-generated plan required

Table 15.2 Algorithm comparison result

Item	Method		Ratio
	Lagrangian	Greedy	
Calc. time (sec)	2,491	3	830.2
Lot lead time (min)	3,604	3,604	1.0
Total slide time (min)	6,654	5,241	1.3
Num of slide machine	2	1	2.0

Table 15.3 Slide volume comparison

Maintenance no.	Machine no.	Slide volume (min)	
		Lagrange	Greedy
1	14	0	0
2	27	0	0
3	40	697	0
4	44	0	0
5	79	5,957	5,241

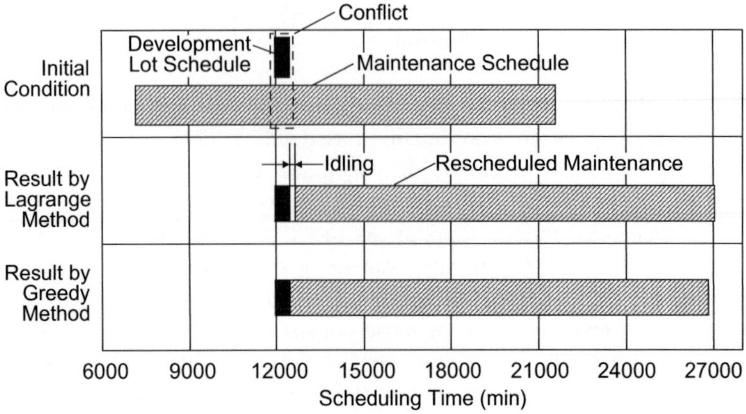

Fig. 15.17 Schedule performance at machine no. 79

rescheduling of two machines, while the multi-start/greedy method-generated plan only required rescheduling one machine. Furthermore, the total required slide time was 30 % longer under Lagrangian decomposition coordination than under the multi-start/greedy method.

Table 15.3 provides a comparison of slide volumes in the maintenance schedule. While the Lagrangian decomposition coordination method rescheduled maintenance for Machine No. 40 and 79, the multi-start/greedy method only rescheduled Machine No. 79. Additionally, the slide volumes were lower in the latter method. Figure 15.17 addresses this, showing how the maintenance schedule for Machine No. 79, in particular, varied with the prototype lot schedule.

The horizontal axis of Fig. 15.17 is the scheduling time. From the top, the initial condition before PSM procedure, the schedule optimized by Lagrangian decomposition coordination, and the schedule optimized by the multi-start/greedy method. In the initial condition, approximately 12,000-min interference exists between the prototype lot schedule and the maintenance schedule for Machine No. 79. Additionally, the Lagrangian decomposition coordination result shows a gap between the prototype lot schedule and Machine No. 79 maintenance schedule, which represents wasted time. In contrast, the multi-start/greedy method result assigns a maintenance schedule to Machine No. 79 just after the prototype lot schedule. This was probably due to the fact that the former method's treatment of the schedule as a Lagrangian relaxation problem means it seeks suboptimal solutions without the constraint of schedule continuity. The multi-start/greedy method moves the schedules forward.

15.3.5.3 Application Trial Result to Actual Production System

Figure 15.18 shows the application trial result of the PSM, developed in this study, in a real semiconductor device production system. Figure 15.18 indicates a maintenance plan optimization result in the system by PMS, to keep the high

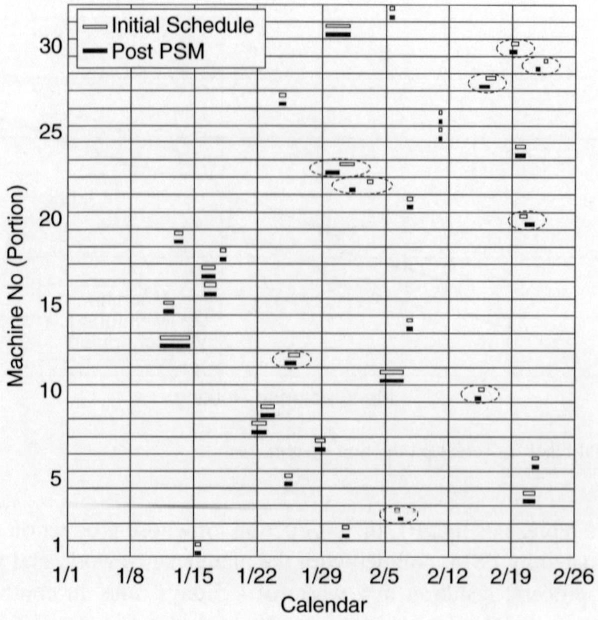

Fig. 15.18 Optimized result in actual production system by PMS

productivity without collisions between manufacturing processes and machine maintenances. The horizontal bar indicates a portion of the manufacturing calendar, the vertical bar shows the machine No (portion), white bars mean initial machine maintenance schedules, and black bars mean post PMS process machine maintenance schedules.

In the figure, dotted circles indicate optimization results of PMS to avoid collisions between manufacturing processes and machine maintenances by PMS calculation. That is, initial schedules were moved by PMS calculation to avoid those collisions.

Also, Fig. 15.19 shows an actual sample of PMS effect. In the figure, the horizontal axis is the process route for the semiconductor device and the vertical axis is the lead time after initiation of manufacture. P1 is the production lot whose production schedule was left un-coordinated with the PSM, whereas P2 is the lot whose production schedule was coordinated by the PSM. The figure contains maintenance work on one machine in a process, and shows the manufacturing lead time for P2, whose schedule was coordinated with that work schedule, and for P1, whose schedule was not coordinated.

In this case, the results of the coordination of the schedules by the PSM were provided to all of the offices in charge of machine maintenance, after which they revised their schedules. The facility automatically collects all process start and finish times for all lots, so these data were collated to obtain the manufacturing lead times for each process.

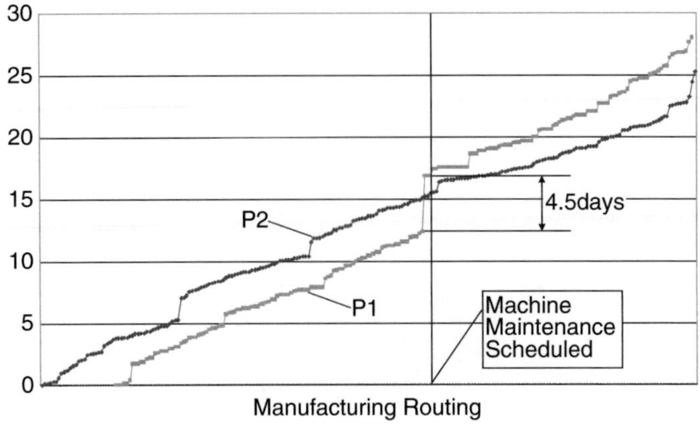

Fig. 15.19 Trial result to actual production system

This figure represents that P1, the production lot whose production schedule was left uncoordinated by PSM, collided with the maintenance work, and was forced to wait for that process, resulting in a waste of 4.5 days time. In contrast, the coordinated schedule of P2 allowed it to pass through this process without conflicts. This indicates that the schedule coordination by the PSM was effective.

15.3.6 PSM Summary

In this subsection, a procedure was proposed in which maintenance schedules are proactively adjusted in order to prevent error conditions while preserving manufacturing system productivity. A solution method incorporating a sub-optimization procedure based on the multi-start/greedy method was proposed and its effectiveness was compared with that of the Lagrangian decomposition coordination procedure in a numerical experiment. Its effectiveness was also validated in an actual production system.

15.4 Summary

In this chapter, the operation and maintenance has been summarized with re-cent technologies trends of integrated maintenance system. Motivated by an importance of facility maintenance scheduling in re-entrant flow shop, proactive scheduling for maintenance (PSM) was developed. The feasibility of PMS was tested in a real semiconductor device production system. The test confirmed the effectiveness of the proposed method.

Acknowledgments We would like to extend our sincere thanks to Ms. N. Yano of KOZO KEIKAKU ENGINEERING Inc. for useful discussions and encouragement. We also would like to thank to Professor S. Takada of Waseda University for useful comments.

References

1. Takata S, Kimura F et al (2004) Maintenance: changing role in life cycle management. Annu CIRP 52:37–40
2. Lee CY (1999) Two-machine flowshop scheduling with availability constraints. Eur J Oper Res 114:420–429
3. Lee CY, Chen ZL (2000) Scheduling jobs and maintenances on parallel machines. Nav Res Logistics 47:145–165
4. Sad S, Penz B, Rapine C, Blazewicz J, Formanowicz P (2005) An improved approximation algorithm for the single machine total completion time scheduling problem with availability constraints. Eur J Oper Res 161:3–10
5. Zribi N, El Kamel A, Borne P (2008) Minimizing the makespan for the mpm job-shop with availability constraints. Int J Prod Econ 112:151–160
6. Sortrakul N, Nachtmann HL, Cassady CR (2005) Genetic algorithms for integrated preventive maintenance planning and production scheduling for a single machine. Comput Ind 56:161–168
7. Xu D, Sun K, Li H (2008) Parallel machine scheduling with almost periodic non-preemptive maintenance and jobs to minimize makespan. Comput Oper Res 35:1344–1349
8. Berrichi A, Amodeo L, Yalaoui F, Chatelet E, Mezghiche M (2009) Bi-objective optimization algorithms for joint production and maintenance scheduling: application to the parallel machine problem. J Intell Manuf 20(4):389–400
9. Cassady CR, Kutanoglu E (2003) Minimizing job tardiness using integrated preventive maintenance planning and production scheduling. IIE Trans 35(6):503–513
10. Aggoune R (2004) Minimizing the makespan for the flow shop scheduling problem with availability constraints. Eur J Oper Res 153:534–543
11. Naderi B, Zandieh M, Ghomi SMTF (2009) Scheduling sequence-dependent setup time job shops with preventive maintenance. Int J Adv Manuf Technol 43(1–2):170–181
12. Shimizu T, Shomura Y, Masukawa H, Takeda Y (2014) Traffic management solutions for social innovation. Hitachi Rev 63(1):51–56
13. Sato J, Katutani Y, Baba T, Koimaru T, Nagasato A, Shirai T (2014) Capturing benefits of M2M in manufacturing. Hitachi Rev 63(1):25–29
14. Pipe K (2008) Practical prognostics for condition based maintenance. In: International conference on prognostics and health management, pp 1–10
15. Viassolo DE et al (2007) Advanced estimation for aircraft engines. In: Proceedings of the IEEE American control conference, pp 2807–2821
16. Goebel K, Saha B, Saxena A et al (2008) Advanced estimation for aircraft engines. IEEE Instrum Meas Mag 33–40
17. Wegerich SW et al (2003) Nonparametric modeling of vibration signal feature for equipment health monitoring. Proc IEEE Aerosp Conf 7:3113–3121
18. Maeda S, Shibuya H (2010) Study of anomaly detection on tracking of residual vector based on subspace method. In: Proceedings of spring meeting of the Japan Society for Precision Engineering, pp 329–330
19. Takata S, Kimura T (2003) Life cycle simulation system for life cycle process planning. Annu CIRP 53:634–655
20. Kim SK, Ishii K, Beiter KA et al (2010) Design for service innovation: a methodology for designing service as a business for manufacturing companies. Int J Serv Technol Manage 13:40–62

21. Moritsu T, Fujishiro T, Koda K, Kutsuna T (2013) O&M Service for sustainable social infrastructure. Hitachi Rev 62(7):370–375
22. Imazawa K et al (2010) Development of potential failure detection system for semiconductor manufacturing equipment. In: Proceedings of spring meeting of the Japan Society for Precision Engineering, pp 223–224
23. Nonaka Y, Lengyel A, Suginishi Y, Sugimoto K, Aizawa R, Yano N, Kato M, Kaihara T, Fuii N, Kurano T (2010) Maintenance planning in re-entrant flow (2nd Report)—application of multi-start/greedy method and verification test. Jpn Soc Precis Eng 76(12):1398–1405
24. Kaihara T, Fujii N, Nonaka Y, Takako K (2010) Maintenance planning in re-entrant flow (1st Report)—application of multi-start/greedy method and verification test. Jpn Soc Precis Eng 76 (4):468–473
25. Xiang Y et al (2011) Optimization of fatigue maintenance strategies based on prognosis In: Annual conference on the prognostics and health management society, pp 1–11
26. Bohlin M et al (2008) Reducing maintenance using condition monitoring and dynamic planning. In: 4th IET international conference on railway condition monitoring, pp 1–6
27. Applied Materials (2008) Xsite computerized maintenance management software (CMMS), Applied XsiteTM
28. Berry W, Rao V (1975) Critical ratio scheduling: an experimental analysis. Manage Sci 22 (2):192–201 (INFORMS)
29. Railway Technical Research Institute (2006) Planning systems, Railway Scheduling Algorithm, NTS
30. Kubo M (2003) Logistics engineering. Asakura Publishing Co. Ltd, Japan, pp 104–108
31. Takahara S Solution algorithm of 3-dimentional packing problem by multi-start/greedy method. In: Autumn conference, the operations research society of Japan, pp 82–83
32. National Research Institute for Earth Science and Disaster Prevention, Ministry of Education, Culture, Sports, Science and Technology (2006) Development of integrated earthquake disaster simulation system. 2006 report of metropolis great earthquake disaster mitigations special project, pp 221–225

Chapter 16
Managing Design Change with Functional Blueprints

Jacob Beal, Aaron Adler, Fusun Yaman, Jeffrey Cleveland, Hala Mostafa, Annan Mozeika, Kyle Usbeck, Gretchen Markiewicz and Benjamin Axelrod

Abstract Long-lived complex electromechanical systems, such as vehicles or industrial machinery, often need to be adapted for new uses or new environments. Adapting the design for such a system is frequently complicated by the fact that they are often tightly integrated, such that any change will have consequences throughout the design, and must take many different aspects of the system into consideration. Functional blueprints simplify adaptation by incorporating the reasons for design decisions and their consequences directly into the specification of a system. This allows a human designer to be supported by automated reasoning that can identify potential conflicts, suggest design fixes, and propagate changes implicit in the choices of the designer. This chapter presents the functional blueprints approach in detail, including both review of prior work and new results.

16.1 Introduction

Complex electromechanical designs, such as robots, vehicles, and industrial machinery, tend to be "brittle," meaning that it is often difficult to modify any significant aspect of the design without triggering a cascade of complex, difficult to predict and often costly changes. Such cascading changes are the result of interlocking constraints between elements that are modified and other parts of the design. An expert engineer, in fact, would likely consider many of these consequential changes to actually be keeping the design the same. When considered from the viewpoint of knowledge representation, a contradiction of this sort, where complex changes are required to keep a property the same, often indicates a critical flaw in representation.

J. Beal (✉) · A. Adler · F. Yaman · J. Cleveland · H. Mostafa · K. Usbeck · G. Markiewicz
Raytheon BBN Technologies, Cambridge, MA, USA
e-mail: jakebeal@bbn.com

A. Mozeika · B. Axelrod
iRobot Corporation, Bedford, MA, USA

L. Redding and R. Roy (eds.), *Through-life Engineering Services*,
Decision Engineering, DOI 10.1007/978-3-319-12111-6_16

If system specifications were more closely aligned with the ways in which human experts conceive of and work with designs, then many consequential changes would be either implicit from the specification or simple to automate. This would facilitate the development of design automation systems capable of making similar judgements about how to best maintain design integration in the face of changes. For complex electromechanical systems, such as aerospace vehicles, such tools might be able to significantly decrease the time and cost of both initial development and through-life upgrades and servicing. At the lower end of the complexity scale, such tools could enable simpler systems such as tactical ground robots to be rapidly modified in the field by operational experts in response to their evolving needs. Functional blueprints [1] are a representation aimed at providing such adapt- ability, inspired by biological development.

Functional blueprints capture expert knowledge by specifying a design as a set of behavioural goals and topological constraints and a method for incrementally adjusting the design when those goals or constraints are not being met. This chapter introduces the concept of functional blueprints and their application to electromechanical design. In particular, the work presented here focuses on robotic design, where systems are typically complex and highly integrated, yet relatively small and inexpensive, using an example robot similar to the iRobot LANdroid, but simpler and less expensive, called the "miniDroid." Section 16.2 introduces the functional blueprint concept and how it can be applied to electromechanical design. Section 16.3 then examines how interacting networks of functional blueprints can be used to generate both parametric and qualitative adaptation of a design. Section 16.4 discusses extensions to these approaches that can enable greater design plasticity. Finally, Sect. 16.5 discusses currently feasible applications and presents key open problems.

16.1.1 Comparison with Alternative Approaches

Design automation for electrical and mechanical systems has a long history, in which many significant results have been attained (e.g., [2–4]). A number of evolutionary methods have also been developed (e.g., [5, 6]). Applications have been limited, however, primarily due to the complexity and lack of smoothness in the design spaces that must be searched. Commercial modeling and simulation tools have thus been generally restricted to parametric exploration of relatively simple subsystems (e.g., [7]). Constraint-based local search methods, such as Kangaroo [8] and Comet [9], attempt to address this problem by using local parameter changes to minimize constraint violations, similar to functional blueprints but without their encoded knowledge of local repair strategies.

Control theory also addresses problems of system integration, but generally has difficulty with large numbers of non-linearly interacting parts. A notable exception may be viability theory [10], but its applicability is limited as it requires a system to be specified completed in terms of differential equations.

Other approaches to adaptive design of functional structures includes work on distributed adaptive construction [11–13], and various projects in self-reconfigurable robotics e.g., [14–16]. These approaches, however, are generally intended for more homogeneous and loosely coupled systems and would be difficult to adapt to electromechanical design. Recent work on "morphogenetic engineering" as proposed by Doursat [17] and for robotics in [16, 18], aims to support more heterogeneous systems. In particular, Doursat has laid out a framework for evolvable pattern formation [19], while Meng et al.'s generate patterns coordinating the configuration of a modular robot [20]. A more formal mathematical model can be found in [21], though the representational consequences are not yet explored.

16.2 Functional Blueprints

Functional blueprints, as defined in [1], specify a design in terms of behavioral goals and a method for adjusting the structure when those goals are not met, rather than a fixed structure. The approach is inspired by animal development, in which feedback processes maintain continuous integrated functionality across diverse subsystems such as muscles, nerves, and blood vessels as the animal grows from an embryo to a mature adult, despite the fact that the relationship and relative sizes of the elements making up these subsystems may change radically over time. Such decentralized adaptation is a critical enabler for the evolution of natural systems [22, 23].

Perhaps surprisingly, engineered systems also often show a family resemblance reminiscent of natural phylogeny, as in the iRobot product family shown in Fig. 16.1, to which also belongs the "miniDroid" robot (Fig. 16.3) used as a running example through the remainder of this chapter. This is due to the preference of human engineers to adapt functioning designs rather than to build from scratch. Functional blueprints aim to build on this parallel to enable engineered designs to exhibit the same power of facilitated change as is exhibited by natural systems.

Based on the feedback model of angiogenesis [24] and similar processes, a functional blueprint is thus defined as a collection of four elements: (1) a system behaviour that degrades gracefully across some range of viability, (2) a stress metric quantifying the degree and direction of stress on the system, (3) an incremental

Fig. 16.1 Families of engineered systems often exhibit "phylogenetic" relationships similar to those of natural organisms. For example, these four iRobot products share a base body plan, including symmetric two-wheel treads, flippers coaxial with one wheel, and a top-mounted sensor/manipulator package. **a** iRobot Warrior, **b** iRobot PackBot, **c** iRobot SUGV, **d** iRobot LANdroid

program that relieves stress through growth, shrinking, or other structural change, and (4) a program to construct an initial viable minimal system.

In essence, this model uses stress as a coordinating signal by which independently developing subsystems are integrated. The stress metric and incremental program combine to shift the design back toward required functionality; graceful degradation ensures that there is a margin for error in the interactions between subsystems, and the minimal system ensures there is a viable place to start. A network of interacting functional blueprints may thus be viewed as a piecewise specification of a parametric model. Moreover, unlike a typical parametric model, this approach does not require on a closed-form relationship between parameters.

16.2.1 Application to Electromechanical Design

The Morphogenetically Assisted Design Variation (MADV) architecture shown in Fig. 16.2a applies the functional blueprint concept to the problem of electromechanical design adaptation [25]. Under this architecture, electromechanical designs are adapted following a three-phase loop: the current model is run through a set of functional blueprint evaluators to determine stress on each functional blueprint; from these stresses come requested adjustments of the design, which are blended together to produce an incrementally updated design. The loop continues to iterate until the design reaches a stable point—at zero stress if adaptation succeeds, and greater than zero if it runs into a contradiction it cannot resolve.

Users control design variation either directly by modifying parameters, or indirectly through modifications of the environment for simulation-based evaluations [e.g., changing the height of an obstacle that a robot is expected to climb over through a user interface like that in Fig. 16.2b]. In either case, this modification injects stress into the system, which causes the values to begin adjusting toward a new equilibrium. The user of the architecture can then observe the ongoing process

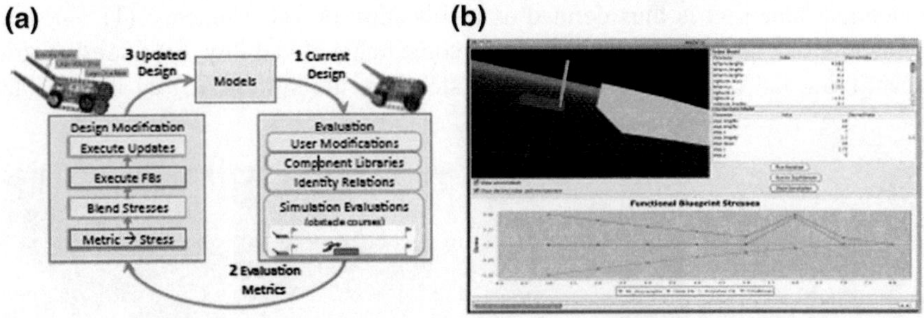

MADV architecture

Prototype user interface

Fig. 16.2 **a** MADV architecture: designs are adapted following a three-phase loop: the current model is run through evaluators to determine stress on each functional blueprint; from these stresses come requested adjustments of the design, which are blended together to produce an incrementally updated design. **b** Screenshot of prototype MADV software

of adaptation to the new requirement, adjusting their specifications if they prove infeasible or to give hints to help the system if it gets stuck. When the user is satisfied, the final design can then be exported for further refinement or for fabrication.

For the work discussed here, the functional blueprints for electromechanical designs fall into four categories (presented in detail in [25]): (1) simulation-based blueprints measuring the ability of the systems as a whole to accomplish a task (e.g., a robot climbing a step), (2) families of COTS components (e.g., a collection of servo motors with various torque limits), (3) closed form relations (e.g., the identity relation between mass, density, and volume, the inequality between the vertical dimensions of a motor and of the robot body that contains it, or a functional specification requiring total mass less than a certain amount), and (4) user modification (e.g., a request to double the height of the step that can be climbed).

The collection of functional blueprints are integrated to form a complete network using a manifold-based representation that mixes geometric and topological elements [26, 27]. This allows the representation to include both architectural decisions (in the form of topological constraints and symmetries), geometric commitments (in the form of parametric values and geometry-based constraints), and functionality (in the form of functional blueprint constraints).

Finally, because parameters may have very different magnitudes, and because multiple functional blueprints may act on the same parameter, functional blueprints as implemented for MADV act on parameters only by expressing a stress in the range of [0,1] and a direction; the combination of stresses then produce value changes according to an adaptive process as described in Sect. 16.3.1.

Discussion and examples of applying the MADV architecture will largely focus on the miniDroid robot, shown in Fig. 16.3. This system was created by iRobot, based on the LANdroid from the robot family shown in Fig. 16.1 and slightly expanded and simplified to be a better target for investigation of adaptive design. Much of the work discussed in this chapter has been carried out using the miniDroid as a driving electromechanical design example, and it will thus be used as a running example. Figure 16.4 shows a functional blueprint network representing the miniDroid, including all design features at least 1 cm^3 in volume. In total, this comprises 23 components, 112 design parameters and 111 functional blueprints. Of the functional blueprints, three (step-climbing, self-righting, and fast driving) are

miniDroid prototype　　CAD design　　Step-climbing sim

Fig. 16.3 miniDroid base robot design, in CAD (**b**) and reality (**a**). **c** Screenshots from the miniDroid ROS simulations

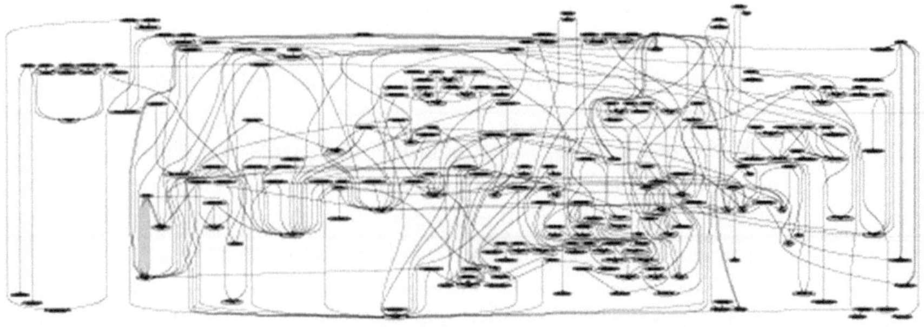

Fig. 16.4 Parameters, constraints, and initial values for the miniDroid: *blue* is parameters with an explicit initial value, *purple* is parameters whose value is inferred, *green* indicates constraint relations, and arrows link constraints to the parameters they affect (Color figure online)

evaluated with simulations implemented using ROS (Robot Operating System) [28] and the Gazebo simulator, while the remainder are either closed-form or represent COTS components.

16.2.2 Diagnosis and Assistive Design

Functional blueprint representations of electromechanical design can also provide useful services at lower levels of automation. In particular, since functional blueprints capture the intentions and requirements of a design, they can be used to assist a human designer with diagnosis of problems and suggestions for their solutions. This is a much less radical change to existing processes than automatic adaptation, yet still addresses many of the key challenges that motivate the investigation of automatic adaptation.

Figure 16.5 shows an example of a transcript from a prototype functional blueprint diagnosis system. The text for the explanations is generated automatically from the structural relations encoded in the functional blueprints and the parameters

```
Assuming fixed value for Instrument Package Mass and Instrument Package Volume

Detected 2 problems in current design (in descending order of importance):
Self Righting: Bot Body Length and Bot Body Mass are too large, relative to Flipper Motor Torque
Layout Geometry: Instrument Package Volume is too large, at 0.0100 Meters^3,
  Should be significantly smaller than Bot Body Volume, which is 9.43e-4 Meters^3

Suggestions to fix design problems:
  Decrease Bot Body Length from its current value of 0.200 Meters
  Increase Flipper Motor Torque from its current value of 1.22 Newton-Meters
  Increase Bot Body Volume from its current value of 9.43e-4 Meters^3
```

Fig. 16.5 Example transcript generated by electromechanical diagnostic system based on functional blueprint representation

they interact with. These explanations may be able to clarify the relations of a design to a human, and could form the basis of "expert advisor" systems for assisting domain experts in adapting designs.

16.3 Adapting Electromechanical Designs

Having captured the relationship between form and function as a network of functional blueprints, integrated adaptation of an electromechanical design may then be carried out as an iterative process of incremental adjustments. Many adaptations can be effectively carried out entirely by parametric variation, adjusting effective step size to optimize convergence speed while maintaining coherent design. Larger scale changes in specification, however, may also require qualitative changes in the collection of components making up the design.

16.3.1 Convergence of Functional Blueprint Networks

Functional blueprints deliberately do not contain any specification of how they are expected to interact with other functional blueprints: to do so would require an exponential number of relationships to be considered and would prevent them from being modular and capable to being reused in new designs. Instead, functional blueprints interact indirectly, by changing parameters such that stress is induced in other functional blueprints. That stress then induces those other functional blueprints to act, which may induce other stress, etc., propagating changes through the design. This raises a critical question, however: under what conditions is it possible to ensure that the stress thus generated will converge and return to zero (meaning that an acceptable adaptation has been found) in a reasonable amount of time?

The graceful degradation property of functional blueprints ensures that it is always possible to maintain the integration of a design. As proved in [1], it is always possible to reduce step-size such that no parameter ever saturates on stress. This does not, however, guarantee progress toward a user's desired specifications. Consider, for example, if one simultaneously requires a miniDroid to climb twice as tall a step and to be small enough to fit in a pocket: this may simply be physically impossible. It is also possible that a path to a solution may exist, but that interactions in the functional blueprint specifications may render it inaccessible to any particular algorithm.

If all functional blueprints are linear in their interactions, then convergence can be guaranteed, as demonstrated in [25]. Furthermore, the speed of convergence is quite rapid, outperforming genetic algorithms by more than two orders of magnitude. In most electromechanical designs, however, many interactions are non-linear; consider for example, that scale typically has a quadratic relation with strength and cubic with mass. Navigation of the stress space may then be treated as a non-convex optimization problem, around which there is already well-established literature. The graceful degradation property of functional blueprints is helpful, as it ensures a relatively smooth stress function with response to any given parameter, and thus smoothness in the joint space as well. Unfortunately, many non-convex methods are still extremely computationally expensive, particularly given simulation-driven functional blueprints that lack a closed-form expression for the relationship between parameters.

Even simple heuristic methods, however, appear to be able to support large-scale variation in a real electromechanical design such as the miniDroid. One such heuristic that has been investigated for balancing convergence speed versus stability is to adaptively modify the size of each incremental adjustment of a parameter based on global and local stress. Specifically, the heuristic for parameter adjustment is:

$$S_p = \max_r \{s_{r,p}\} \cdot \frac{\sum_r s_{r,p}}{\sum_r |s_{r,p}|} \quad S_M = \max_p \{|S_p|\} \quad \Delta_p = \varepsilon \frac{S_p}{S_M} \tag{16.1}$$

where the $s_{r,p}$ is the stress exerted by relation (functional blueprint) r on parameter p and ε is the current incremental scaling factor for step size. The relative stress on a parameter S_p is then proportioned based on the degree of conflict between the stresses exerted on it by relations in the network, and the step size Δ_p proportioned based on the maximum stress S_M in the network and the incremental factor e. Finally, each parameter is changed by multiplying its current value by $1 + \Delta_p$. The size of e is critical to convergence: the larger that e is, the faster that the system will converge, but if it is too large, then parameter values will oscillate and possibly become unstable. The critical value for e may be difficult or impossible to determine statically (and indeed, may vary depending on parameter values). It is therefore useful to adaptively select the value of e based on the observed behaviour of the system. One simple heuristic which performs well is to examine the value of stress over a k-sample window: if stress is steadily decreasing, then multiply e by 2; if more than some threshold amount of oscillation is observed, then divide e by 2. Figure 16.6 shows examples of stress rising and converging during large-scale design adaptation using these heuristics (as well as qualitative design change at local minima, as discussed in the next section). Note in particular the transitions between oscillatory and smooth behaviour, as e is adjusted to attempt to control the rate of descent.

Using lightweight heuristics such as these, the time to execute one iteration of incremental adaptation is driven primarily by the cost of evaluating the functional blueprints, particularly those that are evaluated by simulation. Here, it is possible to greatly accelerate adaptation by exploiting the graceful degradation property of functional blueprints. Typically, only a small fraction of the relations in a system are simulation-driven functional blueprints; most others implement fast-evaluating relations such as specifications (e.g., a system mass limit), identity relations (e.g., total mass being the sum of component mass), physical relations (e.g., mass equals volume times density), or availability of components (e.g., dimensions vs. torque

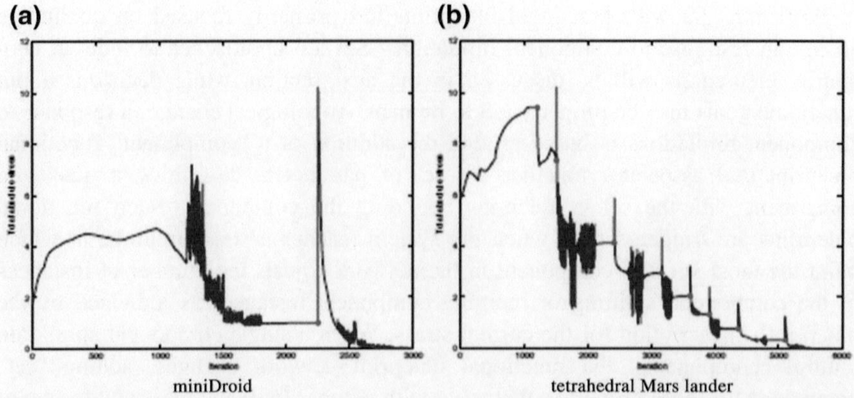

Fig. 16.6 Example traces of stress over time for constraint resolution including functional blueprints for qualitative change. Topological changes are marked with a *red* star; note that stress resolution reaches a stable "stalled" point before each change. **a** Shows a miniDroid adapting as the step size is raised from 10 to 25 cm, including shifting from one to two flipper drive motors to maintain self-righting with increasing mass. **b** Shows a tetrahedral lander adding motors until ultimately there are nine motors per panel as instrument package mass and power demand both increase 10-fold

for a family of COTS servo motors). Rather than compute simulations with every iteration, the stress value of each simulation may be cached and reused, such that there is only one simulation every n iterations. If the simulation has the required graceful degradation property, then its stress should not change significantly in a short time, so the rest of the network can continue adjusting while reducing the number of simulations required (and thus the approximate time) by a factor of n. Experiments have shown that it is likely possible to obtain one to two orders of magnitude speed-up with this mechanism, depending on the particulars of the design.

16.3.2 Parametric Versus Qualitative Design Changes

The discussion thus far has focused on parametric design changes, in which the parameter values of a blueprint network are adjusted, but the network (i.e., the set of components and their key topological relations) remains unchanged. Qualitative changes in design, on the other hand, make changes in the network of interacting parameters. For example, a new component may be added, an existing component removed, or the geometric relations of components significantly changed. Typical reasons for a human designer to make such a change include adding a new type of function, limitations in available components, and indirect geometric interactions of components.

Work thus far with functional blueprints has primarily focused on qualitative change in response to component limitations. Several approaches to indirect geometric interactions will be discussed in the next section, while decisions about functional goals may be properly left to humans. Topological change in response to component limitations is based around the addition of a "component" functional blueprint that associates together the set of parameters describing a quantized component with the set of relations bounding the component. Such functional blueprints are triggered only when the system reaches a stress minima, at which point the most stressed component in the network adjusts the number of instances of the component, splitting or merging component instances as indicated by the blueprint's prescription for the current stress. When a single component splits into multiple components, the functional blueprint network changes, adding "set" parameters for those properties that scale with number (e.g., the mass and torque of motors, but not their maximum rotational speed) and rewiring other relations to connect to the set or original parameters depending on the class. A complementary change occurs when a component set merges into a single component.

Both quantitative and qualitative variation have been tested by specifying large changes to the values of functional parameters of the miniDroid (e.g., the step height to be climbed, the amount of undercarriage clearance), and demonstrating that the network of functional blueprints successfully adapts, returning to a zero-stress state. Figure 16.6a shows an example of a stress trace from a typical such adaptation experiment, in which changing the step height from 10 to 25 cm causes parameter values to change throughout much of the design, as well as causing a change from one to two flipper drive motors in order to maintain self-righting with increasing mass. Notice the local minimum just before the number of flipper drive motors increases, and the rapidly dispersed pulse of stress afterwards, as the new values propagate through the network.

16.3.3 Reusability of Functional Blueprints

It is important to ensure that functional blueprints can often be reused from one design to another design, because it allows the effort required to create a functional blueprint to be amortized across the benefit of its use in many systems—a reuse expected to be further facilitated by the adaptivity inherent to the functional blueprint concept. This has been tested by constructing a design that reuses the self-righting functionality of the miniDroid, but is generally extremely different in form: a simplified tetrahedral Mars Lander based off of the landers for NASA's Spirit and Opportunity rovers.

These landers used a tetrahedral shape to ensure that the rover they delivered would be upright: three of the four panels open away from the fourth, like petals of a flower, so that if the rover lands on any face other than the intended bottom, it will be flipped upright by the opening of the panels. This functionality is similar to the miniDroid using its flippers to right itself, so the functional blueprints associated with this functionality should be reusable in the new design.

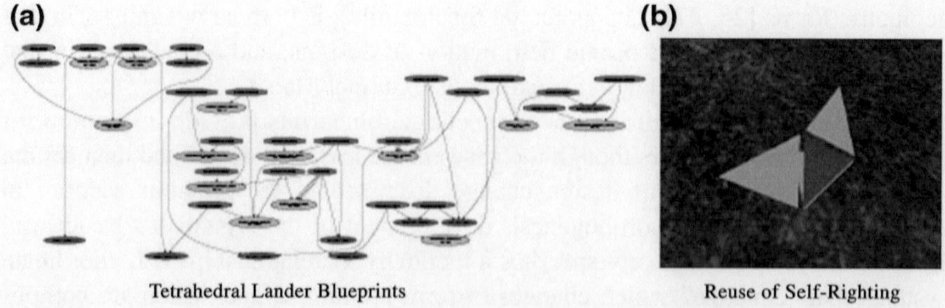

Fig. 16.7 Demonstration of functional blueprint reusability with a simplified design for a tetrahedral Mars lander based off of the landers for NASA's Spirit and Opportunity rovers: **a** shows the functional blueprint network for the lander, **b** shows a screenshot from a simulation carried out for the self-righting functional blueprint, one of a number reused from the miniDroid

Figure 16.7a shows the functional blueprint network for a simplified tetrahedral lander containing an instrument package attached to its base and solar cells on the interiors of the "side" panels, which are exposed when the tetrahedron opens, containing a total of 11 modules (4 panels, 3 solar panels, 3 flipper motors, and 1 instrument package). The symmetries of the design allow these component to be described with only 23 design parameters (abstracting away details of the instrument package contents), which are connected together by 22 constraints. Of these constraints, 18 are reused, including the flipping functional blueprint and the motor library. In many cases, reuse entailed minor modification of parameters [e.g., changing to Mars gravity, linking the flipping blueprint with the alternate simulation shown in Fig. 16.7b], but the main work required for each functional blueprint design was reused. As a result, creation of the tetrahedral lander's initial design and blueprint constraint network took only 2 h.

This design, like the miniDroid, enables large-scale design variation to be driven by changing critical parameters and allowing the rest of the parameters to adjust accordingly. For example, Fig. 16.6(b) shows the result of increasing the mass and power demand of the instrument package 10-fold. The lander grows in size, increases its solar panels to deal with the expected higher demands from a larger package, and undergoes qualitative change in the number of motors per panel, eventually finding that nine motors per panel are required to ensure that the lander can right itself. These appear are fairly regular intervals as the system gracefully moves towards resolution, stalling just before the introduction of each motor.

16.4 Expanding the Plasticity of Design

The adaptations considered thus far have been limited to relatively simple geometric forms and relations. The mixed geometric-topological representation used by MADV, however, is capable of supporting a much broader variety of complex geometric forms [26, 27]. This potential for flexibility is both an advantage, in that it can in principle support plastic deformation of designs, and a challenge, in that the number of such possible deformations is extremely large.

Once again, the biological roots of functional blueprints provide inspiration for how to tackle this problem, though the answers are less well developed than for the more constrained forms of design change discussed in the previous section. In animals, the process of morphogenesis development of the organism's basic form from an initial egg effectively specifies a hierarchy of relationships and coordinate systems that "canalize" which changes of form are simple and which are complicated [22, 29]. As development continues, function-based feedback drives many forms of system integration [23], such as the ramification of blood vessels (driven ultimate by oxygen demand from tissue and by tension in the walls of blood vessels) and the assortation of neurons to control muscles (driven by competition for effective control).

Both of these forces are known to greatly facilitate the resilience and evolution of biological organisms [22, 23], so such developmental processes may also provide a useful model for facilitating greater plasticity in electromechanical designs. Such developmental models are in effect applying functional blueprints on a second and finer level, not just in transitioning from design to design, but also as a feedback process in the translation of a design specification to a geometric form that can then be evaluated by the "higher-level" functional blueprints already considered.

References [26, 30] present a variety of approaches for integrating such developmental models with the MADV architecture. At the coarsest scale [30] presents a "tissue development" model in which the topological and parameter relations of a design are generated by applying a set of developmental rules. Each developmental rule is an independent and asynchronous operation that applies a sequence of operators (the rule's body) on any tissue that matches the rule's preconditions. Executing a set of rules on an initially undifferentiated electromechanical body ("egg"), results in a body plan of a design "fetus" made up of various "organs" like limbs, flippers and wheels—the topological component of the mixed topological/geometric representation. Parameters in the rules become the geometric parameters constraining the topological elements generated by applying the rules, with design symmetries emerging from the repeated application of a single rule. Figure 16.8a shows an example of the basic miniDroid body-plan generated by a set of 12 rules operating in eight conceptual stages. Detailed description of the rules and stages may be found in [30].

The layout of wires, cables, and other such connections is another area of design where plasticity of form is useful, as these often have a great deal of flexibility in how they can be routed from place to place within a design. Making analogy to the chemotactic process by which neurons grow axons to their appropriate targets in biological development, routing of wires and other linear connectors can be accomplished by "seeding" each connector at one end-point and simulating an informational gradient from the other [26]. The connectors are then routed by climbing up the gradient, avoiding one another and fixed elements of the design as they grow.

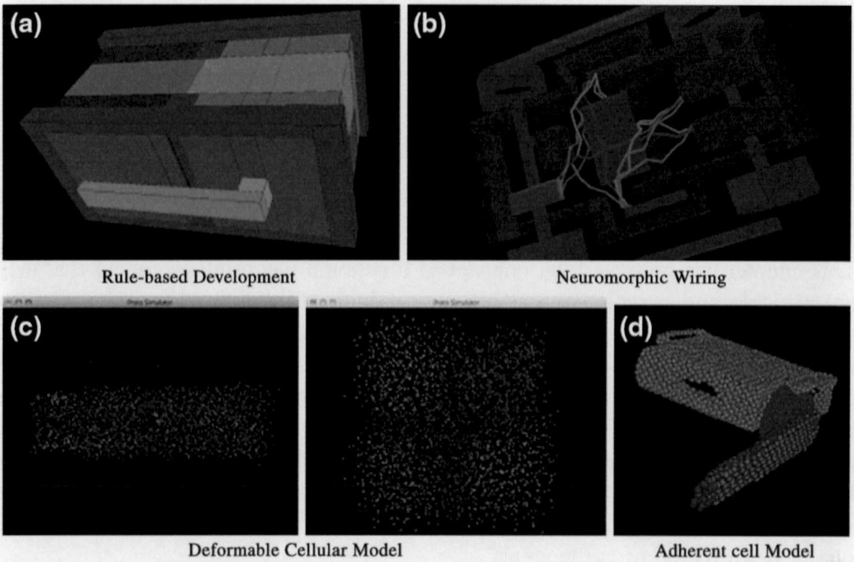

Fig. 16.8 **a** Rule-based development of a miniDroid topology from an initial undifferentiated "egg." **b** Neuromorphic routing connecting electronic components in the interior of a miniDroid design with wires carrying power (*red*), ground (*green*), and signal (*other colors*). **c** Distortion tolerant developmental program using the Proto manifold model to automatically adapt to execution on a modified underlying shape, e.g., shifting from a thin *rectangle* (*left*) to an *square* with a twisted coordinate system (*right*). **d** miniDroid flipper growing from the tip (*green*) in adherent cell simulation using MASON (Color figure online)

Figure 16.8b shows an example of this wiring model, from [26], being applied to lay out power, ground, and signal wires connecting the batteries, CPU, encoder, and motors of a miniDroid.

At a finer grain, Figs. 16.8c, 16.8d show cellular models of development [26]. These cellular models have the advantage that the distribution of "cells" can directly represent arbitrarily complex shapes, but the disadvantage that it is necessary to interpret these shapes to re-connect them with the parametric relationships that higher-level functional blueprints act upon. Another challenge is the tension between the resolution of the cells and the cost of simulation. Figure 16.8c shows how a fine-grained cellular model, implemented using the manifold geometry operations of the Proto [31, 32] aggregate programming language, allows a developmental plan to distort to match the conditions of its execution. In this case, the layout of electronics, wheels, and motors in a miniDroid "body plan," executing on a model of 2,000 cells distributed through a rectilinear 3D volume of space, is changed in proportion (left) and twisted by a change in coordinate system (right), showing the inherent geometric adaptation of the program. Further coherence in adaptation can be enabled by soft-body simulation, which can allow design elements to adhere, compress, deform, penetrate, or otherwise physically interact with one another for co-adaptation during the developmental process. Figure 16.8d shows an example of a computationally inexpensive soft-body model implemented using MASON, a Javabased multi-agent simulator [33], implementing tapering of a miniDroid flipper by means of an adhesion-based growth process.

16.5 Applications and Open Problems

This chapter has presented the concept of functional blueprints, showing that they are a viable and potentially scalable approach to adaptation of complex electromechanical designs. Applied in combination with unified topological-geometrical representation and self-tuning step sizes, functional blueprints can adapt complex designs effectively across a large range of variation, producing both quantitative and qualitative changes that maintain functionality in a changing electromechanical design. Functional blueprints are also composable and reusable, as demonstrated by the transfer of blueprints from the miniDroid design to the tetrahedral lander design, and the approach may be further extended to allow greater adaptability through the plastic deformation of components.

The visionary goals for this approach are two-fold. The first is to allow non-experts to produce design variations that satisfy new requirements, even without a good understanding of subsystems, simply by indicating the critical changes that are needed and allowing the rest of the complementary changes to propagate automatically. The second is to enable a continuous design and manufacturing cycle, in which emerging additive manufacturing technology joins with a functional blueprint driven radically decreased cost of redesign to enable even highly complex electromechanical systems like aerospace vehicles to be updated on the fly, rapidly incorporating new technologies and responding to changing requirements. Realizing these visions will require an investment to develop libraries that associate existing CAD components with corresponding functional blueprints, as well as development of appropriate user interfaces, and additional work to enhance the speed and reliability with which large networks of functional blueprints can be guaranteed to converge.

Because functional blueprints encode knowledge about a design, they can also be used to assist human designers in other ways than adaptation. In particular, functional blueprints can not only detect potential design problems but can be used to generate human-readable explanations of the causes of those problems and suggestions for approaches to fixing them. This holds the potential for improvements in the way that mechanical engineering is conducted and taught.

In the practice of mechanical engineering, functional blueprints can be used to import the software notion of continuous integration into electromechanical design. Continuous integration is an important tool for rapid and reliable software engineering, in which every incremental step in the realization of a design is automatically tested against a suite of "regression tests" ensuring that important existing functionality is not endangered by progress in other areas. By evaluating the ability of a design to satisfy requirements, functional blueprints could be used to effectively implement such regression testing, decreasing the cost of integration and the number of design problems identified after production.

Finally, functional blueprints could also be used in several ways to help educate students on electromechanical design and/or manufacturing. First, functional blueprints could be used as part of an active learning process to give a student instantaneous feedback on the strengths and weaknesses of their current design. Second, since functional blueprints can be used to adapt a design to find an integrated solution, this sort of "look-ahead" could be used either to give students hints to help with design or to evaluate what aspects of design a student is most struggling with, and therefore to adaptively present or recall relevant curriculum elements.

In summary: the engineering of complex electromechanical systems is an important problem that impacts society in myriad ways. Functional blueprints offer the potential to improve the engineering process at every phase: assistance in design, improved diagnosis of potential faults, simplification of through-life adaptation and redesign, democratization of engineering, and even improvement of the education of future electromechanical engineers.

References

1. Beal J (2011) Functional blueprints: an approach to modularity in grown systems. Swarm Intell 5(3):250–281
2. Campbell MI, Cagan J, Kotovsky K (1999) A-Design: an agent-based approach to conceptual design in a dynamic environment. Res Eng Design 11(3):172–192
3. Hoover SP, Rinderle JR (1989) A synthesis strategy for mechanical devices. Res Eng Design 1 (2):87–103
4. Fromherz MPJ, Bobrow DG, de Kleer J (2003) Model-based computing for design and control of reconfigurable systems. AI Magazine 24(4):120–130
5. Fan Z, Wang J, Goodman E (2005) Cutting edge robotics. In: Kordic V, Lazinica A and Merdan M (eds) Exploring open-ended design space of mechatronic systems, Pro Literatur Verlag, Germany, pp 707–726
6. Koza JR, Bennett III FH, Andre D, Keane MA (1998) Adaptive computing in design and manufacture, Springer, Berlin, pp 177–192
7. ANSYS, Inc. ANSYS DesignXplorer (2013) http://www.ansys.com/Products/Workflow+Technology/ANSYS+Workbench+Platform/ANSYS+DesignXplorer
8. Newton MAH, Pham DN, Sattar A, Maher MJ (2011) Kangaroo: an efficient constraint-based local search system. In: Principles and practice of constraint programming 17th international conference, CP, Italy, pp 645–659
9. Van Hentenryck P, Michel L (2005) Constraint-based local search, MIT Press
10. Aubin JP (1991) Viability theory, Birkhauser
11. Werfel J (2006) Anthills built to order: Automating construction with artificial swarms. Ph.D thesis, MIT
12. Werfel J, Nagpal R (2007) In: International conference on intelligent robots and system
13. Estevez N (2007) Functional blueprints: a dynamical systems approach to structure representation. Cornell University
14. Stoy K, Nagpal R (2004) Self-reconfiguration using directed growth. In: International symposium on distributed autonomous robotic system (DARS), Springer-Verlag, New York, pp 3–12

15. O'Grady R, Christensen AL, Dorigo M (2009) SWARMORPH: multi-robot morphogenesis using directional self-assembly. IEEE Trans Rob 25(3):738–743
16. O'Grady R, Christensen AL, Pinciroli C, Dorigo M (2010) Robots autonomously self-assemble into dedicated morphologies to solve different tasks. In: AAMAS, pp 1517–1518
17. Doursat R (2011) Morphogenetic engineering weds bio self-organization to human-designed systems. PerAda Magazine
18. Jin Y, Meng Y (2011) Morphogenetic robotics: an emerging new field in developmental robotics. IEEE Trans syst Man Cybern C Appl Rev 41(2):145–160
19. Doursat R (2008) Organic computing. In: Wurtz R (ed) Springer, Berlin, pp 167–200
20. Meng Y, Zhang Y, Jin Y (2010) A morphogenetic approach to self-reconfigurable modular robots using a hybrid hierarchical gene regulatory network. In: International conference on the synthesis and simulation of living systems, pp 765–772
21. MacLennan B (2010) Models and Mechanisms for Artificial Morphogenesis. In: Natural computing, Proceedings in information and communications technology, vol 2. pp 23–33
22. Carroll SB (2005) Endless forms most beautiful, W.W. Norton and Company, New York
23. Kirschner MW, Norton JC (2005) The plausibility of life: resolving darwin's dilemma, Yale University Press, New Haven
24. Carmeliet P (2003) Angiogenesis in health and disease. Nat Med 9(6):653–660
25. Adler A, Yaman F, Beal J, Cleveland J, Mostafa H, Mozeika A (2013) A morphogenetically assisted design variation tool. In: AAAI, pp 9–15
26. Beal J, Mostafa H, Axelrod B, Mozeika A, Adler A, Markiewicz G, Usbeck K (2012) A manifold operator representation for adaptive design. In: GECCO 2012, pp 529–536
27. Beal J, Adler A, Mostafa H (2013) Mixed geometric-topological representation for electromechanical design. In: Extended abstract at GECCO 2013, pp 105–106
28. Quigley M, Gerkey B, Conley K, Faust J, Foote T, Leibs J, Berger E, Wheeler R, Ng AY (2009) In: Proceedings of the open-source software workshop at the international conference on robotics and automation (ICRA)
29. Waddington CH (1942) Nature 150(3811):563
30. Beal J, Lowell J, Mozeika A, Usbeck K (2011) Using morphogenetic models to develop spatial structures. In: Spatial computing workshop 2011 at IEEE SASO '11, pp 85–90
31. Beal J, Bachrach J (2006) IEEE intelligent systems, pp 10–19
32. MIT Proto software available at http://mitproto.net. Retrieved 16 Sept 2012
33. Luke S, Cioffi-Revilla C, Panait L, Sullivan K, Balan G (2005) Simulation 82(7):517

Part V
Cost, Uncertainty, Risk, and Standards

Chapter 17
Obsolescence Management

John Ahmet Erkoyuncu and Rajkumar Roy

Abstract Rapid technological advances in component electronics driven by the consumer market are forcing acceleration in such components becoming obsolescent. For sectors that employ long life assets the situation is becoming costly, estimated at $750 m per annum for the US Navy alone. In this chapter mitigation and resolution solutions for this problem are explored as well as how to approach estimation of its cost. Skills shortages and cost estimation problems are identified as key issues.

17.1 Introduction

Manufacturers of electronic, electrical and electromechanical (EEE) components since the 1980s, have migrated away from the low volume military market and focused their efforts on the more profitable commercial market (e.g. mobile phones and personal computers) [1, 2]. As a result, the proportion associated with defence and aerospace is becoming a smaller percentage of the global semiconductor market [3, 4]. This implies that nowadays vendors are highly likely to discontinue their components as they launch new ones with more advanced technological features, drastically reducing their life cycle [5, 6]. The mismatch between the duration of the life cycle of the components and the systems they are in is the main cause of obsolescence issues. An obsolescence issue arises when a component is no longer; available from stock of own spares, procurable, nor produced by its original manufacturer at the original specifications [7].

In sectors such as defence, aerospace, oil and gas, railway and nuclear the systems need to be supported for many decades. As illustrated in Fig. 17.1, in these type of systems it is not unusual that 70–80 % of the electronic components become

J.A. Erkoyuncu (✉) · R. Roy
EPSRC Centre for Through-Life Engineering Services, Cranfield University,
Cranfield, Bedfordshire MK43 0AL, UK
e-mail: j.a.erkoyuncu@cranfield.ac.uk

L. Redding and R. Roy (eds.), *Through-life Engineering Services*,
Decision Engineering, DOI 10.1007/978-3-319-12111-6_17

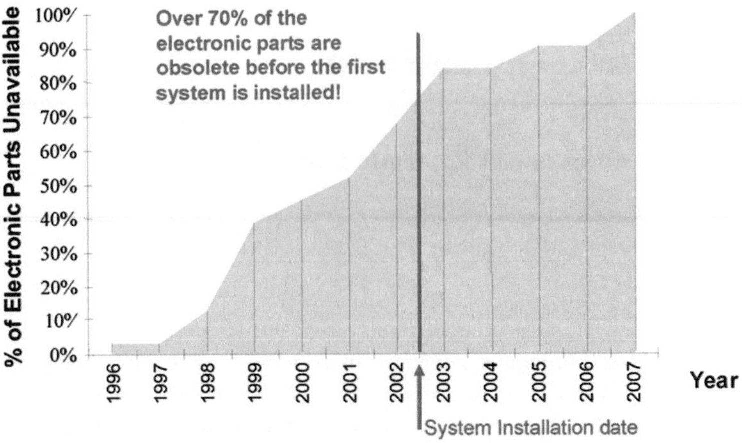

Fig. 17.1 Percentage of electronic parts unavailable versus the first 10 years of a surface ship sonar system's life cycle (Adopted from [9])

obsolete before the system has been fielded [8, 9] Therefore, obsolescence has become a major issue in these sectors [10–12]. For instance, the British prime contractor for the Eurofighter project declared that obsolescence is the No.2 risk to the project and it is taking vast amounts of money to design out obsolescence from one version of the aircraft to the next.

Although the obsolescence problem is commonly associated with EEE components [13], it is not restricted to them. There are many other areas of a system that can become obsolete such as: (Fig. 17.2)

Fig. 17.2 Holistic view of obsolescence

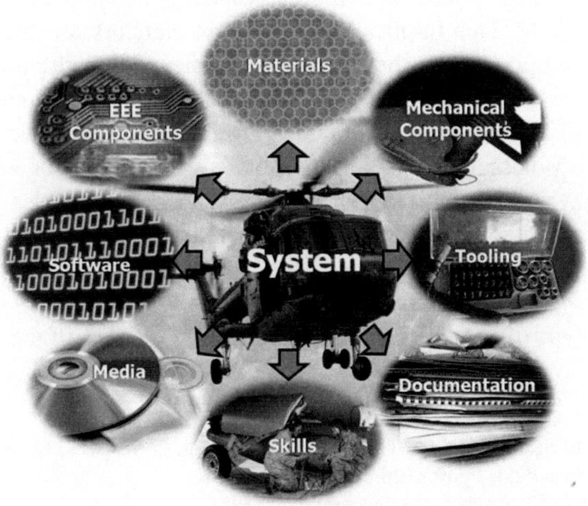

- Mechanical components
- Materials
- Software
- Media
- Tooling and test equipment
- Documentation
- Skills

Among those areas there are many interdependencies; so there is a need to in a holistic manner rather than analysing each one independently [14].

17.2 Cost of Obsolescence

Obsolescence in many sectors is becoming very costly. For instance, the estimate of the total through life obsolescence costs for the Nimrod MRA4 is £782.3 m, according to the Obsolescence Scoping Exercise [15] (QinetiQ ref: D&TS/CS/TR058826, Nov. 2005). In United States, the obsolescence issues cost up to $750 million annually according to the US Navy estimations [11].

In sectors such as defence, aerospace, oil and gas, railway and nuclear, the vehicles (e.g. aircraft) and installations (e.g. refinery) usually need to be sustained over many decades. This is why they are commonly known as sustainment-dominated systems. These systems are generally composed of low volume complex electronics, which are affected by the fast changing market trends and the ongoing technical revolution in the electronics industry. They are characterised by high costs associated with their redesign because of the strict requalification requirements, and little or no control over their supply chain because of their low production volumes. Due to the high costs and long life times associated with technology insertion and design refresh, these systems often fall behind the technology wave [9].

One of the main problems that these systems will definitely face during their lifetime is obsolescence. A component becomes obsolete when it is no longer; available from stock of own spares, procurable, nor produced by its original manufacturer at the original specifications [5].

The fast technological changes caused by the rapid growth of the electronics industry and the diminishing demand for aged components are exacerbating the obsolescence of electronic components. Thus, obsolescence has become one of the main costs in the life cycle of the systems in these sectors [11]. For instance, the prime integrator for the Eurofighter project has declared that obsolescence is the No.2 risk to the project and it is taking vast amounts of money to design out obsolescence from one version of the aircraft to the next.

17.3 How to Manage Obsolescence

In recent years, more awareness has been raised about the problem of obsolescence among researchers and practitioners. This enabled the ongoing shift from reactive obsolescence management to a more proactive approach. In the past, obsolescence was usually dealt with in a reactive way, so that obsolescence issues were only identified when a particular component was required to continue sustaining the system and it could no longer be procured. In most cases, the possibility of making a last-time buy (LTB) is missed by then. This situation implies that the resolution of this issue is likely to become urgent and costly. Currently, most organisations employ obsolescence managers to proactively predict obsolescence issues and plan the best strategy to mitigate them, so that the obsolescence impact can be considerably reduced.

In the literature, the terms 'mitigation' and 'resolution' are frequently used interchangeably. However, it is important to make a distinction between their meanings. The term 'mitigation' refers to the measures taken to minimise the impact or likelihood of having an obsolescence problem, whereas the term 'resolution' refers to the measures taken to tackle an obsolescence issue once it appears [8]. The most common resolution approaches and mitigation strategies are presented in Figs. 17. 3 and 17.4 respectively.

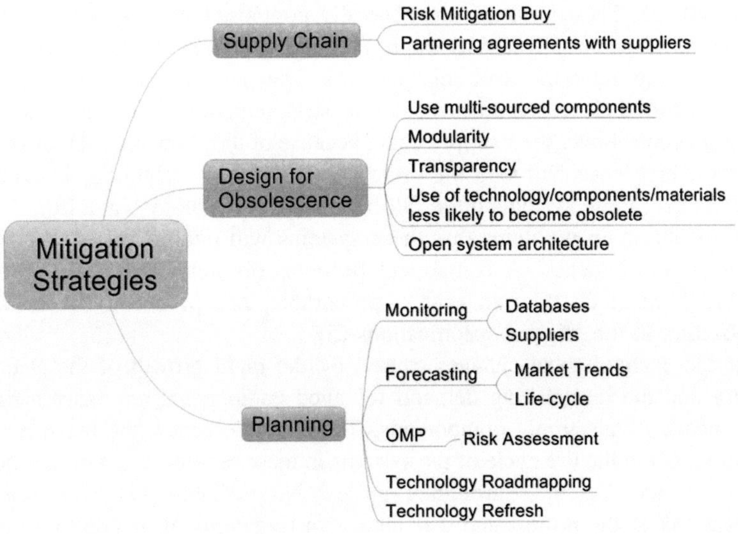

Fig. 17.3 Obsolescence mitigation strategies [8]

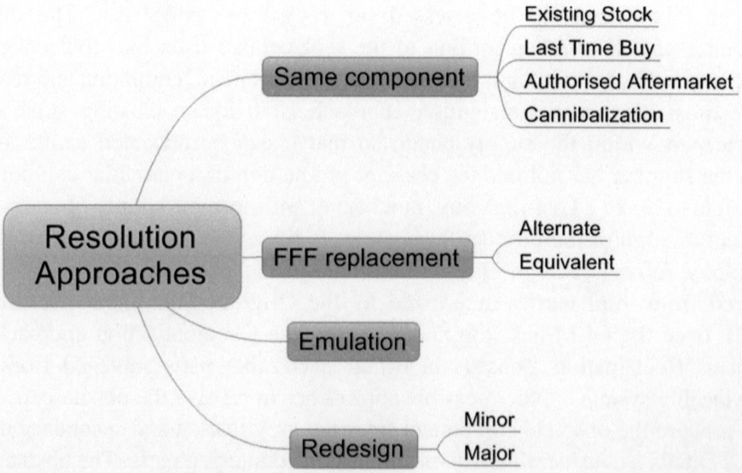

Fig. 17.4 Obsolescence resolution approaches [8]

17.3.1 Obsolescence Mitigation Strategies

Obsolescence risk can be mitigated by taking actions in three main areas: supply chain, design and planning. The mitigation strategies that can be taken in the supply chain are: Life-time Buy and partnering agreements with suppliers. The Life-time Buy approach, also known as Risk-Buy, involves purchasing and storing enough obsolete items to meet the system's forecasted lifetime requirements. It is also suggested to make partnering agreements with suppliers to ensure the continuous support and provision of critical components. Awareness of the obsolescence problem is leading to tackle it at early stages of the project by designing for obsolescence. This may involve the use of open system architecture, modularity, transparency, increase of standardisation in the designs and avoid using single-sourced components. These actions will definitely ease the resolution of obsolescence issues that may arise at the component or line replaceable unit (LRU) level. The third way of mitigating obsolescence is planning. It implies the development of an Obsolescence Management Plan (OMP), the use of technology roadmaps and the use of obsolescence monitoring tools, which are commercially available from different vendors [8].

17.3.2 Obsolescence Resolution Approaches

When an obsolescence issue arises, a resolution approach must be applied immediately to tackle the problem. There are several possible resolution approaches, but their suitability needs to be assessed on a case by case basis. The different approaches are classified according to the replacement used into four categories: same component, form, fit and function (FFF) replacement, emulation and redesign.

The most inexpensive resolution approach is to locate existing stock of the obsolete part within the supply chain, so that it can be allocated to the system. When the supplier has notified the cease of production of a particular component, it is possible to make a Last-time buy, purchasing and storing a supply of components sufficient to support the product throughout its life cycle or until the next planned technology refresh (Bridge Buy). Occasionally, the obsolete component can be procured from third parties authorized by the Original Equipment Manufacturer (OEM), once the OEM has stop producing it. The Cannibalization approach, also known as Reclamation, consists in using serviceable parts salvaged from other unserviceable systems. Other possible approaches to resolve the obsolescence issue are to procure the obsolete component from the grey market and secondary market, but this entails a considerable risk of purchasing counterfeit parts. The obsolescence issue can also be resolved by identifying a form, fit and function (FFF) replacement, which can be regarded as an equivalent or an alternate. An equivalent is a functionally, parametrically and technically interchangeable replacement, without any additional changes; whereas an alternate part may be less capable than that specified for one or more reasons (e.g. quality or reliability level, tolerance, parametric, temperature range). The most expensive resolution approaches involve emulating the obsolete component using state of the art technologies or redesigning it [8].

17.4 Obsolescence Cost Estimating Framework

A three year research project was undertaken by Cranfield University to address this issue in collaboration with the UK Ministry of Defence, BAE Systems, GE Aviation, Thales Aerospace, Rolls-Royce, Selex Galileo, the Component Obsolescence Group and many other organisations. The framework developed for the NRE cost estimation of EEE components obsolescence issues, named Electronic, Electromechanical and Electrical Framework for Obsolescence Robust Cost Estimation (EEE-FORCE), is mainly based on three elements: the estimate of the number of obsolescence issues during the contract period, the Obsolescence Resolution Profiles and the Obsolescence Cost Metrics as shown in Fig. 17.5. Each of these elements is briefly explained as follows (more details are provided in: [7]).

The estimate of the number (and probability) of obsolescence issues during the contract period is based on the analysis of the Bill of Materials (BoM). The characteristics of each component are taken into account to determine the level of complexity. This will influence the probability of using each resolution approach to solve an obsolescence issue. As shown in Fig. 17.6, EEE components can be classified into three categories based on the complexity level: low, medium and high. The use of obsolescence monitoring tools allow forecasting the end of life (obsolescence date) for each component. This information can be combined with

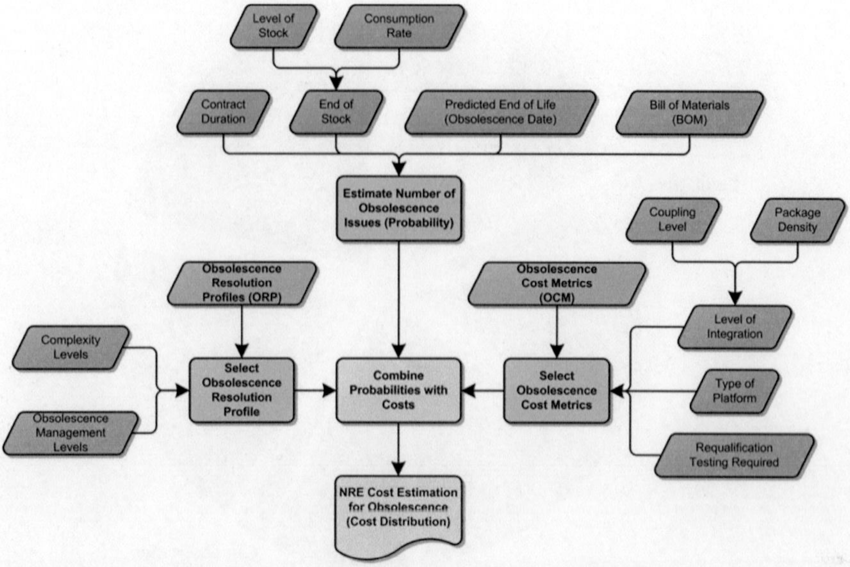

Fig. 17.5 EEE obsolescence cost estimating framework structure [7]

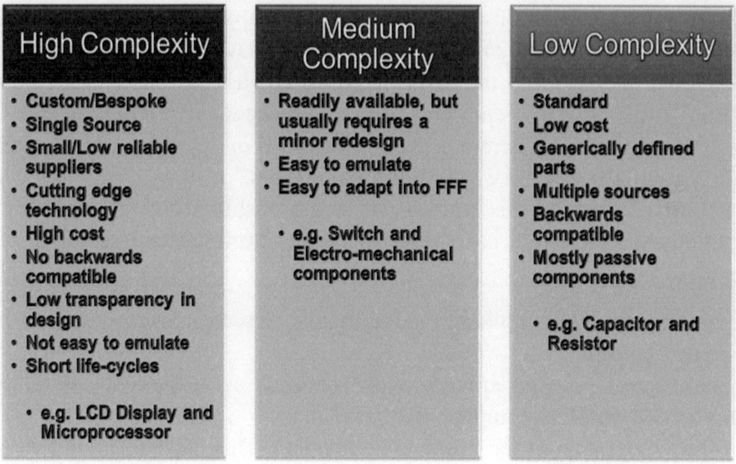

Fig. 17.6 EEE components complexity levels [7]

the level of stock available and consumption rate to assess the likelihood of having an obsolescence issue.

When an obsolescence issue arises, it can be tackled using any of the following resolution approaches as covered in Fig. 17.4. The most suitable approach has to be decided on a case by case basis. The Obsolescence Resolution Profiles represent the

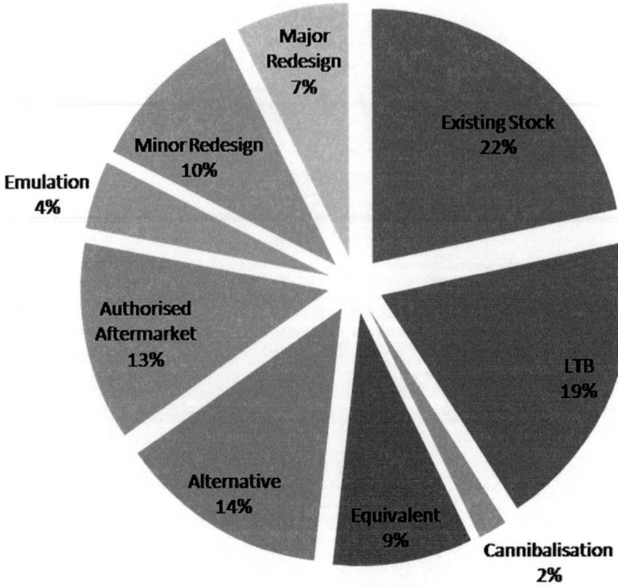

Fig. 17.7 Example of an obsolescence resolution profile

probability of using each obsolescence resolution approach to tackle an obsolescence issue (Fig. 17.7). They mainly depend on the level of component complexity (low, medium or high) and the level of pro-activeness for obsolescence management (from 1 to 5, where 1 represents total reactiveness and 5 represents the highest level of pro-activeness). Therefore, the probabilities of applying each obsolescence resolution approach will vary with these parameters.

The third element of this framework is the obsolescence cost metrics. They define the most likely NRE cost of resolving an obsolescence issue. The key cost drivers identified are [7]:

- Resolution approach applied to solve the obsolescence issue.
- Type of platform.
- Requalification testing required, which depends upon the level of safety/criticality of the obsolete component.
- Level of integration of the obsolete component. This depends upon two factors:
  - The Package Density, which is based on the space available in the product (e.g. Line-replaceable unit (LRU) or assembly) and the level of interaction within the obsolete item.
  - The Coupling Level, which is characterised by the number of interfaces that the obsolete item has with adjacent items (e.g. mechanical, optical, electrical, software or communications protocols) and the characteristics of each interface.

17.5 Discussion and Conclusions

The awareness raised about the problem of obsolescence over recent years has enabled the ongoing shift from reactive obsolescence management to a more proactive approach. The development of mitigation strategies to reduce the probability and potential impact of obsolescence issues is essential to reduce the obsolescence cost over the life cycle of a system. At the same time, it increases the readiness and sustainability of the system.

The contractors and OEMs are considered to be in the best position to manage obsolescence in the most cost-effective way, so that this responsibility has to be transferred to them by the customer. A key challenge for including obsolescence management in support contracts is the difficulty to estimate the non-recurring engineering (NRE) costs of resolving obsolescence issues and to have suitable skills to manage obsolescence.

References

1. Humphrey D, Condra L, Pendse N, Das D, Wilkinson C, Pecht MG (2000) An avionics guide to uprating of electronic parts. IEEE Trans Compon Packag Technol 23(3):595–599
2. Barton DL, Chawla P (2003) Designing systems for future obsolescence. In: Proceedings of digital avionics systems conference, DASC 2003, IEEE, Indianapolis, 12–16 Oct 2003, vol 2, p 10, C.2-101-11
3. Gibbs R (2011) An alternative approach to semiconductor obsolescence management. In: Proceedings of the 7th component obsolescence group international conference: the economics of obsolescence, COG 2011, York, 28–30 June 2011, pp 63–69
4. Josias C, Terpenny JP, McLean KJ (2004) Component obsolescence risk assessment. In: Proceedings of the 2004 industrial engineering research conference (IERC), 15–19 May
5. Pecht MG, Das D (2000) Electronic part life cycle. IEEE Trans Compon Packag Technol 23 (1):190–192. doi:10.1109/6144.833060
6. Solomon R, Sandborn P, Pecht M (2000) Electronic part life cycle concepts and obsolescence forecasting. IEEE Trans Compon Packag Technol Part A 23(4):707–717. doi:10.1109/6144.888857
7. Romero Rojo FJ (2011) Development of a framework for obsolescence resolution cost estimation. PhD thesis, Cranfield University, UK
8. Romero Rojo FJ, Roy R, Shehab E (2010) Obsolescence management for defence and aerospace systems: state of the art and future trends. Int J Adv Manuf Technol 49 (9–12):1235–1250
9. Sandborn P (2007) Designing for technology obsolescence management. In: Proceedings of the 2007 industrial engineering research conference, Nashville, 19–23 May 2007
10. Hitt EF, Schmidt J (1998) Technology obsolescence (TO) impact on future costs. In: Proceedings of digital avionics systems conference, 17th DASC, 31 Oct–7 Nov, vol 1(A33). The AIAA/IEEE/SAE, pp 1–7
11. Torresen J, Lovland TA (2007) Parts obsolescence challenges for the electronics industry. In: Proceedings of design and diagnostics of electronic circuits and systems, DDECS'07. IEEE, 11–13 Apr, pp 1–4

12. Romero Rojo FJ, Roy R, Kelly S (2011) A comparative study between obsolescence cost metrics. In: Proceedings of the 7th component obsolescence group international conference: the economics of obsolescence, COG 2011, York, 28th–30th June 2011, pp 83–92
13. Adams C (2005) Getting a handle on cots obsolescence. Avionics Magazine. Available online http://www.aviationtoday.com/av/categories/military/887.html. Accessed on 13th July 2011
14. Romero Rojo FJ, Roy R, Shehab E, Cheruvu K, Blackman I, Rumney GA (2010) Key challenges in managing software obsolescence for industrial product-service systems (IPS2). In: Proceedings of the 2nd CIRP industrial product-service systems (IPS2) conference, Sweden, 14th–15th Apr 2010
15. Romero Rojo FJ, Roy R, Shehab E, Cheruvu K, Mason P (2011). A Cost estimating framework for electronic, electrical and electromechanical (EEE) components obsolescence within the use-oriented product-service systems contracts. In: Proceedings of the institution of mechanical engineers, part B, journal of engineering manufacture (in Press) doi:10.1177/0954405411406774

Chapter 18
Planning to Extend the Life of Major Assets: Metro Rail Example

Mark Norris

Abstract Purchases of major long-life assets such as rolling stock, planes, wind turbines etc., represent a substantial investment by companies or governments, from which the owner needs to ensure maximum return over the asset's economically viable operational life. A big question facing industry is: "What tools and techniques could be used when planning to introduce a major long life (40+ years) asset, to extend the design life and could this be achieved in a cost effective manner?" This chapter discusses one particular example of how a customer explored the potential to extend the life of their asset purchase by approximately 50 % and the methods they were using to achieve this. Of particular interest is the creation of a decision support tool in the form of a Whole Life Cost Model. What this provided was a method to assess the affect of various factors, such as degradation, inflation, additional load (e.g., 2012 Olympics, extended running hours) to

Detailed in this chapter are the experiences and methodologies used when considering ways to extend the life of a new fleet of trains for London Underground, which included a Whole Life Cost Model (WLCM) to be used as a Decision Support Tool by the Fleet Asset Management team at Metronet BCV and SSL, two of the three companies formed by the Public Private Partnership (PPP) to, "Renew the Tube," a project initiated to bring private enterprise thinking and investment into London Underground. Although the PPP failed, for a variety of reasons, the thinking, tools and techniques introduced carried over to London Underground and continue to have a positive affect on asset management within that organisation. As a result, London Underground became the first train operating company to achieve PAS55 certification for their asset management strategy and regime.

This is, therefore, a case study and not a purely technical discussion. It describes the thought process and background to developing the WLCM, as well as going into the WLCM structure.

I will also describe other tools and techniques, such as Six Sigma, 5S and other elements of Lean Thinking introduced to improve working methods and enhance the availability of the existing and future fleets.

Although commercially sensitive at the time, the performance data can now be shared as it comes from trains that are no longer in service or whose maintenance regimes are unrecognisable from those in place at the time of measurement.

M. Norris (✉)
Asset Reliability Consultants Ltd., 7 Fairdene, Southwick, Brighton BN42 4QN, UK
e-mail: arc.ltd.uk@gmail.com

L. Redding and R. Roy (eds.), *Through-life Engineering Services*,
Decision Engineering, DOI 10.1007/978-3-319-12111-6_18

determine the circumstances under which life extension would be financially viable, or early replacement would be a wiser investment. The reader will also be introduced to Lifecycle Reliability Engineering and how it links to Asset Management through tools such as Reliability Centred and Risk Based Maintenance. How lessons learnt and problem solving processes link to Reliability Growth Plans.

18.1 Introduction

This is a case study taken from my experience working with the Asset Management Team whilst at Metronet on the, "Renewing the Tube," Project. With Metronet in the process of purchasing the first new fleet of London Underground trains for 17 years in 2009, whilst planning to introduce another four fleets from 2012 onwards, the Fleet Asset Managers wished to lay the foundations of a long term investment plan to maximise the return on their investment and insulate that investment from political changes.

The design life of the new fleet would be 40 years, however, existing fleets had already exceeded this design life based on technology and materials developed in the early 1960s. The asset management team at Metronet needed a tool that would help determine the actions required to achieve an extended life.

Significant forward planning was required as the known time interval for extension would be 20 years. This is the life of the bogies. After 20 years, the cast iron develops sufficient fatigue cracking to require replacement, so for an investment to be worthwhile, the fleet would need to achieve 60 years. A 50 % life extension was a challenge but one that had already been met by the Hong Kong metro who's rolling stock had achieved 65 years of effective service and continued to operate efficiently.

With this in mind and a remit to achieve a cost effective life extension, we needed a way to assess the through-life cost of operating the fleet, that could be adjusted according to lessons learnt and changes in circumstances and technology, e.g., the current 67TS (1967 Tube Stock) did not have LED based passenger information systems fitted when it was delivered in 1967. The 2012 Olympic Games and the proposed extension to running hours was not a factor when the 92TS on the Central Line was purchased. Each change would require investigation and assessment to aid decision making on when to refurbish or replace major systems, such as door engines, traction motors, wheels and brakes. This tool would need to take account of a wide variety of inputs from labour rates and inflation to system degradation.

It became clear that in order to encompass the wide variety of inputs a computer based model would need to be developed. After researching market availability, early in 2007, no commercial off the shelf solution would provide the breadth or depth of input required, a bespoke model was necessary in order to assess whole life cost, hence a WLCM was commissioned in Excel. It is worth noting that the bare model was 200 Mb.

There are no startling revelations here, no new technologies just the marriage of different technical and financial ways of thinking to assess ways to achieve a cost effective life extension, or show that such a life extension would only be cost effective under certain circumstances that would enable a decision to be taken after, in the case of rolling stock, 20 years (the life of the bogies) on whether or not the fleet would require replacement, or if a refurbishment would be a worthwhile investment after 40 years to achieve an additional 20 year's service.

18.2 Objectives

I feel it is important to let the reader know what my intensions were in writing this chapter, so I will state them here:

- To provide an insight into how an organisation lays a long term plan that will obtain and maintain Senior Leadership approval in order to maximise the earning potential of the new asset over the longest financially viable life.
- To provide supporting evidence to present to a Senior Leadership team to discuss a long term planning strategy for new or existing major assets.
- To show how Lifecycle Reliability Engineering and Asset Management come together to ensure value for money throughout the asset's life.

This is an opportunity for the reader to consider how this thinking and planning could be applied to your own assets and presented to your senior leadership team to propose a similar approach, citing this as an example of how others have been able to develop tools and methods to aid investment and operational decisions that allow for, and encompass, change.

It is not a script for producing such a tool but does show how much additional effort and thinking is required around the subject and the type of thought process required.

18.3 Lifecycle Reliability Engineering

Lifecycle reliability engineering is closely related to asset management. It looks at all aspects of the asset lifecycle from concept through specification to design, production, commissioning, operating, maintaining, decommissioning and disposal/recycling. In this instance, it looks at both the fleet and infrastructure (depot and equipment) due to the multiple interfaces between the two.

From the beginning of the product lifecycle, it seeks to reduce the overall cost of ownership, supporting higher initial investment costs to yield long term cost reduction benefits.

Where it differs from asset management is the technical delivery. Lifecycle reliability engineering looks at how a design can be made more cost effective by designing for ease of maintenance or replacement. It considers whether or not it would be better to create modular systems that can be removed for service off-line, than allowing easy access in situ, rather than how to minimise the cost of operating what is already in place, or what your selected supplier delivers after interpreting your specifications.

Lifecycle reliability engineering also considers the degree of performance recovery from planned maintenance and how best to plan maintenance interventions through the use of tools such as Reliability Centred and Risk Based Maintenance (RCM and RBM).

Asset management adds an additional layer of investment decisions to ensure technical choices achieve the lowest cost option rather than the best technical or feasibly technical route.

Ideally, both work streams run in parallel from the concept stage, however, not all suppliers have embraced such working techniques and many products were supplied prior to the development of RCM, e.g., the 67TS was in service 11 years prior to the release of Nowlan and Heaps' landmark study of commercial aircraft reliability and maintenance in [1].

18.4 Asset Management Planning

At the time of considering the life extension of the new 09TS, due for delivery to begin in Sep-09 and complete in Sep-11, in 2007, the design, build and testing of the new trains was well advanced, so there was limited opportunity to influence the design.

In 2006 I joined as a RAM Engineer, however, in 2007 the BCV Fleet Asset Manager (Bakerloo, Central, Victoria and Waterloo & City lines) left Metronet, so I was appointed to the role of asset manager. With a reliability engineering background I was able to understand the technical solutions being proposed by fleet reliability engineers. A crash course in corporate finance to understand internal rates of return, nett present values, discount rates etc., provided me with an insight into the investment decisions that were required by a private company necessary to remain profitably operational. This allowed me to calculate the cost of proposed modifications, the performance improvement they would bring and hence whether or not Metronet would make an operating profit on those investments. I then had the unenviable task of telling reliability engineers their technically feasible modifications were unprofitable, so the most cost effective solution was to fix-on-failure.

This proved a successful combination, so Metronet actively recruited another reliability engineer for the position of SSL Fleet Asset Manager (sub surface lines—District, Circle, Metropolitan, Hammersmith & City).

Together we formed an improved asset management plan for all eight Metronet fleets, introducing RCM and RBM as key performance and investment controls, delivering the first annual asset management plan to be approved by the customer, London Underground, in 2008, the ninth year of operation.

At this point we began to work on ways to carry over the new tools and techniques being introduced to the new fleets and ensure maximum life and return on investment could be achieved.

The design life of the new "09TS" Victoria Line trains was 40 years (Metronet contractual requirement), however, the asset management team at Metronet studied other Metro systems and determined it would be possible to extend the life considerably if it were planned early and applied from first introduction.

This is Asset Management Planning, which Metronet took very seriously whilst aware that there were few resources available with the correct skills set, hence reliability engineers became asset managers.

The planning cycle began in 2007 with a view to increasing the supplier's proposed fleet life of 40 to >60 years. To support this, a range of Decision Support Tools (DST) were developed, the key one being a Whole Life Cost Model (WLCM).

18.5 Preparation

In order to determine what would be required to extend the life of a fleet due for delivery 2 years later, it was necessary to understand the current situation with the existing fleets and what had been learnt over nearly 150 years of operating the London Underground.

The initial stage would be to gather data and lessons learnt from the existing fleets. This would be used to determine the range of inputs required for the WLCM and allow the trial of processes new to London Underground, i.e., RCM and RBM.

The WLCM would be trialled on the 92TS of the Central Line as its performance was below that expected of a new fleet, so poor, in fact, that early replacement was being considered after just 15 years in service.

18.6 Data Gathering

The Six Sigma Team were engaged to carry out the lessons learnt study and gather data. Initially in 2007, Lessons Learnt were investigated that could be applied to the new design. The 67TS in use on the Victoria line was 40 years old at this point and clearly life expired, although it should be noted that the Fleet Asset Manager and Fleet Manager worked closely with the Fleet Reliability Engineer from 2007 to 2009 to improve reliability to $\sim 20\ \%$ above the level achieved when new, whilst simultaneously reducing costs.

Failure data was gathered from the eight Metronet fleets in order to identify common failures, which would allow each fleet to learn from improvements made on other fleets as well as provide feedback into the design process.

Although much of the data collected is commercially sensitive, and hence cannot be discussed here, certain findings are generic and recognisable to other industries. One such common issue was the breakdown of wiring insulation as it hardened, becoming impossible to service, as any movement would result in the insulation crumbling away.

Fortunately, most of the fleets concerned have now been removed from operation and the other data is out of date and unrepresentative of current operation, so can be reproduced here. The Six Sigma team gathered failure data and produced two simple pareto analyses.

The first looked at the relationship between system and failure rate, the second looked at the relationship between system and cost of failure, measured in Lost Customer Hours (LCH), which was the performance penalty system used by London Underground to determine payments made to Metronet for operating the fleets.

The LCH abatement was created to account for lost revenue should there be a significant delay (above 2 min) or a train removed from service. Under these circumstances it could be necessary to close a station to new customers whilst clearing those left on the platform either due to being detrained or building up as there was no train for them to board. If people were still allowed to enter the station, there could be an incident where those on the platform were endangered by those trying to access the platform.

Under such circumstances, those customers unable to gain access to the station would have to find alternative ways to continue their journey, which would represent lost revenue to London Underground and reduce the payment to Metronet. A sliding scale was used according to the time of day and location, e.g., Oxford Circus in either rush hour would carry a severe penalty as three lines meet there, so a 2 min delay could take 20 min or more to correct.

The initial analysis showed a correlation between failure rate and LCH, the highest being Doors, followed by Traction then brakes, as can be seen in Figs. 18.1 and 18.2.

18.7 Data Analysis

The case study initially examined the Doors as this system failed most frequently at the highest cost, closely followed by traction and brakes. These three systems accounted for over 40 % of failures by both frequency and cost, refer to Figs. 18.1 and 18.2.

An investigation was initiated into the differences between the door system maintenance regimes on the different fleets to identify, and then roll out, best practice. This investigation showed that the Bakerloo and Victoria line trains had the fewest number of incidents per 28 day period, approximately 30 % of the incidents on the District Line (D Stock), as shown in Fig. 18.3.

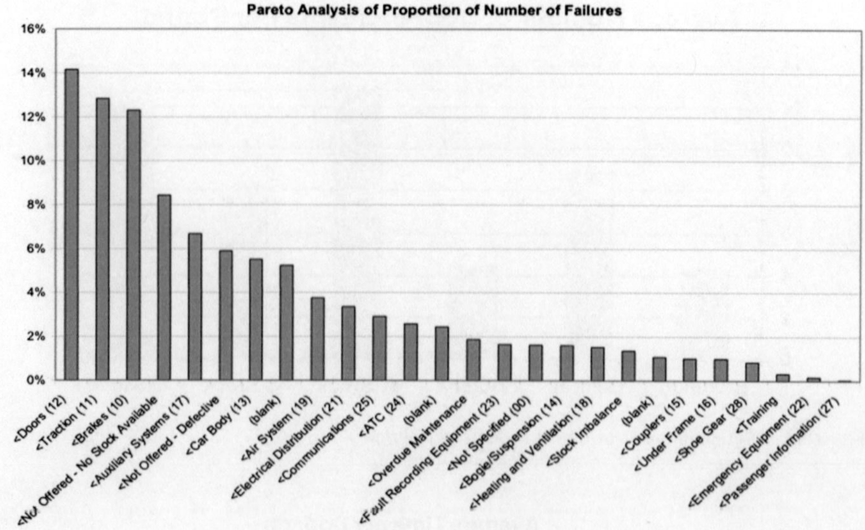

Fig. 18.1 Pareto analysis of proportion of number of failures

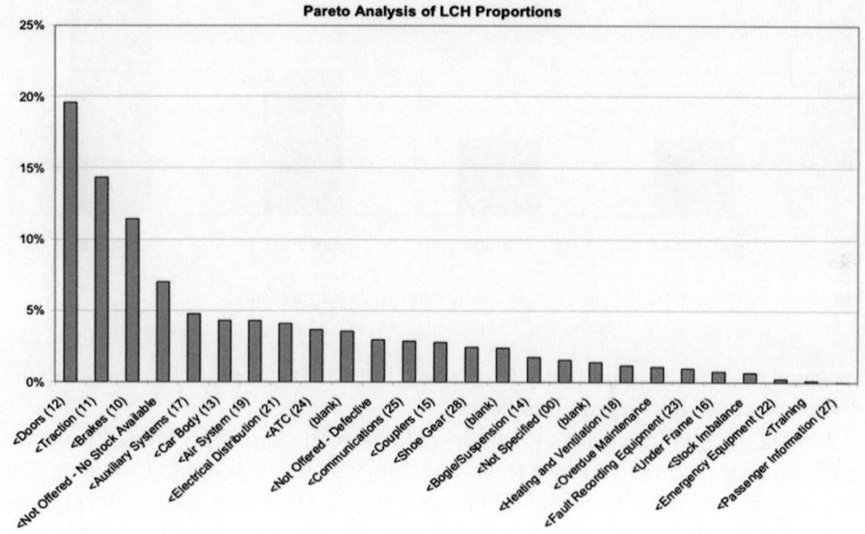

Fig. 18.2 Pareto analysis of LCH proportions

Furthermore, the investigation showed that maintenance effort in service time during the 38 week door service on each fleet varied significantly, with the Victoria line spending 300 % of the time taken by the District line and 150 % of the time taken by the Bakerloo line, see Fig. 18.4. This warranted further investigation as it

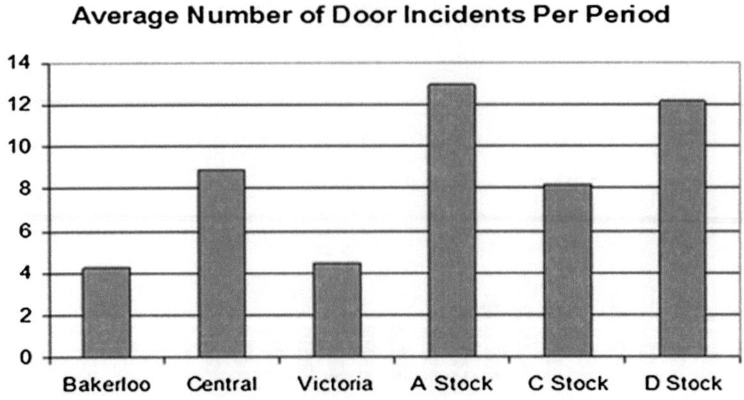

Fig. 18.3 Average number of door incidents per period

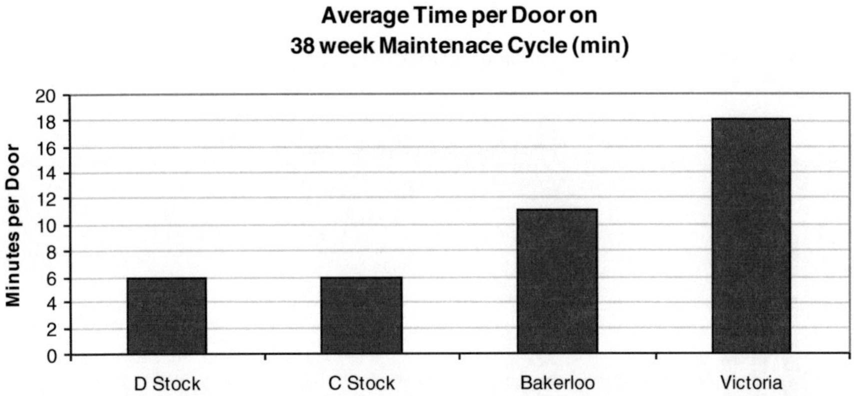

Fig. 18.4 Average time per door on 38 week maintenance cycle (min)

appeared to show an opportunity to reduce effort and hence cost of maintaining the door system on the Victoria line fleet.

I asked the Six Sigma team to continue their investigation by visiting the depots to observe the door maintenance teams and note the process being followed in order to highlight and understand the differences. By observing the measurement techniques, the cause of the difference was found, i.e., the time taken to retrieve parts wasn't included by the Bakerloo or District line, which distorted the figures but highlighted an issue across the fleets, as shown in Fig. 18.4.

Further investigation and discussion about the observations of the Six Sigma team showed the maintenance process to be poorly controlled with each fleet's maintenance technicians behaving differently following their own locally produced process. None of the maintenance teams had a bespoke tool kit or parts kit to facilitate door maintenance, requiring frequent interruptions to the flow of work to locate either tools or parts.

Based on my previous experience in automotive and blue chip manufacturing, I suggested the use of additional Six Sigma techniques associated with Lean Manufacture. One such technique is 5S, which, in very simple terms, is a place for everything and everything in its place. What was clearly required was a set of tools and parts required in order to service all the door systems on a train.

A set of three trolleys were purchased which had shadow boards on top, like a pitched roof, to hold the tools, and trays below with foam cut outs to hold the parts required and troughs at each end where damaged tools could be placed. In this way, it was obvious at a glance when tools were missing or damaged and if parts were missing. By having three trolleys, one would be in use for planned maintenance, one would be available for unplanned (known as casualty) maintenance and one in its designated storage area (a painted box on the floor labelled, “Door Trolley”).

This was successfully implemented for the Door Service Teams, with the store man tasked with replenishing the trolleys in between Door services by checking the contents of the shadow boards and trays in the trolleys in the designated storage area. This reduced the time taken and improved the quality of servicing.

What the Six Sigma Team also reported was the dissatisfaction amongst the maintenance technicians with the rotation system implemented whereby they would change roles every six months. This had been implemented following discussion with the fleet managers in order to keep their jobs fresh and improve the flexibility of the maintenance workforce. Unfortunately, this did not suit everyone. Some preferred to remain within a team and on a given zone or system. Following this feedback, the rotation system was made optional so team members were invited to change every six months.

The combination of better, easier access to tools and parts coupled with stability of their daily routine resulted in reduced maintenance time and improved door performance as the Doors Teams took ownership of the Door System performance and pride in making improvements.

The Six Sigma investigations did not stop at the depot maintenance routines.

In parallel an investigation was carried out into the location of failures on the network, as shown in Fig. 18.5. Some were understandable, e.g., at Oxford Circus where the track is on a bend and a crest with a twist, so the carriages at the rear tilt up and lean into the platform but down and away from the platform at the front. With three lines meeting at the busy station, coupled with the oddly shaped track, issues with successfully closing doors can be understood, particularly those resulting in No Defect Found (no fault found is a more popular terminology but this a 150 year old system with it’s own idiosyncrasies) when checked in the depot, as there could be sufficient distortion of the chassis and body to cause sticking or otherwise uneven movement.

Meanwhile, there were an unusually high number of failures at Seven Sisters in Zone 3, a relatively benign area.

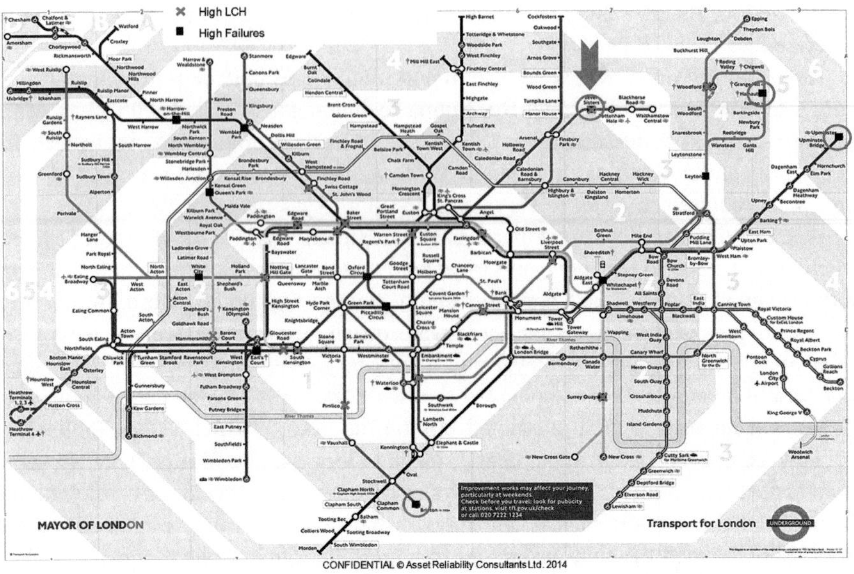

Fig. 18.5 Six sigma doors project

18.8 Root Cause Analysis

The Six Sigma team initiated an investigation into potential causes of door failure, brainstorming with the fleet reliability engineer, depot duty manager, maintenance team leaders and maintenance technicians. For this, they used an Ishikawa or Fishbone, diagram, shown in Fig. 18.6.

I do not favour this approach. Mr Ishikawa developed this technique as a quick and dirty method of analysing issues affecting the world's largest ship yard in Japan. It has much in common with 5 Whys which was banned from use by the company that created it due to poor results and poor repeatability.

Highly skilled practitioners with access to a team having excellent technical and domain knowledge may be able to find the correct root cause this way but the results vary enormously and it is highly unlikely that a similar team would be able to repeat their result.

In this case, it can be seen that the analysis failed to recognise spares availability as a potential issue under Materials.

(My preference is Fault Tree Analysis. Software tools can include probabilities of component failure, which is ideal for the electronics industry, however, the basic logic process naturally leads to a wide range of potential causes that can then be ranked by the likelihood of occurrence according to engineer's judgement.)

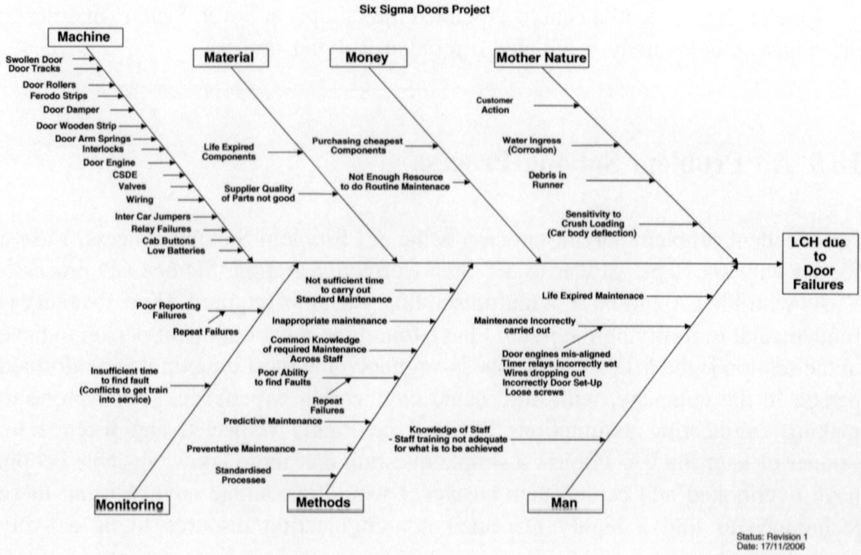

Fig. 18.6 RCA—course and effect diagram

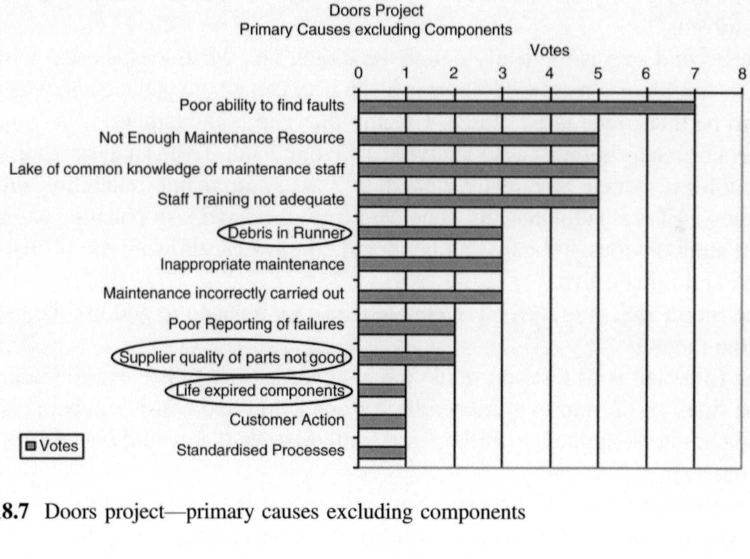

Fig. 18.7 Doors project—primary causes excluding components

Following the creation of the Cause and Effect/Fishbone/Ishikawa diagram, the Fleet's maintenance and Reliability Engineering teams voted on which they thought to be the most likely cause based on experience, the results are shown in Fig. 18.7.

Note: of the 12 primary causes, only 3 are physical faults with the doors, all the others are people or process related, i.e., what is now know as Human Factors.

None of these potential causes explained the failures at Seven Sisters. In order to investigate this anomaly, I initiated a problem solving process.

18.9 A3 Problem Solving Process

An excellent problem solving process is the A3 Problem Solving Process. I use a format with six steps, similar to a 7 Step Corrective Action, 8D or G8D process.

As with FMEA and FTA, a multidisciplinary team is required. Here, diversity is fundamental to resolving the issue. I have found the most important person to have in the session is the MD's PA. He/she is, arguably, the most important and informed person in the company, with little or no engineering experience, so not prone to making engineering assumptions but will be highly respected and listened to. Sooner or later the PA will ask a simple question that more knowledgeable people have overlooked and cannot then answer. I would encourage anyone using these techniques to find a highly respected non-engineering resource to be actively present.

Firstly, define the problem which is an important step. Where I have found teams struggle to subsequently identify a root cause is with a vague, wooly description of the issue they are trying to address, so the first step is to clearly define the problem to be solved.

The second step is to clearly define the target, i.e., what does success look like when the problem is resolved? There is a gap between where you are and where you wish to be that needs to be closed. Closing that gap is the target.

The third step is root cause analysis. I favour Fault Tree Analysis as a robust methodology, which is directly linked to FMEA in major reliability software packages—FTA is automatically generated from the FMEA so potential causes are immediately obvious and gaps can be identified quickly, allowing the FMEA to be revised and remain live.

The fourth step is to develop a countermeasure, planned to address the gap and meet the target.

The fifth step is to test and refine the countermeasure to achieve the target.

The final, sixth, step is to assess the lessons learnt and consider where else this solution could be applied, or if there are similar issues that would benefit from this solution.

One element of the A3 Problem Solving process is Genchi Genbutsu, "Go and see," so a member of the Six Sigma team, the fleet reliability engineer and I, the Asset Manager, went to find out what was unusual about Seven Sisters.

The platform at Seven Sisters could not have been more benign, straight and level without any interconnections to other lines, overground or mainline rail. We then rode the train to the end of the line to review the remaining stations. At the last stop, Walthamstow Central, the drivers were found sitting on a platform bench having their lunch. When I asked why they were having lunch there, they explained that their mess room, with coffee and snack machines, had been removed so that the room could be used for storing rubbish—the result of the distribution of free newspapers, magazines etc., at station entrances. In the morning rush hour it simply took too long to take bags of discarded newspapers out of the station, so a storage are was required. One of the performance measures used by London Underground was, "Ambience," which included cleanliness as well as lighting etc., so these papers not only created a safety hazard and potentially blocked doors, they created a mess which had to be cleared.

The fleet reliability engineer reminded us that Seven Sisters was not only a station, it was the point where the track divided between the Victoria line and its Northumberland depot. He went on to explain that a train suffering a door failure here would be taken out of service immediately and then checked whilst the driver went to the depot staff canteen.

We talked to the distributor of the newspapers and arranged for collection of the bagged newspapers from the platform, so that I could reinstate the mess room at the end of the line. The door failure rate on the Victoria line fell by 30 %. This type of event is one of the drivers behind investigations into the influence of Human Factors on product reliability.

18.10 Reliability Growth Plan (RGP)

Reliability growth planning was introduced in order to better control proposed modifications to the aging Victoria line fleet. Modifications could be overlaid effectively in this way and predicted performance uplift monitored to ensure the expected return on investment was realised.

In order to initiate reliability improvement activities, a simple flow diagram was created. Each new process introduced had the potential to be rolled out to other fleets and support the introduction of the new fleet in autumn 2009.

Two routes for reliability growth were introduced, see Fig. 18.8, one for procedural changes to maintenance, the other for physical changes to the rolling stock, as procedural changes would not incur the significant cost of a physical change required on every car on each train ($8 \times 85 = 680$ changes).

In both cases the process began with investigation, data collection and root cause identification. From this, the need for a modification could be established.

A procedural change would be trialled to see if it were effective and adjusted as necessary after assessing the effectiveness of the trial. If a physical modification was required, the cost for the materials and labour time to implement were assessed in order to determine whether or not the Lost Customer Hours (LCH) would be reduced, increasing earnings from London Underground so that the modification's affect on profitability could be assessed.

If a reduction in failures over time could be shown to be profitable prior to decommissioning, a Business Case was created and presented to the Metronet board in order to gain approval to invest in the improvement.

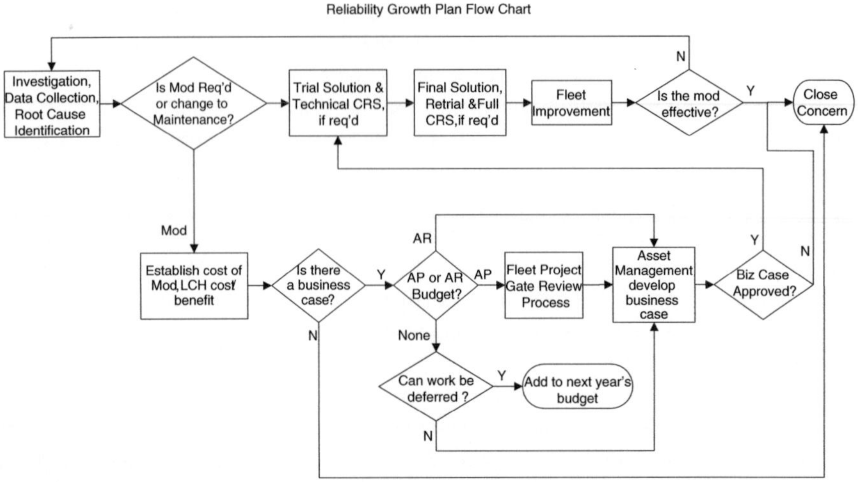

Fig. 18.8 Reliability growth plan (RGP), *AR* asset management regime, *AP* asset performance management, *CRS* change to rolling stock

The proposed modifications were then overlaid on a chart showing the previous 13 periods' data to indicate when programs would begin and end (Fig. 18.9). This also shows the expected improvement, which was reported each week on the senior leadership team's visualisation board, cascaded down to the fleet visualisation board, from the fleet reliability engineer.

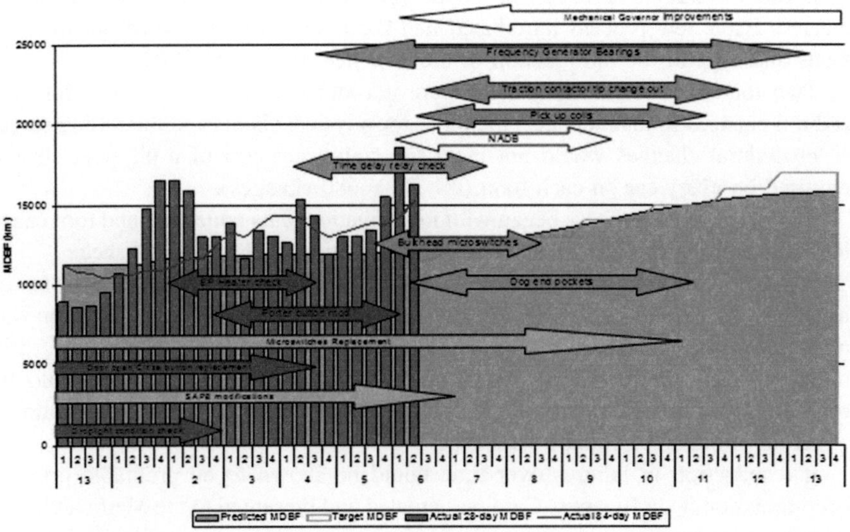

Fig. 18.9 Reliability growth plan—67TS example

It also shows the measured performance and trend in order to ensure the expected benefits were achieved, allowing early review and revision if the performance uplift was below expectations. The predictions were necessarily conservative so were very much a limit to trigger further action.

18.11 Asset Management Planning

A benchmarking exercise was carried out by the Six Sigma Team in order for to Metronet to observe, learn and determine what best practice should be by visiting other organisations involved in fleet maintenance, such as Bombardier and Thales, not just other London Underground depots.

In addition, the Fleet VP and Fleet Managers went on a fact-finding trip to tour Hong Kong Metro who had successfully extended the life of their rolling stock to 65 years.

One of the key drivers behind HK Metro's success was a very stable management structure with the CEO having been in place for over 20 years—almost unheard of in large corporations in Europe. This gave consistency of management philosophy and steady development of processes and procedures within a well executed asset management framework, where Reliability Centred Maintenance was core to cost effective use of resources.

It was this stability that Metronet wished to replicate by implementing a set of decision support tools that would withstand changes in leadership, (political and within Metronet) allowing a 65 year plan to be laid.

18.12 Reliability Centred Maintenance (RCM)

RCM is a common sense approach to maintenance. Rather than blindly follow generic maintenance intervals, the timing is tailored to the application. Even now I still see Bathtub curve referred to as a pattern for reliability, despite the fact we have know since Nowlan and Heap's research on commercial airlines in the 1970s that only 2 % of failures follow this pattern, so here is what really happens.

What this shows is most systems fail at random after an initial period of instability, so, servicing systems tends to reintroduce that period of infant mortality and is to be avoided until it's necessary.

Planning RCM for a new untried fleet, would have provided more questions than answers due to the number of assumptions that would have to be made, so the focus was put into existing fleets. This would improve reliability whilst reducing costs, introducing the tools and processes, preparing to apply them to the new fleets when rolled out from 2009 onwards. This also allowed learning and fine tuning within the boundaries of just one of the eight lines and depot environments under Metronet's control. Two years before the retirement of the existing fleet, it was an excellent

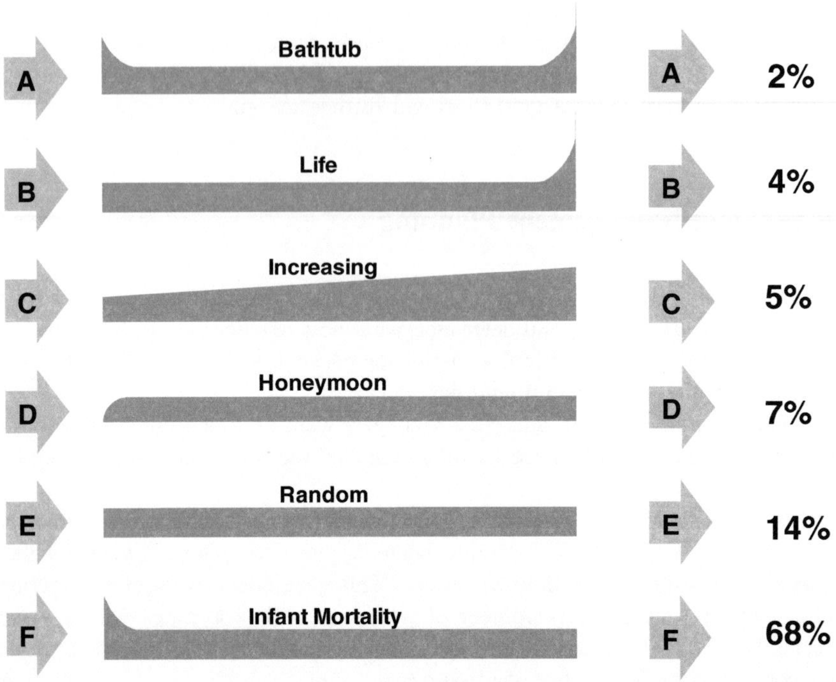

Fig. 18.10 New working methods

opportunity to prepare and low risk as the fleet was representative of the way decades of underinvestment had affected London Underground's ability to deliver the performance necessary to support a 21st century capital city.

Predicting the level of unplanned maintenance provided a significant challenge for the new fleets. The eight existing Metronet fleets were assessed to provide guidance on the probable ratio between planned and unplanned maintenance loading as provisions needed to be made in terms of maintenance equipment, staff, training, tooling, depot roads etc. With further fleets to be delivered and a 10 year breathing space until heavy lifting facilities would be required, there would never be a lower risk opportunity to introduce new working methods (Fig. 18.10).

As the intention was to push out the end of life for the new fleet, the RCM regime was expected to stretch maintenance intervals to ensure such life extension would not only be possible but cost effective. The condition of 8 (out of 85) × 8 car trains at 50, 100 and 150 % of the proposed scheduled maintenance interval was to be investigated, noninvasively without maintenance, to see if such stretch targets were achievable without reducing availability. It also allowed the probable trend (linear, progressive, or regressive) of component degradation to be established, ideally prior to failure. Note that systems behave differently to components. With far higher degrees of freedom the mix of components results in a random failure frequency for the system.

Note: I can retrospectively add that the time for regular maintenance was stretched from 39 to 59 days. This was found to be the limit due to the contamination of grease by the dust in the tunnels, i.e., after this time it became a grinding paste, not a lubricant.

18.12.1 Asset Management Planning, a Reprise

As discussed earlier, following the employment of reliability engineers as Fleet Asset Managers, Metronet began planning the life extension of the new fleets. Metronet required a tool to assist in decision making, with an opportunity to influence the supplier of the new rolling stock. If Metronet could articulate their requirements and the potential impact of missing the targets created, a better design could be achieved and an understanding of the impact of failing to achieve targets.

The model itself would be built in Excel to match business requirements, rather than attempting to adapt generic software to those requirements.

(It is worth noting that subsequent conversations with finance managers have highlighted an issue with COTS software. When being audited, the auditor may ask for visibility of finances at any given point in time. This has resulted in the software supplier having to be called into facilitate the cut required in the midst of complex calculations. Excel models are far more accessible should this become necessary.)

There were many degrees of freedom, interdependencies and variables to build into the model, which will be discussed later. A mathematician and computer scientist began creating the spreadsheet which would have multiple degrees of freedom, creating a large matrix for each year of operation with data manually entered into the year one sheet flowing through calculations to year 65.

This allowed Metronet to build a base model, making adjustments to maintenance intervals, performance degradation and recovery, residual life, refurbishment intervals, replacement intervals, resource levels, material and labour costs etc., and to begin to define unacceptable degradation rates and cost increases that would act as an early warning that the cost effective life target required action to achieve.

18.12.2 Whole Life Cost Model (WLCM) Methodology

Note: What is discussed here is not intended to be a definitive matrix for Metro or any other asset, it merely describes what we felt would help us at the time in 2007/2008, before any of the five new fleets now operating in London had been introduced.

The basis of the WLCM would be a matrix of systems versus performance, with lookup tables that could be used as variables, such as the performance risk degradation curve and the safety risk degradation curve. It would contain cost data as well, including forecasted cost of failure as well as maintenance costs.

A great deal of the data would need to be entered manually for the first year, after which the model would generate subsequent years according to the various multipliers built in.

18.12.3 WLCM Inputs

The initial stage of development was to define the inputs and outputs for the WLCM. The purpose of the model is to compute material, labour, Opex and Capex costs based on forecast; planned maintenance volumes, corrective maintenance volumes, enhancements and renewals, headcount and labour rates.

Here the model parameters to achieve this are reviewed.

The majority were time based:

- Input TB (Time Based):
- Opex costs per intervention, i.e., Unit Labour hours, rates of pay, material costs entered in current prices.
  - Corrective volumes for the base year
  - Planned Volumes (annual volumes by planned activity)
  - Corrective Hours (annual hours by corrective activity)
  - Deterioration Rates for Corrective activities (assuming no intervention)
  - Capex cost per annum by project with a single date for recognising the improvement benefits
- Input NTB (Non-Time Based):
  - Corrective intervention improvement percentages entered manually

The Corrective Intervention Improvement as a percentage are an estimation and one of the many parameters whose impact is assessed by trial and error, experimenting with the level of degradation and recovery that can be tolerated as cost effective.

18.12.4 Asset Definition

The asset needs to be broken down into systems and sub systems. Here, some guidance is provided on how the rolling stock was broken down for the purpose of the model as an example for others to carry out the same exercise on their company's major assets.

- Consider ways to describe the asset and the sub systems.
- Does it make sense to break the system down too far?
- How would these systems be replaced?
  - As a unit? As sub systems?

- Ensure there are sufficient degrees of freedom to create a representative model but not so that it becomes unwieldy.
- By leaving spare capacity (blank lines) in the model it allows for the option of adding subsystems if required—there is no guarantee the same modeller will be available, so better to have spare capacity than risk corrupting the model by inserting additional capacity.

Even though this simplified the systems, the basic unpopulated Excel model was nearly 200 Mb, as mentioned previously.

18.12.5 WLCM Asset Definition Example

Figure 18.11 shows an example of the breakdown of the train systems by zone and items—the zone information has been included to show the complexity beneath each of the sub system titles. Note the additional blanks built into the model, which were intended to allow the model to be used on any and all fleets where future developments could increase complexity. In the case of Metro rolling stock, Doors, Traction and Brake Systems are the highest failure rates. Additionally there is a generic lead item where NDF (No Defect Found) can be attributed. It is also an unavoidable fact that on Metro Rolling Stock some failures are the result of passenger action, not defects but there is a limit to the number of times a driver can attempt to close the doors before he has to assume there is a fault and take the train out of service for investigation (Fig. 18.12).

In Fig. 18.13 is the breakdown of the Braking and Air system as well as the Bogies and Underframe Equipment (non-structural). The brake system is a high use and wear rate system requiring frequent checks and servicing, governed by safe operating regulations.

Note that we have not attempted to break this down to individual components and hence over complicate the model and calculations. This is a good example of

Sub Systems	Zone Reference	Components/Functional Units
1.0 Traction & Propulsion	1199	1.0 Traction & Propulsion OTHER + NDF
	1101	1.1 Motors (Stator, Rotor, Carcus)
	1132, 1133, 1134	1.2 Brushes & Brushgear
	1102, 1103, 1104, 1105, 1106, 1107, 1108, 1112, 1113, 1114, 1115, 1116, 1117, 1118, 1119, 1121	1.3 Electromech (Switchgear)
	1109, 1110, 1111, 1123, 1124, 1125, 1130, 2121	1.4 Electronics (Control, Solid State)
		1.5
		1.6

Fig. 18.11 Whole life-cycle cost model 92TS example

2.0 Doors	1299, 1224	2.0 Doors OTHER + NDF
	1201, 1211, 1212, 1213	2.1 Passenger Doors / Sliding
	1206, 1207,	2.2 Passenger Door / Pneumatic
	1209, 1210, 1214, 1215, 1219, 1220, 1221, 1222, 1223,	2.3 Passenger Door / Electronics
	1208, 1216, 1218	2.4 Passenger Door / Electromech (switches and actuation)
	1201, 1202, 1203, 1204, 1205	2.5 Other Doors / Mechanical
		2.6

Fig. 18.12 Whole life-cycle cost model 92TS example

3.0 Braking & Air System	1099, 1999	3.0 Braking & Air Control OTHER + NDF
	1001, 1011, 1012, 1014, 1016, 1017, 1018, 1019, 1020, 1021, 1022, 1023, 1026, 1027, 1028,	3.1 Braking Control / Pneumatic (Valves, Gauges, Filters)
	1907, 1908, 1909, 1910, 1911, 1912, 1913, 1914, 1915	3.2 Air Control / Pnumatic
	1006, 1007, 1008, 1010, 1013, 1015, 1024, 1025, 1029, 1030, 1031	3.3 Controls (incl. Adhesion, electronic and switching)
	1009	3.4 Brakeblocks
		3.5
		3.6

Fig. 18.13 Whole life-cycle cost model 92TS example

how the complexity of such a system is reduced for the purpose of the model, balanced against the high proportion of Opex cost in this system.

When carrying out this exercise I would encourage you to be pragmatic about the way you breakdown a complex machine, even where the system is safety critical.

These examples are sufficient for the purpose of providing the reader with an insight into the methodology.

Other systems assessed were:

Bogies and Underframe Equipment.

Car Body, Bogie and Underframe Structure—this included the cast iron bogies known to have a life of $\sim$20 years. What we did not attempt to include was vandalism against the car body. We did include graffiti as this was a measure of our ability to apply coatings such as scratch glass film and paint film/anti graffiti coatings, the application and replacement of which represented a significant investment and ongoing cost.

Auxiliary Systems and Electrical Distribution—here we were concerned about wiring due to aging and embrittlement of the insulation on existing fleets, power cables etc. As many industries have found, any disturbance of existing wiring leads to the insulation crumbling away, which is why new cables are laid alongside existing ones so as not to disturb them. The long term performance of new materials is an unknown factor, hence we requested that wiring on the new trains be contained in removable, replaceable cartridges to minimise the cost of replacement.

Ambience—this is a measure of performance used by LUL's, "Mystery Shoppers," who rove the Underground riding trains and scoring them against guidelines for the internal and external condition, so includes graffiti, heating, ventilation, lighting, seat condition, litter etc.

Depot—Maintenance is the driver for Opex spend, which includes the infrastructure necessary to carry out the maintenance, so this was captured here. The depots are complex working environments, so there was much debate over whether or not a separate WLCM should be created for the depot as a system. Bespoke lifting and maintenance equipment is required for the new fleets, however, as no heavy lifting was required for the first few years of operation, the build of these facilities was delayed to spread the cost.

Interfaces—This section captured interactions with other systems that had a trainborne element of cost, e.g., Signalling would include SPADs (Signal Passed At Danger) which had a cost impact on operation (Lost Customer Hours). It was also used to capture costs due to train operator or passenger behaviour as well as items that did not easily fit elsewhere. This is where the costs attributed to the Wheel Rail Interface were captured, i.e., the cost of track maintenance associated with damage caused by operation of the train, such as heavy braking, over enthusiastic acceleration etc. This is an area hotly contested within any operational rail environment.

18.12.6 Maintenance Costs

Next to gain an understanding of the Opex costs required to populate the model, preventative and corrective costs were examined by task (Figs. 18.14 and 18.15 respectively).

It should be noted that although the highest failure rates are Doors, Traction and Brakes, the repairs required to the doors are quick and cheap, whilst the traction, car-body (graffiti) and suspension are far more time consuming and expensive. This underlines the importance of looking at the overall cost and not just the frequency of failure. Likewise, door maintenance is relatively quick and easy but takes up another 5 % showing the level of attention paid during annual door maintenance, interlock checks and bond circuit testing. It is also important to note that the majority of the cost (85 %) is incurred carrying out regulatory safety checks and tests.

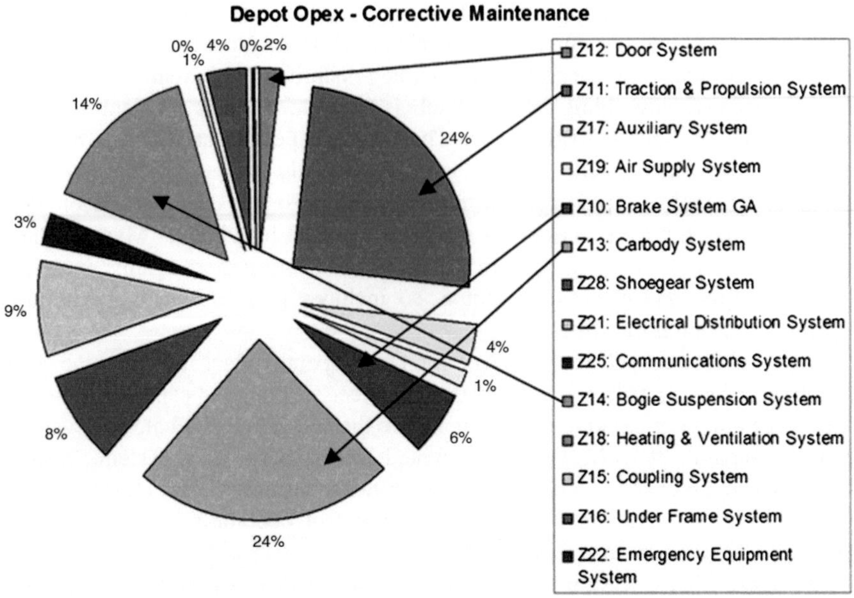

Fig. 18.14 Depot opex—corrective maintenance

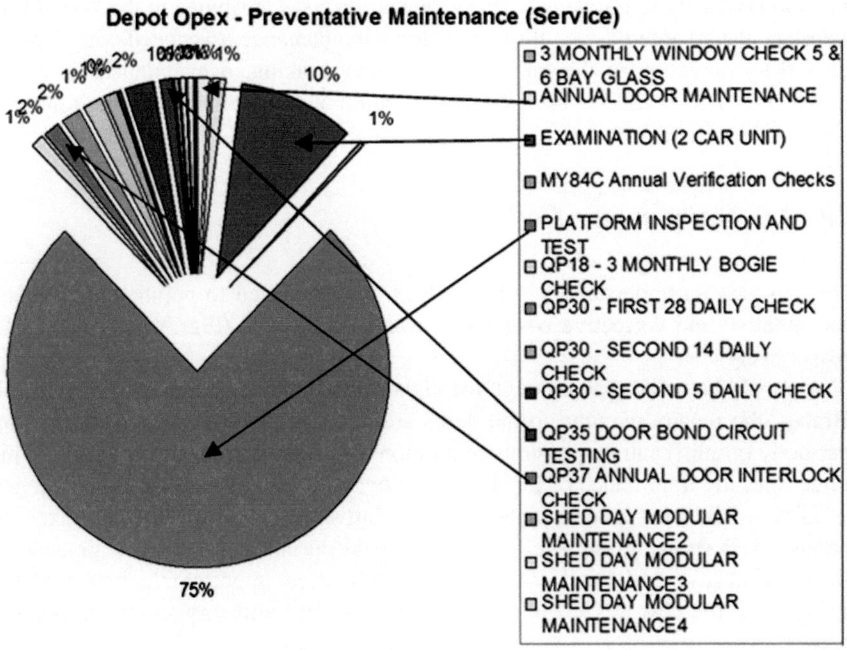

Fig. 18.15 Depot opex—preventative maintenance (Service)

18.12.7 Whole Life Cost Model—Input Data

A sample of the model input data is shown in Fig. 18.16, although this is not a definitive list for Metro or any other Major Asset. I have listed the full breakdown with the rationale behind the inclusion of each.

This is intended to prompt organisations into thinking about what is important for their assets, as supplier or owner. Each organisation must determine what is important for the application of their Major Asset. Some guidelines include:

- Consider the degrees of freedom for the Asset—Systems, Subsystems, components, infrastructure etc.
- Understand the cost drivers—labour, parts, upgrades etc.
- Review maintenance activities, equipment and spare parts
- Take account of unplanned activities/maintenance/repairs
- Create a plan to monitor degradation in order to create/update/modify lookup tables

The lookup tables contain details of our variables, where each variable represents a different level of risk, e.g.,

- Performance—this will grow and improve after initial introduction, then deteriorate in wearing systems e.g. door system contamination, brake wear, traction motor brush wear, with a cost associated with maintenance to recover performance.
- Residual Life—allows the assessment of the impact of refurbishments to extend life. This, Duty and Condition were entered manually between 0.05 and 10 to experiment with the affects.

Components/Functional Units	1.0 Traction & Propulsion OTHER + NDF	1.1 Motors (Stator, Rotor, Carcus)	1.2 Brushes & Brushgear	1.3 Electromech (Switchgear)	1.4 Electronics (Control, Solid State)	1.5	1.6
Performance Risk Degradation Curve	Not Used	Not Used	Not Used	Not Used	Not Used	Not Used	Not Used
Safety Risk Degradation Curve	Not Used	Not Used	Not Used	Not Used	Not Used	Not Used	Not Used
Initial Population at Start of Year	340	340	340	340	340	340	340
Component Duty Multiplier Override (For This Year)	None	None	None	None	None	None	None
Component Duty Multiplier (For This Year)	1.00	1.00	1.00	1.00	1.00	1.00	1.00
Average Residual Life (Years) at end of this Year	-1.00	-1.00	-1.00	-1.00	-1.00	-1.00	-1.00
Average Duty Performed (KM) at end of this Year	1,700,000	1,700,000	1,700,000	1,700,000	1,700,000	1,700,000	1,700,000
Average Cumulative Duty Performed (KM) at end of this Year	1,700,000	1,700,000	1,700,000	1,700,000	1,700,000	1,700,000	1,700,000
Update to Condition Assessment (This Year)	None	None	None	None	None	None	None

Fig. 18.16 Whole life cost model—data input

- Duty performance—allowed the assessment of the impact of increased duty during Olympics.
- Condition assessment performance refers to earnings through Ambience measurement.
- Life Safety Risk—is there an increased risk over time?
- Duty Safety Risk—is the risk higher for increased duty during Olympics?
- Condition Assessment Safety—refers to potential for tripping hazards etc., over time.
- Performance failures per year—assessed from averages for other current fleets, to be updated with data from new trains.
- Cost of failure—assessed from average LCH, current versus future.
- Assess frequency and cost of safety risk failures.
- Assess Opex repair and Capex maintenance costs.
- Enhancement and renewal costs—these occur at irregular intervals—the model accounts for cost and workload peaks.
- Availability—calculated from time required to maintain and impact of longer duration renewals or enhancements.
- Residual Life—between enhancements or renewals, e.g. Bogies will be replaced every 20 years.
- Average Duty—allows for increased duty cycle if enhanced signalling (DTG) equipment is installed.
- Condition—allows for cosmetic enhancements.
- Population—allows for an increase in number of trains if DTG and enhanced duty cycle indicates too high a wear rate, or reduction during major refurbishment/enhancement.

In addition to the above the Interventions were defined for each two carriage unit:

> Ambience, Inspection, Testing, Overhaul, Refurbishment, Component Change, Fault, Repair, Technical, Testing, Warranty, Audit, Capital Works, External Party, Modification/Renewal, Safety, Specialist Cleaning.

The time taken, equipment, resources, material costs etc., were also included in this setup process for Preventative and Corrective maintenance as well as Renewals or Enhancements—as mentioned previously, the LED passenger information displays were not a feature in 1967 and we could foresee enhancements such as WiFi and mobile phone aerials becoming features of the train in future.

18.12.8 Whole Life Cost Model—Output Data

The output runs for 65 years. The model's complexity is still only adequate for a modest number of reports, which combine into a Cumulative Net Cost of Operation, see Fig. 18.17. Note the additional capacity built into allow for additional cost not anticipated at the time of building the model.

Model Year	1	2	3	4
Year	2008	2009	2010	2011
Overheads (Cost £)	-	-	-	-
Preventative (Planned) Maintenance (Cost £)	0	0	0	0
Corrective (Unplanned) Maintenance (Cost £)	0	0	0	0
Maintenance Other (Cost £)	-	-	-	-
Initiatives and Changes (Cost £)	-	-	-	-
One-off Renewal Costs (not condition) (Cost £)	-	-	-	-
Enhacnements and Renewals (Cost £)	0	0	0	0
Total Cost of Maintenance & Renewal (Cost £)	0	0	0	0
LCH Abatements (Cost £)	0	0	0	0
Service Points (Cost £k)	0	0	0	0
ISC JTC Uplift (Benefit £k)	0	0	0	0
Total Performance Impact	0	0	0	0
Safety Risk (Cost £)	0	0	0	0
Capital (Cost £)	-	-	-	-
Depreciation (Cost £)	0	0	0	0
Total Hours required for Maintenance (Hours)	0	0	0	0
Availability Budget (Hours)	1,000,000	1,000,000	1,000,000	1,000,000
Availability (Cost £)	0	0	0	0
Total Other Costs (£)	0	0	0	0
	-	-	-	-
	-	-	-	-
	-	-	-	-
	-	-	-	-
	-	-	-	-
	0	0	0	0
	0	0	0	0
Net Cost (£)	0	0	0	0
Cumulative Net Cost of Operation (£)	0	0	0	0

Fig. 18.17 Whole life cost model reports

Each iteration of the WLCM provides additional guidance from the user and repeated application with different parameters to understand the different influences rates of degradation and improvements bring.

18.13 Conclusions

The case study has shown an example of Lifecycle Reliability Engineering and Asset Management Planning using a Whole Life Cost Model. This has provided an insight into how an organisation can lay a long term plan that will obtain and maintain Senior Management approval.

Lessons Learnt from existing assets such as Door Maintenance were shown to highlight the usefulness of this methodology. Furthermore the use of the Early Intervention Planning method was shown to assist in the prediction of degeneration and help revise the WLCM. Here the WLCM was used to determine whether or not to purchase a new fleet vs. a mid-life refurbishment.

Finally both RCM and RGP were used as a Continuous Improvement aids for the new fleet based on current knowledge, developed as the new fleet entered service. The methodology also highlights the importance of looking at the overall cost of maintenance, not just the frequency of failure.

Reference

1. Nowlan FS, Heap HF (1978) Reliability centred maintenance. Report number AD-A066579

Chapter 19
Identification of Risks Related to Integrated Product Service Offerings of Rail Infrastructure: A Swedish Case

Sofia Lingegård and Mattias Lindahl

Abstract Risk-averse actors, in combination with a lack of trust between the actors, hinder development. An increased level of trust between them could make the process more effective. Allocation of risk, lack of information, renegotiations and supply chain management are all aspects that affect risk. Most of the risks and uncertainties identified are due to lack of experience; other uncertainties are inherent in long-term contracts. Obstacles for change also lie more in the culture and attitude of the actors than in technical complications. The current contracting forms for rail infrastructure in Sweden have low incentives for development at the same time as large amounts of materials are used, resulting in high costs and significant environmental impacts. The concept of the Integrated Product Service Offering (IPSO) has the potential to increase cost efficiency and quality from a life cycle perspective by providing incentives to optimize the use of energy and material. This chapter aims to identify if IPSO contracts could improve the management of rail infrastructure, and to identify potential risk aspects for both the provider and buyer. The empirical data is based on interviews with actors in the industry.

19.1 Introduction

The predominant part of the Swedish rail infrastructure is controlled by one actor, the Swedish Transport Administration, and the number of rail infrastructure suppliers is few. From the government, there is an increased focus on getting more goods and passenger transports transferred from road to train traffic 9 [1]. At the

S. Lingegård (✉) · M. Lindahl
Department of Management and Engineering, Linköping University,
581 83, Linköping, Sweden
e-mail: sofia.lingegard@liu.se

M. Lindahl
e-mail: mattias.lindahl@liu.se

L. Redding and R. Roy (eds.), *Through-life Engineering Services*,
Decision Engineering, DOI 10.1007/978-3-319-12111-6_19

same time, the rail infrastructure in Sweden is suffering from many years of neglected maintenance, which has damaged the reputation of the railway and the train traffic [1]. Reinvestments are not realized to the extent needed and old infrastructure, in combination with an increased traffic volume, is a major problem [2]. Passenger transportation and freight traffic use the same tracks, making the infrastructure heavily-used, and in some areas even overburdened. It has also been shown that the corrective maintenance is increasing, and not the preventive [3]. In the UK, the actor in charge of the rail infrastructure has weak incentives to make efficiency improvements due to the monopoly situation, with no competition in combination with debt guaranteed by the government [4]. This resembles the situation in Sweden, where the STA holds a monopoly position for infrastructure management.

Among construction companies in Sweden, incentives for development, increased efficiency and raised competence are low [5]. Construction contracts are currently used to a large extent, and have shortcomings concerning weak incentives for development of procedures [6]. The over-detailed specifications cause major obstacles for the rail infrastructure industry in terms of technical development [7]. Benefits with the traditional contracts are that all actors can relate to and calculate the tender, and the buyer knows what will be realized since everything is specified in detail.

A buyer who wants to reduce variation in possible outcomes and is willing to sacrify expected returns to do so is risk averse in its decision making [8]. The construction industry is mostly risk averse [9–11]. Risk averse decision makers have a tendency to overestimate possible losses [12]. This could be a reason for contracts specified in detail and unwillingness to reform the industry. Construction risks are in general activities that affect the cost, time and quality of a project, and formal risk analysis and management are not used frequently due to lack of knowledge [9].

The question is how to change the contracting in a conservative industry to deal with the problems stated above regarding availability and accumulated maintenance. An additional matter to consider is the fact that large amounts of materials are used when building and maintaining rail infrastructure, resulting in significant environmental impacts [13].

Performance contracting has been highlighted as a potential way to increase the drivers for change within an industry. Using a product-service mix with more durable materials and other designs may prolong the lifetime of the product, and potentially optimize maintenance and operations [14]. This type of contracting is also known as Product Service Systems, or PSS, and implies that one actor has the responsibility to deliver a result, and therefore has incentives to optimize the use of energy and material [15, 16]. In this chapter, the term Integrated Product Service Offering, IPSO, will be used.

An IPSO gives the provider the opportunity to increase the value of the solution for the customer by integrating components in new ways, as well to provide incentives for the supplier to realize better economic and environmental development when considering the entire life cycle. Previous research for road infrastructure projects has concluded that integration of the different stages in the life cycle seems to be a logical step towards sustainable performance [17]. This is why a similar development seems feasible for rail infrastructure.

In line with the motivation above, the objective of this article is to identify if IPSO contracts could improve the management of rail infrastructure and determine potential risk aspects for both the provider and buyer perspectives.

This chapter first provides a background to the IPSO concept in Sect. 19.2, followed by a presentation of the methodology used for this research project in Sect. 19.3. Section 19.4 presents the practice for rail infrastructure procurement in Sweden, followed by Sect. 19.5, where IPSOs for rail infrastructure are introduced. Thereafter, potential risk aspects are identified in Sect. 19.6, while there is a discussion of contractual IPSO contracts in Sect. 19.7. This is followed by a discussion of organizational aspects of IPSO contracts in Sect. 19.8. The chapter ends with conclusions and suggestions for further research needs in Sect. 19.9.

19.2 Integrated Product Service Offerings

An IPSO has a life cycle perspective, and the combination of products and services can be combined into an optimized solution for the customer; it can also give the manufacturing company the possibility to have control over the product throughout its entire life cycle [18, 19]. This provides more degrees of freedom for the provider to design a system and the way it will be delivered, resulting in the provider trying to achieve the most cost-efficient way to do so [20]. The provider needs to be competitive, something which requires a minimum use of resources for a maximum utilization of the element in the offering [21].

The importance of making decisions early in the product development process—when there is still freedom to make changes—has been stressed in earlier research [22]. This is also true for the infrastructure industry, where earlier involvement of the provider in the planning process creates better opportunities to adapt the content and realization of the project to its specific conditions and requirements [23]. The further along in the process the more modifications cost, due to the difficulty in making the changes. More money spent on the construction, and thereby improved quality, could result in reduced cost for maintenance work. On the other hand, too high a cost for construction can never be motivated by future savings for the maintenance cost [23]. Additionally, the provider has the possibility to gain knowledge during the use of the offering to reconfigure or redesign it [21]. New business models, such as IPSO contracts, create challenges like uncertainty concerning forecasting costs at the bidding phase of the contract [24]. With a business model focusing on delivering a result, a lot of the risk previously carried by the user is now assumed by the provider, and it can be difficult rededicating and controlling the risks and uncertainties [16]. In this chapter, uncertainty is defined as "*any deviation from the unachievable ideal of completely deterministic knowledge of the relevant system*" [25], meaning the deviation from the ideal outcome and the actual outcome. Uncertainty implies an inability to plan, but can also be seen in terms of trade-offs between risk and reward [26]. Risk assessments, including forecasting and economic development, are very important for these long-term contracts, and both the supplier side and the buyer side have to be considered [27]. For long-term performance contracts, risks caused by uncertainties arise in the bidding stage [24]. Key uncertainties for an IPSO contract are performance, operation, training, engineering, affordability and commercial uncertainties [24].

The IPSO contracts discussed for railway infrastructure in this chapter include part of the design phase, as well as building and maintaining the infrastructure for a longer period of time. The contract includes all the parts of the infrastructure and is e.g. not only limited to the tracks.

19.3 Methodology

The respondents were asked to consider the design, building and maintenance phases of the rail infrastructure, which are all included in IPSO contracts for a rail infrastructure project. The concept of triangulation has been used throughout the research, using several different sources of information [28]. Individual interviews were first performed, followed by a group interview to gain deeper knowledge. Finally, the information collected was validated using a questionnaire.

19.3.1 Individual Interviews

The initial steps of this research project were an introductory interview with a well-informed employee at the STA, followed by a literature study. Subsequently, more in-depth interviews with actors in the industry were performed, where a total of seven respondents from the STA and seven respondents from the contractors participated.

The choice of respondents was made to get an overall picture of the industry, and to gain knowledge of both the buyer and the providers' perspectives and their interaction. The criteria for the respondent selection at the STA were to include representatives from both the Investment Division and the Traffic Division of the organization, as well as to focus on those holding positions at a managerial level. This was a conscious choice, since an overview of the organization and an understanding of the strategy and market was preferred to contribute to the research. The respondents themselves suggested others as potential respondents during the course of the interview study. A similar approach was used for the respondents at the contracting companies. The respondents at the STA provided contact information to their contacts within the contractors' organization. Almost all of the respondents from the contractors' organizations worked in the marketing or business divisions of the companies.

19.3.2 Group Interview

To gain more knowledge and deeper understanding on the topic, a group interview was conducted [29]. Those chosen represented important areas within the STA: the Business Developer for maintenance contracts and the Procurement Manager for maintenance contracts from the Traffic Division, and a manager from the Investment Division. The three respondents had knowledge on the topic and they had shown interest in the topic during the interviews, and were outspoken with their beliefs.

19.4 Rail Infrastructure Procurement in Sweden

In this section the rail infrastructure procurement in Sweden is presented to provide a background to the current praxis. The STA is responsible for 80 % of the total rail system in Sweden [30], and since [29] the contracts have been procured through competition [31]. This has resulted in a cost reduction, but nevertheless costs are still increasing. The other two main actor groups in the industry are the contractors that realize the construction and maintenance work, and the design consultants that assist both the STA and the contractors in planning and designing the infrastructure.

The most common contracts within the infrastructure construction industry in Sweden are Design-Bid-Build contracts, where the procurer specifies what to build, how to build it and how much of each element is required [32, 33]. The contractor is obliged to realize the project within the set time, price and standard level based on detailed specification realized by consultants on behalf of the STA [34]. The tender is in most cases chosen based on the lowest price [35].

The maintenance contracts used in Sweden since 2005 are based on functional requirements. These requirements are set on a detailed level, far from an overall function. An examples of a functional requirement is [36]. "*The snow depth at the railway yards (…) is not to exceed 200 mm over the top edge of the sleepers.*" The length of the performance contracts is 5 years, with an additional 2-year option.

A newer type of contracting for building rail infrastructure provides an opportunity for the contractor to influence the construction. The contracting form is called a Design-Build contract, and both the more detailed design phase as well as the construction phase are the responsibility of the contractor [23]. This provides an opportunity for the contractor to influence the construction. The overall design has already been determined by the STA with the help of design consultants, and performance requirements are set for the technical standards.

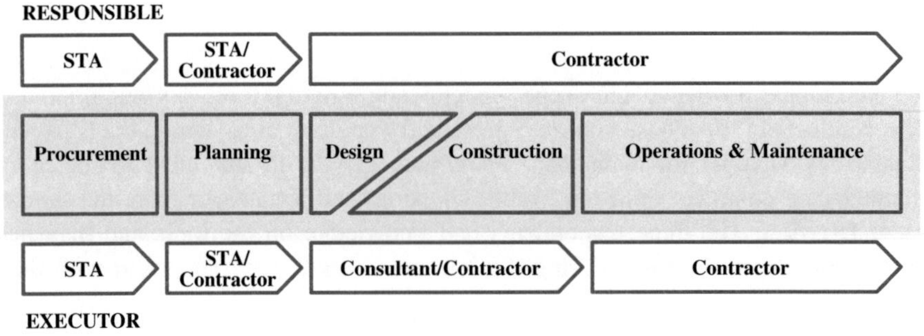

Fig. 19.1 Schematic figure illustrating IPSO contract for rail infrastructure [37]

19.5 IPSOs for Rail Infrastructure

An IPSO contract could be described as a Design-Build contract with a long-term maintenance commitment, where both the design concerning the construction and the maintenance were taken into account and integrated into the initial design phase. The functional requirements are on a higher level than in the Design-Build and performance contracts described in Sect. 19.4. An IPSO contract would include design and construction as well as operations and maintenance, as illustrated in Fig. 19.1. The overall functionality could be described as "build a railway from A to B with capacity C and maintain it for X years. After the contract period the railway should have Y required capacity." The initial planning and environmental impact assessments that could be crucial factors in the approval and realization of the project would still be performed by the STA, according to the respondents. This is not a risk the contractors are willing to take. The design of the construction and the maintenance, however, is the responsibility of the contractor.

19.6 Identified Risk Aspects

In this section, the risks have been divided into three main groups to provide a more comprehensive view of the content.

19.6.1 Size, Duration and Quality of the IPSO Contracts

The IPSO contract is not suitable for all types of projects. The respondents almost unanimously stated that it would be best if the new project was large enough that the contractors could invest in a maintenance organization for the specific contract, and for it to be administratively feasible. Every project requires administration e.g. related to the regulations in the industry, and if the project were not large enough the cost for the administration compared to the project volume would not be justified. The current legal framework is so resource-demanding that, according to the contractors, a large overhead cost is required to match it. If the contract involves a reinvestment, both the STA and the contractors concluded that there would be too many unknowns concerning the status of the construction. It would not be impossible to realize such a project, but it would probably involve a high degree of uncertainty for the contractor and thereby be too expensive for the STA. The contractors are of the opinion that the information e.g. concerning the status of the facilities lacks sufficient detail for using IPSOs for reinvestments.

The respondents did not have a common opinion about the time span of the contracts; the interval of 10–45 years was mentioned. The contractors acknowledge that the longer the contract, the longer payback on machines, etc. On the other hand, a longer contract implies higher uncertainty and risk. Some of the respondents mentioned that long-term contracts could potentially make the involved actors lose commitment and become trapped in work procedures, which would not contribute to development. Renegotiations were mentioned as a way to reduce the risk of these long contracts. The contractors found it unrealistic to believe that there would not be any changes affecting the conditions of the contracts during the time span of the contract; changes in the regulations set by the STA were mentioned as an example.

Another issue related to the long-term contracts is the evaluation of the quality, both during the contract and after it. Respondents from the STA believe this will be difficult, especially since it will take many years before the entire contract can be evaluated. Both sides emphasize the importance of finding appropriate values to measure, and that this could be complicated since many different factors will have to be considered, and that the starting level has to be established as well regarding how to measure the deviation.

19.6.2 Risk Allocation in Contracts

Even though IPSO contracts give the provider more freedom to design the construction and to balance the costs between the construction and maintenance phases, there are still several issues that need to be regulated.

In general, opinion is that the part that has the best ability to calculate the risk should carry it. The start of the contract would be when the design phase starts, and the process before, including e.g. redemption of land and environmental issues, would be the responsibility of the STA. The contractors emphasize that the risk has to be possible to calculate. Since the contractors are not used to putting a price on risk, depending on the method chosen, the price will be set too low or too high. The problem is that they do not have any references; the outcome is unknown, since they have not used this type of contracting before. The worst-case scenario for the contractors is if the calculation does not match reality and they lose money during the duration of the contract. The STA thinks that this could imply higher prices in the tenders, since the contractors might want to protect themselves from this risk. However, the contracts would not only imply a higher risk; they would also generate opportunities, as one contractor describes:

> If a contractor obtains a contract, this implies full responsibility including higher risk and larger opportunities. As long as these two elements balance each other, this type of contract will not cause any problem. *Market Coordinator, Contractor.*

The contractors were in general positive to more transparency concerning risks and costs if that provided better risk-sharing:

> Transparency is a good way to provide a common picture of what the activities cost since this is an area where the procurer many times is greatly uninformed. *Business Area Manager, Contractor.*

Another contractor states that if all the risk is shuffled over to the contractor, no contractor would be interested in a long-term contract where the risk is difficult to estimate. A third contractor talks about the legal framework already existing for regulating risk that could be continuously used. This respondent, however, emphasizes that an IPSO contract would increase the risk, and to some extent this will be expensive, but that the price in total would be lower for the procurer. Furthermore, the price would be controlled by the competition.

When discussing pricing of these contracts, the respondents seem to prefer one fixed and one adjustable part. The contractors in particular prefer an adjustable part as a way to decrease risk-taking, e.g. if and when traffic volumes change. STA, on the other hand, is of the opinion that if a contractor has the overall responsibility for the project, the price should be as fixed as possible.

19.6.3 Supply Chain Management in IPSO Contracts

An issue that was raised by the STA was the survival of the national contractors and their subcontractors as a potential risk for the long-term contracts. In the bidding process, the STA has to take under consideration the long-term survival of the contractor as well. A respondent from one of the major general contractors did not think that the number of contractors would decrease, but did believe the constellation would be different. The smaller contractors will not be participating in the first bidding step, but will take part as subcontractors. The opinion of the general contractor is that they are best suited to take on the IPSO responsibility since they have more financial power and project management competence, while the railway-specific contractor believes they are fitting since they have the technical competence.

19.7 Discussion of Contractual Aspects of IPSO Contracts

19.7.1 Allocation of Risk in IPSO Contracts

There is an unspoken belief from the contractors that the STA would try to shuffle all the risk to the contractors' side of the table. An IPSO contract would entail much more risk for the contractor, an uncertainty that seems to frighten them. The contractors need to acquire new skills for understanding long-term risk as well as being able to identify, evaluate and manage risk, according to previous research [18]. The risk carried by the contractor has to be reasonable, since this limits the possibility for a gambler to send in the lowest tender, and instead the experienced and well-informed contractor wins [38].

Different actors all have different tolerance to risk, and the willingness for an actor to take on a risk is related to the return period of the activity and the size of the risk [12]. Several factors affect the actor's willingness to bear risks, e.g. general attitude toward risk, ability to manage risk and ability bear consequences of risk [39]. Furthermore, Ward states that any additional payment the contractor can charge to bear the risk will influence its willingness to bear the risk [39]. This is why the pricing structure is another precaution for risk [40].

The buyer and the provider in this case have different positions concerning the pricing of the contracts. The buyer prefers a fixed price, probably because this would prevent any unwelcome extra costs during the contract, while the provider prefers a more variable price to reduce the risk. Previous research has concluded that the risk has to be identified, and that the contractor should be supplied with all available information before the contract to have an opportunity to price the risk [39]. In this case, there is not enough information concerning the current state of the infrastructure for the contractors to evaluate the risk in a good way. Considering the limited margins in the industry, a contractor cannot be expected to price risks that cannot be quantified in a proper way [39]. Barnes and Ayers [38] suggest that as a principle, all risks that rise externally should be carried by the buyer, since contractors would charge too much to bear them due to their risk-averse attitude. In this case, such risks could be changes of traffic volume or risks concerning ground conditions that cannot be properly investigated. On the other hand, some of the risks associated with cost and time should be carried by the contractor, since risk and incentives go together, and because the actor who bears the risks has reasons to minimize the impact [38].

If the contracts would include pay per time, material and labor the incentives for finding new solutions would not exist, since the provider would focus more on delivering activities. On the other hand, with a fixed price the contractor would improve efficiency and effectiveness since it would be a self-motivating situation to do so [41]. According to the interviewed contractor, no contractor would ever take on the task of delivering a long-term IPSO contract with a fixed price and no room for adjustments. Therefore, an alternative would be preferable: contracts primarily regulated with a fixed price, but with flexibility for smaller renegotiations and adjustments for factors that the contractor could not affect, such as traffic volume and changes in regulations. This means that the price would be a function of different parameters such as traffic volume, weather etc.

19.7.2 The Need for Renegotiations

As mentioned above, renegotiations would be preferable. Renegotiations are held for similar contracts for the UK defense industry, where the price is reviewed every fifth year [42]. According to the contractors, increased responsibility equals increased risk. This is interesting, since the IPSO literature states that an IPSO offering reduces unpredictability and variability of demand during the contract time, which makes risk reduction a driver for the business model [43, 44]. For the contractors a longer payback time and more freedom, versus a higher degree of risk for the supplier, are the main issues.

By adding flexibility to the contract using soft elements such as renegotiation, uncertainties may be managed [40] this was discussed during the interviews as a possibility. A renegotiation could also be a way for the provider to improve the quality of the service [41]. Renegotiations could then reduce some of the uncertainty, such as changes in traffic volume, which would have a great effect on the wear and degradation of the tracks, but it is a factor that the contractor cannot influence. At the same time, renegotiations could be a way of minimizing the risk of the contractor losing commitment and getting trapped in work procedures, as described by some of the respondents.

Innovation and constant improvement can sometimes be dampened by long-term contracts, since the provider is protected from competition for a longer period of time [41]. On the other hand, the majority of the innovations would probably be created in the design phase, but technical development could be used to improve the maintenance procedures and methods during the duration of the contract. Since the contractors are likely to be paid a relatively fixed price, it is in their interest to prolong the lifetime of the product, and if possible optimize maintenance and operations (see e.g. [14]). This means that long-term contracts could actually increase the innovation and development, as long as the contractors have incentives to do so.

The renegotiations could be seen as a fresh start within the contracts to maintain quality. This would of course all have to be done within the regulations of public procurement; it is not supposed to be a new public procurement process, but rather seen as a degree of flexibility within the contract.

19.7.3 Lack of Information

Currently, there are no incentives for information transfer along the life cycle of the infrastructure, since the life cycle is broken down into several different contracts. Different contractors execute different contracts, and there is no continuity. The STA holds information concerning the function of the rail infrastructure that would be very useful for the contractors when calculating offers. Information concerning the infrastructure also has to be shared throughout the length of the contract within the contractor organization. This information, gained through knowledge and experience, could potentially be used to reconfigure the design of the offering during the contract [21]. Furthermore, the contractor has to be able to show the value of the facility at the end of the contract as well as have measures for evaluation during the contract; this would provide incentives for documentation. These types of incentives are lacking in the form of contract used today, where there is no information transfer between e.g. the maintenance contracts. This is a probable cause to why the actors do not believe reinvestment is appropriate for IPSO contracts, since there are too many uncertainties concerning the condition of the facility. The contractors' reluctance to use IPSOs for reinvestments is confirmed in the literature, where it is stated that lack of historical data causes unpredictability [45]. Lack of information about e.g. ground conditions, in combination with lack of time to prepare the tender, affect the contractors' ability to evaluate the risks [39]. Information gathering has been shown to be a reducing factor for risk, but it is important to not only search for information, but also to focus on how to use it to reduce risk [46]. The trust in a relationship can be facilitated by information sharing, and access to resources can reduce the risk of unpredictable costs [45].

19.7.4 Evaluation During and After the Contract

Other precautions for risk are good performance indicators [40]. Both parties have indicated that the evaluation, both during the contract and afterwards, will be difficult. Operationalization of the functional result of the contracts needs extra attention; one important part of the contracts is to specify precise parameters so that it can be determined whether or not the IPSO is satisfactorily delivered [27, 16]. The actors are new to this kind of thinking, and it is therefore expected they will require a long learning curve. This is the reason flexibility in the contract is so useful, since it would provide opportunities to correct non-functioning conditions naturally. Due to the length of the contract, it will take time before the entire project can be evaluated, and a residual value has to be determined so that the two parties can work towards the same goal. The majority of these considerations for uncertainties have to be dealt with as early as in the design phase [11], and major challenges are assumptions concerning equipment failure, prediction of maintenance routines and communication problems with the customer [47]. This indicates how important information sharing and transparency will be for the IPSO contracts. Managing uncertainties for the entire life cycle at the bidding stage is challenging; the major inputs to calculate the cost are e.g. historical data, supplier inputs and user requirements [21].

19.8 Discussion of Organizational Aspects of IPSO Contracts

19.8.1 Supply Chain Considerations

During the course of these long-term contracts market conditions will change, and with them changes within the supply chain are likely as well. Thus, one of the major uncertainties for the supplier is actually supply chain disruption, which is something the suppliers have to accept and consider [48]. For the partnership to be successful there is a need to coordinate the profit incentives between them [49]. This becomes even more important when considering the fact that a railway infrastructure IPSO contract likely requires both a general contractor and a technical contractor. If a general contractor takes the IPSO responsibility, that company would be dependent on a technical contractor to build the actual tracks, etc., as well as to maintain the construction. This means that the incentives for the technical contractor have to be strong enough to realize the work to the required standard. It would therefore be advantageous if the technical contractor was part of the design phase, since they would have the knowledge and experience needed to achieve technical development. If instead the technical contractor was to be the main contractor, the dependency on the general contractor would not be as significant as in the opposite case. This is because the technical competence is already within the organization, and the general competence needed for the preparation before laying down the tracks can be found more easily. Contractors in general do not have the project management competence needed or the financial strength to pull off such a project. With this said, most of the technical contractors do have corporate groups behind them which could support such a constellation, even though the organization is not formed like that today.

As the market is today, the competence of designing is with consultants; even so, the parties still need competence to work with them. A multi-skilled and cross-functional team is needed to produce the proposal [18], which in this case goes for both provider and customer. Also, the service cost estimations require cross-functional thinking to be able to make effective cost estimations early in the development of the offering [47]. Therefore, it is important that both parties work with a long-term perspective, and especially that the Investment and Operations Divisions at the STA increase their cooperation. This type of cultural challenge has been observed within the defense industry in the UK, where the customer and provider had different ways of thinking about maintenance routines, resulting in extra costs to make up for the difference [42]. This implies the importance of working together and understanding the other parties' perspective. For the Swedish rail infrastructure industry, as for the defense industry in the UK, a massive cultural change is needed [42].

19.8.2 Change of Mindset

Relational issues are required and determined by the business needs, which implies that only changing the business model and contracts would be insufficient [50]. The organization and culture at the STA seems to be a significant hurdle. A major part of the challenge when large companies deal with change in the business model is related to the change of mindset within the organization and the need for internal marketing [51]. The STA lacks a long-term overall perspective, in combination with an internal reluctance to change. Since the rail infrastructure industry in Sweden has one dominating customer, and since this customer sets the rules for the market, the providers in this case cannot change their business model unless this has been initiated by the STA.

The contractors are dependent on the decisions made by the STA in everything from planning to execution; even some decisions have been delegated to the contractors in the past few years when the Design-Build contracts were introduced. Power asymmetry between large buyer firms and their suppliers has been studied before, and it has been concluded that this situation affects partner behavior and operational performance [52]. The same study points out supply chain partners only make adaptations if they are forced to. In the Swedish case, the contractors have to completely adapt to the requirements of the STA that sets the market conditions, the contracting form and the technical and functional requirements. Nyaga et al. [52] also indicate that by adapting to each other the firms can improve their operational performance and reduce misunderstandings and conflicts.

A good relationship also affects the uncertainty and vulnerability for the weaker party, and can as well work to even out the power balance [52]. Considering the fact that the contractors believe there is a risk that the STA will shuffle over all the risk to them indicates a lack of trust between the parties. To achieve a good relationship there needs to be a certain level of trust between the parties, which is an issue that is discussed in the Sect. 19.8.3.

19.8.3 Lack of Trust

The actors have indicated that they see themselves as parties with contradictory interests to some extent. This is in line with previous research stating that, in general, there is a low level of trust between the contracting parties within the construction industry [53]. For a long-term cooperation, however, common interests, shared risks and flexibility are needed, i.e. not making one side take all the risk

[40], as discussed in allocation of risk. Furthermore, the quality of the relationship between buyer and contractor affect the outcome of the project. In high-trust relationships there is a tendency for the contractor to use a lower risk premium, and the effectiveness of the exchange increases [54, 53]. One thing that should improve the contractors trust for the STA is the fact that the STA is state-owned, and therefore always can pay the contractors for their work. This implies an assurance which should at least make the contractors feel safer than working with privately owned-companies.

The STA have so far managed the projects in detail using specifications and limiting the flexibility of the contractors. This could be due to tradition, but it is also likely that the relationships between the STA and the contractors are not based on a high level of trust, which has been indicated by the respondents. The fact that the industry works within the regulations of public procurement hinders the actors in forming long-term relationships. On the other hand, the market in Sweden is limited and there are few contractors active for these types of projects.

This implies that forming better relationships on a general level should not be out ruled.

Within the defense industry in the UK, where similar contracts are used, risk-sharing and transparency are explicitly encourage by the buyer [55].

The same research concludes, on the other hand, that this is not easily implemented in practice due to the lack of trust between the actors. The actors in the UK defense industry suggested open-book relations as a solution to gain trust, but this type of relationship requires high levels of trust to be implemented [55].

The literature emphasizes that there is a cost for mistrust in relationships in the construction industry [53]. In addition, the interpersonal trust between buyer and contractor has more impact on the relational consequences than interorganizational power and dependency structures [54]. This could imply that the lack of trust in the Swedish rail construction industry is in the long run hindering the development of more resource-efficient infrastructure.

19.9 Concluding Discussion

IPSO contracts could provide conditions for innovation and technical development, as well as contribute to a more efficient procurement process and a lower total cost for a rail infrastructure project. However, this business model is new to the actors in the rail infrastructure industry, and therefore involves risks and uncertainties for all parties. The industry and the actors are generally risk-averse, and this in combination with a lack of trust between the actors hinders development. An increased level of trust between buyer and contractors could make the process more effective, since less detailed specifications and control of the process would be needed.

The lack of information affects what type of projects are suitable for IPSO contracting, since too much uncertainty will force the contractors to add high-risk premiums to the project cost. Adequate information and documentation is needed to apply IPSO contracting to reinvestment projects. Another important issue to reduce the risk is the operationalization of the result, and to determine precise parameters for evaluation and control. A fixed-price contract with flexibility, such as renegotiations and adjustment for instance in traffic volume, would be one way to arrange the payment according to the majority of the actors.

The contracts would require a significant cultural change in the industry, and the actors would have to form new constellations to cope with the expanded system that an IPSO contract would encompass. Most of all, it seems like the obstacle to change lies more in the business culture and attitude of the actors than in technical complications. The railway sector in Sweden has one actor dominating the procurement of railway, meaning that initiatives for change have to come from this actor.

The market is far from using IPSO contracts, an implementation would have to be realized in several steps starting with smaller projects, both in terms of geography and components of the infrastructure. The development could probably benefit from a joint evaluation with the contractors and the design consultants to grasp all sides of the story. It is obvious that the actors lack sufficient knowledge about how to plan and calculate the risk, implying that an implementation must occur gradually. Most of all, it seems like the obstacle to change lies more in the business culture and attitude of the actors than in technical complications.

To conclude, most of the risks and uncertainties identified by the actors are due to their lack of experience as well as the working procedures used today. In fact, it is only the uncertainty connected to the time dimension of a long-term contract that will always be an inherent uncertainty factor. If IPSO contracts were to be used in Sweden, it would depend on how the contractors foster this opportunity. They must either embrace the fact that more uncertainties are involved and develop the necessary skills and competence needed to identify and handle these and the potential subsequent risk in a strategic manner, or take the problems as they come in more of an ad hoc way. It is a matter of developing a competitive advantage, or perhaps having to deal with costly miscalculations.

19.9.1 This Work in a Larger Context

There are similarities between road and railway projects; to mention a few, both types of facilities use public procurement, have long-term life cycles and are maintenance-intensive. However, there are differences as well which hinder a straightforward transfer of the concept used for roads to the railway sector. Railways are more complex technical systems, including e.g. signal systems and electricity supply for the transport. Furthermore, the time pressure when maintaining the facilities is significantly larger than for roads, since there are few or no alternative routes for the traffic to take while the track is closed. Therefore, there is less flexibility and tolerance for delays, since they create consequences noticeable in the system far from the closed-down area. Another sector with similar characteristics as the railway and road sectors is the defense sector. This sector, however, is even more maintenance-intensive, and the obsolescence of parts is fast [56]. On the other hand, if one aircraft is broken it does not affect all the other aircraft; compare this to rail infrastructure, where a malfunctioning stretch affects the whole system since there might only be one track to use.

The fact that the empirical data is collected in Sweden does not limit its relevance for other geographical areas since the discussion is on a general level where the focus is on incentives, actors' cooperation and long-term contracting, not on specific geographical conditions.

19.9.2 Further Research Is Needed

This research has had IPSO contracts with rail infrastructure systems as the main focal point. It would be interesting to investigate other large technical systems using the same approach to compare the similarities and differences. Additionally, further investigation and analysis of the functional requirements, e.g. the appropriate level of detail and adequate measurements, are needed for an IPSO contract.

Acknowledgments The authors would like to thank the Swedish Transport Administration for financing the research, and the respondents for contributing with their knowledge.

References

1. Government Offices of Sweden (2012) Press release: 800 miljoner kronor till förbättrat underhåll av järnvägen
2. The Swedish Transport Administration (2011) Situationen i det svenska järnvägsnätet
3. Government Offices of Sweden (2010) Förbättrad vinterberedskap inom järnvägen
4. National Audit Office (2011) Regulating Network Rail's efficiency. National Audit Office, London
5. The Swedish Agency for Public Management (2009) Sega gubbar? En uppföljning av Byggkommisionens betänkande "Skärpning gubbar!". The Swedish Agency for Public Management, Stockholm
6. Nilsson J-E, Ihs A, Leif S, Wiman LG, Wågberg L-G (2006) Funktionsupphandling. Sammanfattning av kunskapsläget och rekommendationer för fortsatt forskning, vol 506. Swedish National Road and Transport Research Institute, Linköping
7. Stenbeck T (2004) Incentives to innovations in road and rail maintenance and operations. Royal Institute of Technology, Stockholm
8. March JG, Shapira Z (1987) Managerial perspectives on risk and risk taking. Manage Sci 33 (11):1404–1418
9. Akintoye AS, MacLeod MJ (1997) Risk analysis and management in construction. Int J Project Manage 15 (1):31–38. doi:10.1016/S0263-7863(96)00035-X
10. Lyons T, Skitmore M (2004) Project risk management in the Queensland engineering construction industry: a survey. Int J Project Manage 22(1):51–61. doi:10.1016/S0263-7863 (03)00005-X
11. Uher TE, Toakley AR (1999) Risk management in the conceptual phase of a project. Int J Project Manage 17(3):161–169. doi:10.1016/S0263-7863(98)00024-6

12. Cha EJ, Ellingwood BR (2012) Risk-averse decision-making for civil infrastructure exposed to low-probability, high-consequence events. Reliab Eng Syst Saf 104(0):27–35. doi:10.1016/j.ress.2012.04.002
13. Svensson N, Eklund M (2007) Screening of environmental pressure from products in the Swedish railway infrastructure: Implications for strategic environmental management. Resour Conserv Recycl 52(2):248–265
14. White AL, Stoughton M, Feng L (1999) Servicizing: the quiet transition to extended product responsibility. Tellus Institute, Boston
15. Goedkoop MJ, Halen CJGv, Riele HRMt, Rammens PJM (1999) Product Service systems, ecological and economical benefits. PricewaterhouseCoopers, Pre Consultants, Netherlands
16. Tukker A, Tischner U (2006) Product-services as a research field: past, present and future. Reflections from a decade of research. J Clean Prod 14(17):1552–1556
17. Lenferink S, Tillema T, Arts J (2013) Towards sustainable infrastructure development through integrated contracts: experiences with inclusiveness in Dutch infrastructure projects. Int J Project Manage 31(4):615–627. doi:10.1016/j.ijproman.2012.09.014
18. Brady A, Davies A, Gann DM (2005) Creating value by delivering integrated solutions. Int J Project Manage 23(5):360–365
19. Lindahl M, Sundin E, Sakao T, Shimomura Y (2007) Integrated product and service engineering versus design for environment—A comparison and evaluation of advantages and disadvantages. In: 14th CIRP International Conference on Life Cycle Engineering. Springer, Tokyo
20. Tukker A (2004) Eight types of product-service-systems: eight ways of sustainability? experiences from suspronet. Bus Strategy Environ 12:246–260
21. Meier H, Roy R, Seliger G (2010) Industrial product-service systems—IPS^2. CIRP Annuals—Manuf Technol 59(2):607–627
22. Lindahl M (2005) Engineering designers' requirements on design for environment methods and tools. KTH Royal Institute of Technology, Stockholm
23. Nilsson J (2009) Nya vägar för infrastruktur. SNS Förlag, Stockholm
24. Erkoyuncu J, Roy R, Shehab E, Cheruvu K (2011) Understanding service uncertainties in industrial product–service system cost estimation. Int J Adv Manufact Technol 52 (9):1223–1238. doi:10.1007/s00170-010-2767-3
25. Walker WE, Harremoes P, Rotmans J, van der Sluijs JP, van Asselt MBA, Janssen P, Krayer von Krauss MP (2003) Defining uncertainty: a conceptual basis for uncertainty management in model-based decision support. Integr Assess 4(1):5–17
26. Eisenhardt KM (1989) Agency theory: an assessment and review. Acad Manag Rev 14 (1):57–74. doi:10.5465/amr.1989.4279003
27. Alonso-Rasgado T, Thompson G (2006) A rapid design process for total care product creation. J Eng Des 17(6):509–531
28. Merriam SB (1994) Fallstudien som forskningsmetod (Case Study Research in Education). Studentlitteratur, Lund
29. Lekvall P, Wahlbin C (2001) Information för marknadsföringsbeslut, 4th edn. IHM Publishing, Göteborg, Sweden
30. Banverket (2008) Banverkets årsredovisning 2008. The Swedish Rail Administration, Borlänge
31. Banverket Produktion (2009) Banverket Produktion Årsrapport 2009. Banverket Produktion
32. Nilsson J-E, Pyddoke R (2007) Offentlig och privat samverkan kring infrastruktur- en forskningsöversikt, vol 601. Swedish National Road and Transport Research Institute, Linköping
33. Nilsson J, Ihs A, Sjögren L, Wiman LG, Wågberg L (2006) Funktionsupphandling: Sammanfattning av kunskapsläget och rekommendationer för fortsatt forskning. VTI, Linköping
34. Pakkala P (2002) Innovative project delivery methods for infrastructure. Finnish Road Enterprise, Helsinki

35. Hedström R, Ihs A, Sjögren L (2005) Funktionsupphandling av väg- och banhållning, Problem och möjligheter., vol 971. Swedish National Road and Transport Research Institute, Linköping
36. Banverket (2009) Entreprenadbeskrivning: Drift och underhåll av järnvägsanläggning Västra Götaland
37. Lingegård S (2012) Integrated product service offerings for rail infrastructure: potential benefits and challenges. Linköping University, Linköping
38. Barnes J, Ayars WB (1977) Reducing new product risk through understanding buyer behavior. Ind Mark Manage 6(3):189–192. doi:10.1016/0019-8501(77)90017-7
39. Ward SC, Chapman BC, Curtis B (1991) On the allocation of risk in construction projects. Int J Project Manage 9(3):140–147. doi:10.1016/0263-7863(91)90038-W
40. Nystén-Haarala S, Lee N, Letho J (2010) Flexibility in contract terms and contracting processes. Int J Managing Projects Bus 3(3):462–478
41. Panesar SS, Markeset T (2008) Industrial service innovation through improved contractual realtionship: a case study in maintenance. J Qual Maintenance Eng 14(3):290–305
42. Datta PP, Roy R (2011) Operations strategy for the effective delivery of integrated industrial product-service offerings: two exploratory defence industry case studies. Int J Oper Prod Manage 31(5):579–603
43. Mont O (2004) Product-service systems: Panacea or Myth? Lund University, Lund
44. Oliva R, Kallenberg R (2003) Managing the transition from products to services. Int J Serv Ind Manag 14(2):160–172
45. Ng ICL, Nudurupati SS (2009) Outcome-based service contracts in the defence industry-mitigating the challenges. J Serv Manage 21(5):656–674
46. Mitchell VW (1995) Organizational risk perception and reduction: a literature review. Brit J Manage 6:115–133
47. Datta PP, Roy R (2010) Cost modelling techniques for availability type service support contracts: a literature review and empirical study. CIRP J Manufact Sci Technol 3(2):142–157
48. Erkoyuncu JA, Roy R, Shehab E, Wardle P (2009) Uncertainty challenges in service cost estimation for product-service systems in the aerospace and defence industries. In: 1st CIRP Industrial Product-Service Systems (IPS2) Conference, Cranfield University, UK, 1–2 Apr 2009, p 200
49. Lockett H, Johnson M, Evans S, Bastl M (2011) Product service systems and supply network relationships: an exploratory case study. J Manufact Technol Manage 22(3):293–313
50. Thompson I, Cox A, Anderson L (1998) Contracting strategies for the project environment. Eur J Purchasing Supply Manage 4(1):31–41
51. Sundin E, Ölundh Sandström G, Lindahl M, Öhrwall Rönnbäck A Industrial Challenges for Product/Service Systems: Experiences from a large company netowork in Sweden. In: CIRP Industrial Product-Service Systems (IPS^2) Conference, Cranfield UK, 2009
52. Nyaga GN, Lynch DF, Marshall D, Ambrose E (2013) Power asymmetry, adaptation and collaboration in dyadic relationships involving a powerful partner. J Supply Chain Manage 49 (3):42–65. doi:10.1111/jscm.12011
53. Zaghloul R, Hartman F (2003) Construction contracts: the cost of mistrust. Int J Project Manage 21(6):419–424. doi:10.1016/S0263-7863(02)00082-0
54. Jiang Z, Henneberg SC, Naudé P (2012) Supplier relationship management in the construction industry: the effects of trust and dependence. J Bus Ind Mark 27(1):3–15
55. Johnsen T, Howard M, Miemczyk J (2009) UK defence change and the impact on supply relationships. Supply Chain Manage: Int J 14(4):270–279
56. Romero Rojo FJ, Roy R (2009) Obsolescence management for long-life contracts: state of the art and future trends. Int J Manuf Technol 49(9–12):1235–1250

Chapter 20
Managing Obsolescence Risk

Peter Sandborn

Abstract Many systems that must be manufactured and supported for long time periods lack control over critical portions of their supply chains; these systems include: military, avionics, industrial controls, and rail infrastructure. During the long lifetimes of these systems the components, technologies and resources that the systems depend on become obsolete (and potentially unavailable) before the system's demand for them is exhausted. The life-cycle (or through-life) cost associated with managing obsolescence can be prohibitive if the problem is ignored. This chapter describes the obsolescence problem, methods of forecasting obsolescence and management solutions applied to hardware, software, and human skills.

20.1 Introduction

The definition of obsolescence is the "loss or impending loss of original manufacturers of items or suppliers of items or raw materials," [1]. The form of obsolescence described in this chapter is known as Diminishing Manufacturing Sources and Material Shortages (DMSMS), which is caused by the unavailability (unprocurability) of the technologies or parts required to manufacture or sustain a system.[1] The unavailability of parts from their original manufacturer means the end of production of the part and the end of support for fielded parts.

[1] "Sudden" or "inventory" obsolescence, which is prevalent in the engineering management literature, is the opposite of DMSMS-type obsolescence. Inventory obsolescence describes product design or system specification changes that cause existing inventories of parts to no longer be needed, e.g., [2].

P. Sandborn (✉)
Center for Advanced Life Cycle Engineering, University of Maryland,
College Park MD 20144, USA
e-mail: sandborn@umd.edu

L. Redding and R. Roy (eds.), *Through-life Engineering Services*,
Decision Engineering, DOI 10.1007/978-3-319-12111-6_20

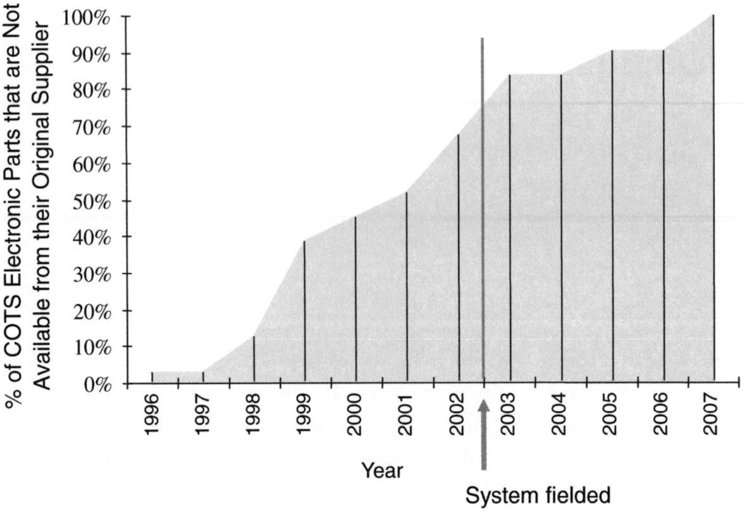

Fig. 20.1 Commercial off the shelf (*COTS*) electronic parts that become un-procurable in the first 10 years of a surface ship sonar system's life cycle [5]

Although many parts will remain available from various aftermarket sources, the use of aftermarket sources for parts creates potentially unacceptable risks for many types of safety, mission and infrastructure critical systems.[2] Electronic parts have the most serious DMSMS problems because the electronic part supply chain is driven by high-volume, short-life consumer products, and creating *the ability to manufacture* parts specifically for low volume, long-life markets is impractical give the nature and cost of semiconductor manufacturing. In the case of electronic parts, the length of time a part is available from its original manufacturer (i.e., its procurement life) can be less than a year in some cases.

The DMSMS obsolescence problem is most problematic for "sustainment-dominated" systems in which the cost of sustaining (maintaining) the system through its life significantly exceeds the cost of procuring or manufacturing the system, [4]. Often the design cycles for sustainment-dominated systems are long enough that many of the system's electronic parts become obsolete prior to the system being introduced to the field, e.g., Fig. 20.1. Once these systems are in the field, their support can last for 20 or more additional years. Matters are made worse because the end of support date for systems like the one shown in Fig. 20.1 are not well known and are often extended one or more times before the system is retired.

In fact some infrastructure-critical systems become "ageless" systems for which there is effectively no end of support date or the end of support dates are extended ad infinitum [6].

[2] For example, sourcing from the aftermarket creates the risk of obtaining counterfeit parts, e.g., [3].

Simply replacing obsolete parts with newer non-obsolete parts is often not a viable solution for many safety, mission, and infrastructure critical systems due to the prohibitively high cost of system re-qualification and re-certification. As an example consider a failed electronic part in the control system of a 25 year old nuclear power plant; replacement of the failed part with anything other than an instance of the original part may initiate a re-qualification of the subsystem (or whole system) that is impractical.

Sustainment-dominated products generally have no control over their supply chain for electronic parts (because their demand for parts is relatively small). DMSMS-type obsolescence occurs when long field life systems are forced to depend on supply chains that have been created to support high-volume products, [7].[3]

The next section describes methods used to forecast the obsolescence of electronic parts. This is followed a discussion of reactive, proactive and strategic approaches for managing obsolescence. Sections 20.4 and 20.5 address the obsolescence of software and critical human skills.

20.2 Forecasting Obsolescence

The key enabler for management of DMSMS-type obsolescence is the ability to forecast the obsolescence events for important parts in the system.

Two types of obsolescence forecasting strategies exist. Long-term (also called model-based) forecasts that are used when obsolescence is a year or more into the future to enable pro-active management of obsolescence events and strategic life-cycle planning for the sustainment of systems. The second type of forecasting is short term. Short-term (or data-driven) forecasting searches the supply chain for precursors to a part's discontinuance. Precursors can include a reduction in the number of sources that a part is available from, reductions in distributor inventories of the part, price increases that may in some cases accompany a reduction in the availability of the part, and announcements made by the part manufacturer that either directly or indirectly indicate that the part is being phased out.

A common strategy is to use long-term obsolescence forecasting while continuously monitoring the supply chain for precursors to obsolescence. This strategy abandons the long-term forecast when a particular combination of subjective indicators of discontinuance associated with a part are observed. In itself, the lack of precursors only indicates that discontinuance of the part may not be immediate; therefore, the supply chain indicators (precursors) are not generally useful for forecasting the long-term availability of the part. The remainder of this section focuses on the long-term forecasting of obsolescence.

---

[3] The dynamic nature of an industry is referred to as "clockspeed," [8]. DMSMS-type obsolescence problems are usually encountered by slow clockspeed industries.

Most long-term electronic part obsolescence forecasting is based on the development of models for the part's life cycle. Traditional methods of life-cycle forecasting are ordinal scale based approaches, in which the life-cycle stage of the part is determined from a combination of technological and supply chain attributes such as level of integration, minimum feature size, type of process, number of sources, etc. [9, 10] and those available in various commercial databases. The ordinal scale based approaches work best as short-term forecasts, but their accuracy in the long-term is difficult to quantify. For ordinal scale approaches the historical basis for forecasts is subjective and confidence levels and uncertainties are not generally evaluatable. For pro-active and strategic obsolescence management approaches (see Sect. 22.3) to be successful accurate long-term forecasts or at least forecasts with a quantifiable accuracy are required.

20.2.1 Data Mining Based Obsolescence Forecasting

Electronic part obsolescence forecasting is easier than forecasting for other technologies because there is a very large and well documented history of part introductions and discontinuances. For example, most commercial electronic parts databases contain over 100 million parts (obsolete and non-obsolete). As a result, data mining based forecasting approaches have been successfully utilized to perform obsolescence forecasting.

Two types of data mining approaches have been used. The first is applicable to parts that have evolutionary parametric drivers [11, 12] and the second applies to parts without evolutionary parametric drivers [7].

20.2.1.1 Sales Curve Forecasting for Parts with Evolutionary Parametric Drivers

An evolutionary parametric driver is a parameter (or a combination of parameters) describing how the part evolves over time. For example, for memory chips the evolutionary parametric driver is memory size, traditionally for microprocessors it has been clock frequency (although this has given way to power consumption).

Product life-cycle curves describe the stages a product goes through in the marketplace and can be used to measure a product's maturity with respect to the marketplace [13]. Product life-cycle curves usually divide the life cycle of a product into the six stages: introduction, growth, mature, decline, phase-out, and obsolescence. For parts with evolutionary parametric drivers, the life-cycle curve of the product is a function of the driving parameter. As an example, Fig. 20.2 shows the life-cycle curves for flash memory chips. Memory capacity is the evolutionary parametric driver of the product in this case.

For flash memory, the life-cycle curves for parts can be parameterized by curve fitting of the sales data (such as that shown in Fig. 20.2) for each memory capacity.

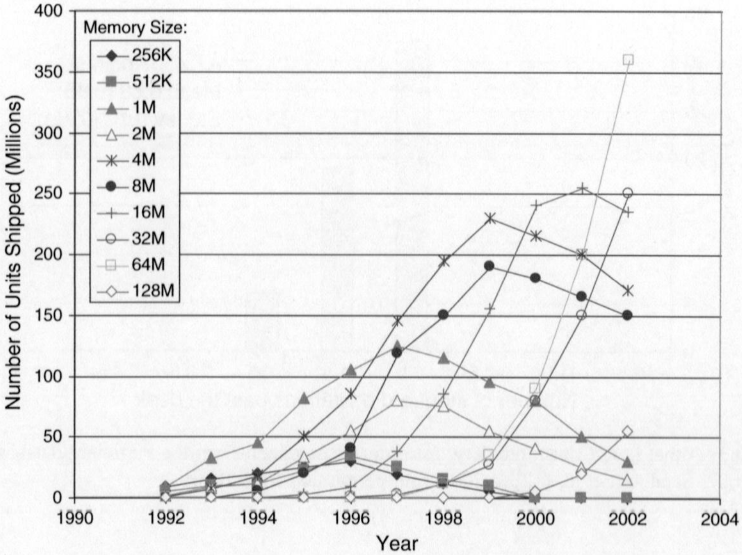

Fig. 20.2 Historical sales data for monolithic flash memory from [14] supplemented with additional data. Sandborn et al. [12] © 2007 with permission from IEEE

The values of the mean and standard deviation that resulted from the best fits to the data sets such as those in Fig. 20.2 can be determined. Using this approach, trend equations for the part's peak sales year (mean) and standard deviation in number of units shipped can be formed. The resulting trend equations can be used reproduce the life-cycle curve for the parts that were used to create the relationships and for parts that are introduced in the future (assuming that the future part exists).

This approach provides a way to create or re-create the life-cycle curve for a part type given its evolutionary parametric driver. In the original implementations of this forecasting approach [11] the "window of obsolescence" specification was defined to be at 2.5–3.5 standard deviations after the peak sales date. Bartels et al. [15] provides the market-based trending based on [11] for many different types of parts. This model is independent of the specific variable plotted on the vertical scale of the life-cycle curve shown in Fig. 20.2; the variable used need only represent some type of quantity measure that reflects the market for the part, i.e., it could be sales volume or even revenue (however, if revenue is used, it needs to be adjusted for part cost changes).

In reality, the window of obsolescence specification is not a constant but depends on manufacturer-specific and part-specific business practices. Sandborn et al. [12] extends the work of Solomon et al. [11] by mapping the actual obsolescence dates of all parts in a database to the number of standard deviations after the mean. The results can then be sorted by vendor and plotted as histograms (e.g., Fig. 20.3) from which vendor-specific forecasting algorithms can be developed.

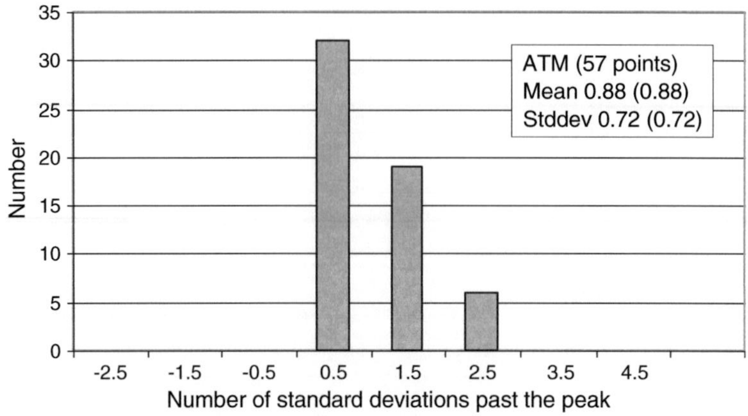

Fig. 20.3 Atmel (*ATM*) flash memory last order dates mined from the Partminer database [16], circa 2007. Sandborn et al. [12] © 2007 with permission from IEEE

20.2.1.2 Procurement Life Modeling

The previously described data mining method for forecasting obsolescence works well when there are identifiable evolutionary parametric drivers. Unfortunately, for the majority of electronic parts, there is no simple evolutionary parametric driver that can be found.

Obsolescence forecasting algorithms that do not depend on the identification of an evolutionary parametric driver for the part can be developed based on predicting the part's procurement life, which is defined as [7],

$$L_P = D_O - D_I \tag{20.1}$$

where L_P is the procurement life, amount of time the part was (or will be) available for procurement from its original manufacturer; D_O is the obsolescence date, the date that the original manufacturer discontinued or will discontinue the part; and D_I is the introduction date, the date that the original manufacturer introduced the part. If a procurement lifetime can be forecasted and the introduction date for the part is known, the obsolescence date can be forecasted from Eq. (20.1).

Figure 20.4 shows procurement life versus introduction date for obsolete linear regulators (a common electronic part). The mean procurement lifetimes for parts can be analyzed using the statistical framework for failure time analysis. The event of interest is the discontinuance (obsolescence) of an instance of a part. The data used includes the introduction dates of all the parts of a particular type and the obsolescence dates for the parts that have occurred through the analysis date (D_A = 2008 in Fig. 20.4). An obsolescence event is not observed for every part in the data set since some of the introduced parts have not gone obsolete as of the analysis date, i.e., the observations are right censored (in the case of linear

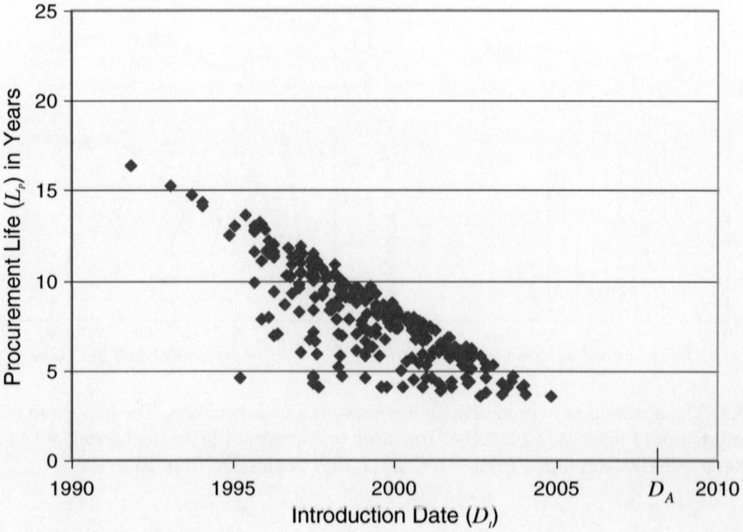

Fig. 20.4 347 obsolete linear regulators from 33 manufacturers mined from the SiliconExpert database [17], circa 2009. D_A = 2008, the analysis date (the date on which the analysis was performed). Reprinted from [7], © 2011 with permission from Elsevier

regulators, there are an additional 500 parts that have been introduced but had not gone obsolete as of the end of 2008). To determine the probability density function, the procurement life data can be fit and the parameters were estimated using maximum likelihood estimation (MLE) assuming right censoring and that the censoring mechanism is non-informative (the knowledge that the observation is censored does not convey any information except that the procurement lifetime exceeds the censoring date, which is the analysis date in our case). Figure 20.5 shows the probability density function and for the procurement life of linear regulators

The mean procurement lifetime can be determined as a function of time by only considering 1 year slices with and without right censoring, e.g., Fig. 20.6. Sandborn et al. [7] analyse several different part types using the procurement life modelling approach.

20.3 Obsolescence Management

Obsolescence is managed in systems at three different levels: reactive, pro-active and strategic, Fig. 20.7. Reactive management resolves the part obsolescence problem after the part has been discontinued and tracks the action(s) taken. Pro-active management is performed on parts that: (a) are at risk of becoming obsolete, (b) will have an insufficient quantity available after obsolescence to satisfy expected

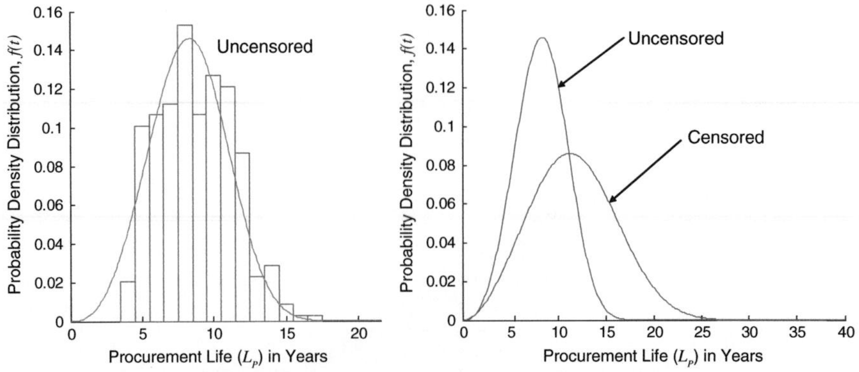

Fig. 20.5 The distribution of procurement lifetimes for linear regulators. The histogram on the *left side* corresponds to the data in Fig. 20.4. The mean procurement lifetime (censored) = 11.63 years, $\beta = 2.84$, $\eta = 13.06$. Reprinted from [7], © 2011 with permission from Elsevier

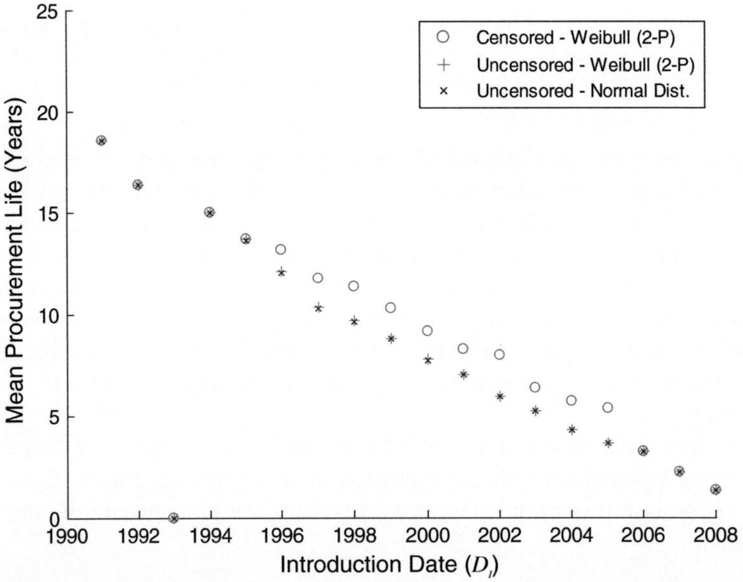

Fig. 20.6 Mean procurement lifetime for linear regulators as a function of time (parts introduced on the date). Note there were no linear regulator parts introduced in 1993. Reprinted from [7], © 2011 with permission from Elsevier

demand, and (c) will represent a problem to manage if/when they become obsolete, [18]. Once critical parts for pro-active management are identified the process of managing obsolescence is initiated prior to their actual obsolescence event. Strategic management of obsolescence refers to the use of part-specific obsolescence

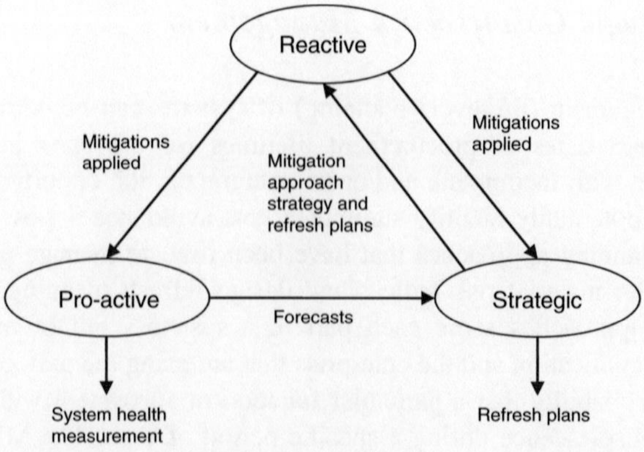

Fig. 20.7 Three obsolescence management levels [18]

forecasts, logistics data, technology forecasting, and business trending (demand forecasting) to enable life-cycle optimization (life-cycle planning). Both pro-active and strategic management depend on an ability to forecast the obsolescence of parts (discussed in Sect. 22.2).

20.3.1 Reactive Obsolescence Mitigation

When parts become obsolete various mitigation approaches can be used [19]. Obsolete parts can be replaced with non-obsolete substitute[4] or alternative parts if there is little or no requirement for re-qualification of the system. Lifetime buys of parts are also commonly used; lifetime buys involve buying and storing a sufficient quantity of parts to last through a system's remaining manufacturing and sustainment life. There are also aftermarket sources for electronic parts that range from sources that are authorized by the original manufacturer that fulfill part orders using finished parts (manufactured by the original manufacturer) and new fabrication in original manufacturer qualified facilities (e.g., Rochester Electronics). Non-authorized aftermarket sources that range from brokers to eBay can also be used—however non-authorized sources introduce counterfeit part and other risks. Electronic part emulation foundries such as GEM and AME [21] use newer technologies to fabricate replacements for obsolete parts that meet the original part qualification standards.

All reactive management approaches are applied after obsolescence has occurred and generally applied on a piece-part specific basis. Rarely do reactive resolutions manage more than one part at a time.

[4] One common type of substitute part is a commercial temperature range part that has been "uprated" to meet the extended temperature range requirements of an obsolete Mil-Spec part, [20].

20.3.2 Strategic Obsolescence Management

Strategic management (life-cycle planning) of systems can be performed if obsolescence forecast dates or procurement lifetimes for the parts in a system are available. Even with incomplete and/or uncertain data, the opportunity to plan the life cycle and potentially identify significant cost avoidance is possible.

Strategic planning approaches that have been used to manage obsolescence in systems include: material risk indices and design refresh planning. Material Risk Index (MRI) approaches score each part in a system's bill of materials in the context of the application and the enterprise that are using the part, e.g., [22]. MRIs evaluate the probability that a particular function or subsystem within a system is impacted by obsolescence during a specific period of time. The MRI also defines the specific actions that must be taken to resolve the issue for the function or subsystem in distant time periods. The MRI-defined risks can be mapped to through-life cost using activity-based cost analysis. MRIs are easy to use, but the initial development of the MRI is non-trivial requiring the creation of a catalog of relevant functions for an organization and the probability of obsolescence precipitated actions (obsolescence risks). MRIs must also be periodically of calibrated against historical actions and their costs.

Sustainment-dominated systems often have very long manufacturing and field support lives during which they are often refreshed or redesigned one or more times to manage obsolescence and to update functionality. Technology "refresh" refers to changes that "Have To Be Done" in order for the system functionality to remain useable. Redesign or technology insertion is a term used to identify "Want To Be Done" system changes that include the insertion of new technologies to accommodate system functional growth and the replacement and improvement of existing functionality, [23]. While the redesign of high-volume commercial products is driven by improvements in manufacturing, equipment and/or technology, the design refresh of sustainment-dominated systems is usually driven by obsolescence and done primarily to keep the system producible and/or sustainable. The objective of design refresh planning (DRP) is the determination of design refresh timing and content, i.e., when should refreshes be done and what obsolete parts should be replaced at each design refresh. DRP seeks to find the optimum balance between expensive refreshes and the management of individual obsolescence events with a reactive obsolescence mitigation approach, [5].

Several DRP modeling approaches exist. The simplest model for performing design refresh planning for electronic part obsolescence was developed by Porter [24] and generalized by Sandborn [22]. Porter's approach calculates the Present Value (PV) of last time (bridge) buys[5] and design refreshes as a function of future date. As a design refresh is delayed, its PV decreases and the quantity (and thereby

[5] A last time or bridge buy means buying a sufficient number of parts to last until the part can be designed out of the system at a design refresh. Last time buys become lifetime buys when there are no more planned refreshes of the system.

cost) of parts that must be purchased in the last time buy required to sustain the system until the design refresh takes place increases. Alternatively, if design refresh is scheduled relatively early, then last time buy cost is lower, but the PV of the design refresh is higher. By balancing the last time buy cost against the design refresh cost, an optimum (lowest life-cycle cost) refresh date can be found. The optimum refresh year from the Porter model is given by [22] as,

$$Y_R = \frac{1}{-r} \ln\left(\frac{P_0 Q}{r C_{DR_0}}\right) \tag{20.2}$$

where r is the weighted average cost of capital (WACC), P_0 is the purchase price of the obsolete part (in year 0), Q is the number of parts needed each year, and C_{DR_0} is the design refresh cost in year 0. The Porter model is a variation of the well know Economic Order Quantity (EOQ) problem from operations research [25].

The porter DRP model is simple to use, but also limited to analyzing a single part and determining a single refresh to manage it during the life of the system. For complex systems with many parts, design refresh planning is usually done using discrete event simulation. The best known discrete event simulator is MOCA [5]. Figure 20.8 shows and example result from MOCA. In the left side of Fig. 20.8, each point represents a unique refresh plan. The vertical axis on the graph is life-cycle cost and the horizontal axis is time. The data points corresponding to the plans are plotted at the mean of the group of refresh dates they represent (note, one plan is expanded in the graph to show the actual two refresh dates it comprises).

The right side of Fig. 20.8 shows a portion of the business case analysis performed by MOCA where the optimum solution is compared to solutions with no refresh (all reactive management), all refresh (refreshing for every obsolescence

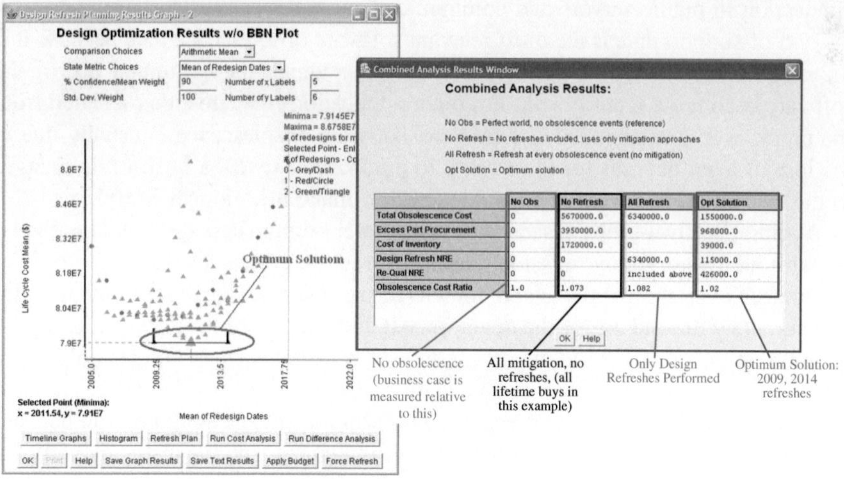

Fig. 20.8 Example MOCA design refresh planning result [18]

event), and an idealized solution where nothing ever becomes obsolete. An alternative version of the MOCA algorithm has also been implemented within integer programming solutions, e.g., [26].

Performing design refreshes on a system for the sole purpose of managing obsolescence is often not practical so refreshes may be coupled with other required system changes. It is common for systems to have technology insertion roadmaps that are developed to define changes in the system's functionality and performance over time. Technology roadmaps may reflect an organization's internal goals and/or budgeting cycles, or may be driven by the needs of key customers. By integrating technology roadmap information into design refresh planning one can ensure the formation of refresh plans that meet timing and budget constraints. The definition of refresh planning constraints created to accommodate roadmapping information is discussed in [27].

20.4 Software Obsolescence

While there are numerous methodologies, databases, and tools that address status, forecasting, risk, mitigation, and management of electronic parts obsolescence, virtually everything that is available today focuses exclusively on the hardware life cycle. In most complex systems, software contributes significantly to the life-cycle costs, and the concurrent sustainment of hardware and software is required.

Definitions of software obsolescence vary depending on the software and the application of the system. Commercial (i.e., COTS—Commercial Off The Shelf) software often has end-of-sale and end-of-support dates separated by long periods of time. For mass-market COTS software (such as operating systems), software vendors may publish the end-of-sale and end-of-support dates. For software with a connection to public servers and communications systems (e.g., through the web), the end-of-support date is the most relevant software obsolescence date because it is the date of termination of security patches after which the continued use of the software becomes a security risk. For other applications that are either isolated from the public web or embedded in hardware, software obsolescence is usually due to the loss of legal licenses (or that ability to purchase or renew a license) or changes to the system the software is embedded within make it no longer useful.

Avoiding software obsolescence in sustainment-dominated systems is unlikely. As with hardware, safety, mission, and infrastructure critical systems have little or no control over the supply chain for COTS software or the development infrastructure they depend on for the development and support of their own software. In addition, in the COTS world, hardware and software have a symbiotic supply-chain relationship where hardware improvements cause software obsolescence, and new software causes hardware obsolescence. This will not change anytime soon because from the viewpoint of commercial PC manufactures and mainstream software vendors, this is a win-win strategy: "The only big companies that succeed will be those that obsolete their own products before someone else does" [28]. Besides

COTS software, systems also depend on organic (custom in-house developed) software, the infrastructure for hardware and software development and testing, and software that enables interoperability between system components—these are subject to obsolescence as well.

Some actions can be taken to proactively manage software obsolescence including: improving the portability of code, the use of open-source development platforms and toolkits, and escrowing source code with a third-party (if the original software developer terminates support, a predetermined third-party takes possession of the source code). Unfortunately, these proactive solutions are often not effective because most require the cooperation of the original software developer and the actual human resources that performed the original development.

Relatively little work exists treating software obsolescence. An overview is given by [29], but besides digital preservation, the termination of sales and support of software comprises the primary focus of the remainder of the literature [30–35].

20.5 Human Skills Obsolescence

Obsolescence is not limited to only hardware and software; it also impacts the human skills necessary to maintain systems. The obsolescence of critical human skills is a significant and growing problem for organizations that must support safety, mission and infrastructure critical systems.

System support organizations need a dependable pool of skills so that they can provide the required level of availability of critical systems. Workforce planning is commonly defined as getting the right number of people with the right competencies in the right jobs at the right time. Workforce planning is complicated by mismatches between the skills that the workforce has and the skills required by the job. These mismatches are classified into the following three categories: skills obsolescence, skill shortage, and critical skills loss.

Skills obsolescence (also called human capital obsolescence) refers to workers that do not have the skills that they need—this includes workers that have obsolete skills and required retraining, De Grip and Van Loo [36] define various types of skills obsolescence. Skill shortage refers to insufficient current skill competences, which requires identification, training and retaining workforce to fill current and expected future skill needs.

The third category is critical skills loss, which is most relevant to the type of obsolescence discussed in this chapter. Critical skills loss refers to the loss of skills that are either non-replenishable or take very long periods of time (many years) to reconstituted [37]. Critical skills loss is a special case of "organizational forgetting," which describes organizations that lose knowledge gained through learning-by-doing due to labor turnover, periods of inactivity, and/or failure to institutionalize tacit knowledge [38]. The majority of the existing literature on organizational forgetting assumes that the interruption is brief and forecast the recovery period and the disruption in productivity and schedule. Alternatively, critical skills loss is

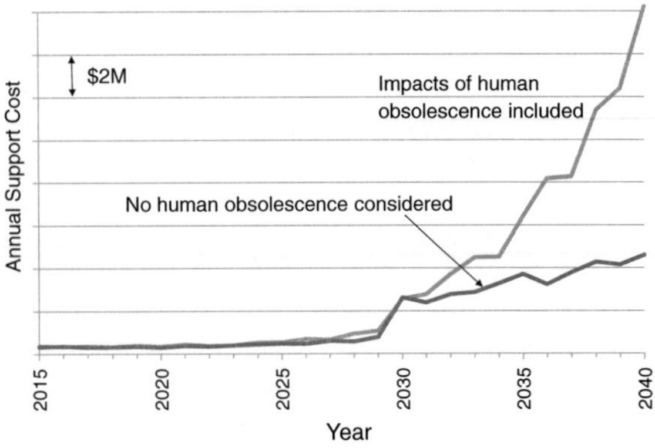

Fig. 20.9 Annual cost to support a legacy system with and without considering skills obsolescence [37]

usually the result of long-term attrition where the workforce with the required skills is lost via retirement and there are not enough younger workers to replace them. In the case of the support of safety, mission and infrastructure critical systems, five or more years of on-the-job experience may be required to become competent, and younger workers often leave to pursue other opportunities. Similar to hardware and software DMSMS-type obsolescence problems, critical skills loss is the inevitable outcome of and organization's dependence on highly specialized skills for which there is relatively little demand.

Critical skills loss is particularly problematic for software support where having access to the software source code post software obsolescence may be of little value without the original programmers. Problems associated with the loss of critical human skills have been reported in industries ranging from aerospace and nuclear to IT and healthcare.

As an example of the impact of the loss of critical skills, Fig. 20.9 shows the annual cost of supporting a legacy system with and without consideration of human skills. The result in Fig. 20.9 indicates stability until approximately year 2029 and then a jump in costs (due to exhausting lifetime buys of hardware) occurs, followed by a steep increase as workers with the necessary skills to support the system become unavailable.

20.6 Discussion

Failure to manage the technological and human content of safety, mission, and infrastructure critical systems is not an option. Obsolescence of technology, software and the humans who are needed to support systems is unavoidable, so solutions are needed in order to keep critical systems operational.

One simple key to successful system management is understanding the state of management of your system (or organization). A very simple, yet effective measure of the state of a system is to form the following index, sometimes referred to as DMSMS a progress indicator, [39],

$$\text{Obsolescence Management Index} = 100\frac{G+V_1}{G+V_1+V_2+R+B} \tag{20.3}$$

where G is the number of parts with multiple qualified suppliers and no problems on the horizon, V_1 is the number of parts with a single supplier and/or suspected problems but a solution is in place, V_2 is the number of parts with a single supplier and/or suspected problems with no solution in place, R is the number of problem parts with no solution in place, and B is the number of unmanaged parts. Unmanaged parts refers to parts for which the status (obsolete or not) is not known and no resolution solution is in place. Applying the metric in Eq. (20.3) to individual boards or cards (or bills of materials) provides a current risk assessment for the supportability of the board or card. Applying the metric in Eq. (20.3) to your whole database of active parts is a measure of the maturity of your entire parts management process.

Too often, sustainment organizations become caught up in addressing obsolescence events as they occur. While it will always be necessary to use reactive management of obsolescence problems, strategic DMSMS management is possible and can lead to substantial cost avoidance for many systems. The difficulty often encountered by when managing obsolescence is that engineers lack the ability to make appropriate business cases to support strategic management efforts.

References

1. Sandborn P (2008) Trapped on technology's trailing edge. IEEE Spectr 45:42–45
2. Song Y, Lau H (2004) A periodic review inventory model with application to the continuous review obsolescence problem. Eur J Oper Res 159(1):110–120
3. Pecht M, Tiku S (2006) Electronic manufacturing and consumers confront a rising tide of counterfeit electronics. IEEE Spectr 43:37–46
4. Sandborn P, Myers J (2008) Designing engineering systems for sustainability. In: Misra KB (ed) Handbook of performability engineering. Springer, London, pp 81–103
5. Singh P, Sandborn P (2006) Obsolescence driven design refresh planning for sustainment-dominated systems. Eng Econ 51:115–139
6. Konoza A, Sandborn P (2014) Evaluating the end of maintenance dates for electronic assemblies composed of obsolete parts. ASME J Mech Des 136:031006
7. Sandborn P, Prabhakar V, Ahmad O (2011) Forecasting technology procurement lifetimes for use in managing DMSMS obsolescence. Microelectron Reliab 51:392–399
8. Fine C (1998) Clockspeed: winning industry control in the age of temporary advantage. Perseus Books, Reading, MA
9. Henke A, Lai S (1997) Automated parts obsolescence prediction. In: Proceedings of the DMSMS conference, San Antonio

10. Josias C, Terpenny J (2004) Component obsolescence risk assessment. In: Proceedings of the industrial engineering research conference (IERC), Houston, TX
11. Solomon R, Sandborn P, Pecht M (2000) Electronic part life cycle concepts and obsolescence forecasting. IEEE Trans Compon Packag Technol 23(4):707–713
12. Sandborn P, Mauro F, Knox R (2007) A data mining based approach to electronic part obsolescence forecasting. IEEE Trans Compon Packag Technol 30(3):397–401
13. EIA (1997) EIA-724—product life cycle data model (ANSI/EIA-724-97). Electronic Industries Alliance, Arlington, VA
14. Matas B, de Suberbasaux C (1997) Chapter 5. The flash memory market. Memory 1997. Integrated Circuit Engineering Corporation. http://smithsonianchips.si.edu/ice/cd/MEMORY97/SEC05.PDF. Cited 2 June 2014
15. Bartels B, Ermel U, Sandborn P, Pecht M (2012) Strategies for the prediction, mitigation and management of product obsolescence. Wiley, Hoboken, NJ
16. PartMiner, Inc., www.partminer.com
17. Silicon Expert Technologies, Inc., www.silconexpert.com
18. Sandborn P (2008) Strategic management of DMSMS in systems. DSP J, 24–30 (April–June)
19. Stogdill RC (1999) Dealing with obsolete parts. IEEE Des Test Comput 16(2):17–25
20. Pecht M, Humphrey D (2006) Uprating of electronic parts to address. Microelectron Int 23:32–36
21. Johnson W (2000) Generalized emulation of microcircuits. In: Proceedings of the DMSMS conference, Jacksonville, FL
22. Sandborn P (2013) Chapter 16: cost ramifications of obsolescence. In: Cost analysis of electronic systems. World Scientific, Singapore, pp 307–328
23. Herald TE (2000) Technology refreshment strategy and plan for application in military systems—a how-to systems development process and linkage with CAIV. In: Proceedings of the national aerospace and electronics conference (NAECON), pp 729–736
24. Porter GZ (1998) An economic method for evaluating electronic component solutions. Boeing Company White Paper
25. Harris FW (1913) How many parts to make at once. Factory Mag Manag 10(135–136):152
26. Zheng L, Terpenny J, Sandborn P, Nelson III R (2012) Design refresh planning models for managing obsolescence. In: Proceedings ASME international design engineering conferences & computers and information in engineering conference, Chicago, IL
27. Nelson R III, Sandborn P (2012) Strategic management of component obsolescence using constraint-driven design refresh planning. Int J Product Life Cycle Manag 6:99–120
28. Gates W (2003) The Bill Gates Method. APT News, 21 Jul 2003
29. Sandborn P (2007) Software obsolescence: complicating the part and technology obsolescence management problem. IEEE Trans Compon Packag Technol 30(4):886–888
30. Rickman T, Singh G (2002) Strategies for handling obsolescence, end-of-life and long-term support of COTS software. COTS J, 17–21 (Jan)
31. Merola L (2006) The COTS software obsolescence threat. In: Proceedings of the international conference on commercial-off-the-shelf (COTS) based software systems 2006
32. Mittelstaedt T (1999) Network community, managing software (and system) obsolescence, Computer Bits, 9 Nov 1999
33. Sheppard R (2005) Software obsolescence, COG workshop Presentation, 19 May 2005
34. Gowland A (2002) Managing software obsolescence in the COTS environment. Component obsolescence group (COG) workshop presentation, 27 Nov 2002
35. CIE (2006) Does software go obsolete? Components in electronics
36. De Grip A, Van Loo J (2002) The economics of skills obsolescence: a review. In: De Grip, Van Loo, Mayhew (eds) Research in labor economics, 21 Elsevier, Amsterdam, pp 1–26
37. Sandborn P, Prabhakar VJ, and Kusimo A (2013) The analysis and modeling of the obsolescence of critical human skills necessary for supporting legacy systems. In: Proceedings of the COG international conference, York, UK

38. Besanko D, Doraszelski U, Kryukov Y, Satterthwaite M (2010) Learning-by-doing, organizational forgetting and industry dynamics. Econometrica 78:453–508
39. SD-22 (2010) Diminishing manufacturing sources and materials shortages: a guidebook of best practices and tools for implementing a DMSMS management program. Defense Standardization Program Office, Fort Belvoir, VA

Chapter 21
TES Service Innovation and the Role of Standards

Paul Tasker, Andy Shaw, Ben Sheridan and Sarah Kelly

Abstract The chapter explores those areas of activity where service innovation can most directly contribute to the design and implementation of engineering services where a number of actors within the service eco-system must collaborate to "co-create" value in use from long-life physical assets. This perspective on through-life engineering services (TES) is then used in order to consider how formal standards may be tailored and used in order to promote performance improvement and innovation in these engineering services. Standards can help promote innovation by codifying and communicating best practice, but have generally been developed to date to suit the conventional, transactional business logic rather than the collaborative, adaptive, outcome-based approach needed for a service-dominant logic. This means that there are gaps in the availability of framework and process standards suitable for engineering services, and that those extant technical standards that are applicable in areas such as dependability and reliability engineering need some tailoring to suit the new business paradigm.

21.1 Introduction

There has been considerable research in recent years on service and service systems enabled by maintaining complex engineering assets [10]. This research has provided insights into the need for a paradigm shift in the way teams behave and resources are integrated in order to focus on the common purpose of creating value in use, for the user, from the engineering assets developed, deployed, updated, replaced and retired during the ongoing delivery of the "complex engineering service" or "product-service

P. Tasker (✉) · A. Shaw
EPSRC Centre for Through-life Engineering Services, Cranfield University, Cranfield, Bedfordshire MK43 0AL, UK
e-mail: p.tasker@cranfield.ac.uk

B. Sheridan · S. Kelly
BSI, 389 Chiswick High Road, London W4 4AL, UK

L. Redding and R. Roy (eds.), *Through-life Engineering Services*,
Decision Engineering, DOI 10.1007/978-3-319-12111-6_21

system". Some large companies have been "servitizing" for some while: IBM and Rolls-Royce, BAE Systems and others have been supporting and adopting into use service research whilst developing their own new practice. Increasingly many large and small manufacturing companies are looking to servitize to maintain or improve market share and profitability as conventional manufacture becomes more competitive but these more recent protagonists struggle to realize the benefits sought [9] let alone mange innovative growth. It is hypothesized that this is largely because research and emerging best practice, demonstrating the need for a new organizational and behavioural paradigm, are inaccessible as learning to date is obscure and has not yet been adequately codified: it is suggested this is exacerbated for complex engineering services by the difficulty of transforming arcane, transactional systems engineering practice into that of a collaborative resource integrator based on the principles of service-dominant logic [20].

Standards promote innovation [2] by providing a clear foundation of best practice and a clear basis for cooperative working and development across teams. Because we know that successful resource integration is critical to effective service operations, it is arguable that the establishment of standards should be critical to service provision by a multi-organisation eco-system. Some more experienced practitioners are getting over this by using in-house or private standards, largely dealing with the processes involved, but this approach struggles to support resource agility (across an agile supply network) and customer variety. There are a number of existing technical and process standards available that are clearly applicable to engineering services, albeit with some tailoring to allow for the new organisational and behavioural paradigm, but new framework standards are needed to support innovative development of the developing market for servitized manufacturing. Until these are developed practitioners should consider developing their own framework standards bespoke to individual applications to ensure resource integration and alignment across the enterprise engaged in delivering engineering services, and as a contextual basis for tailoring available technical and process standards. Throughout this chapter we are using informal BSI definitions for different types of standard:

- Technical Standards deal with technical and material attributes such as material properties and standards for calculation. These are the oldest type of standards
- Process Standards articulate best or recommended practice for business processes such as quality management (ISO-9000)
- Framework Standards and beginning to emerge that will support best practice, enterprise-level operational performance, organization and behaviour.

However, in order to understand how best to use existing and new standards to support engineering services in the context of generic servitized manufacture it is necessary to clarify the scope of TES, or engineering services, as they might apply to a range of manufacturing sectors and a range of scale of application from small to medium sized companies (SMEs) to large enterprises. It is also necessary to develop a language that is sufficiently meaningful across such a wide range of applications, at least to the extent that terms can be recognized sufficiently to allow tailoring to specific sectors.

Another need is to identify, as far as possible where and on what activities to focus—what are the activities and basic processes that most drive value in the TES? To help with this a generic value map for engineering services has been proposed by Shaw et al. [16] that provides a view of those activities or attributes of engineering services that most contribute to realization of the overall value proposition based on the need to improve asset availability at minimum cost, although does not preclude other values.

21.2 Scope of TES, and Importance of the Service Context

What are we trying to innovate? What are we looking to standardize to support this innovation? To answer these questions we need to consider what the generic "system" boundary is for TES—can TES be defined so that it is meaningful across different sectors of manufacturing? This presents a potential problem with the term "through-life engineering services" because it is a term unfamiliar to many potential users, and which is difficult to separate from related terms in more common usage such as maintenance, MRO (maintenance, repair and overhaul) degradation or many others.

Roy et al. [15] define TES as: "the technical services that are necessary to guarantee the required and predictable performance of a complex engineering system throughout its expected operational life with the optimum whole-life cost" and focuses on those services that enable effective maintenance and feedback to new design in the context of integrated product service systems (IPS^2). Roy et al. then go on to illustrate this by describing seven possible "types of TES":

- Application of advanced information technology (IT)
- Optimise component/system life
- Managing degradation
- Autonomous maintenance
- Obsolescence management
- Cost engineering
- Uncertainty modelling and simulation.

Redding et al. [14] note that TES is rooted in the manufacturing trend towards product service systems (PSS) and the need within PSS to consider the life-cycle management of products to reduce their overall cost of ownership. Meier et al.'s discussion [6] expands on IPS^2 describing the impact of, and some frameworks to assist in, the integration of the design of products for service with the design of the delivery service system focused on "the customer and IPS^2 provider working together in order to achieve the highest value from the IPS^2". This does perhaps provide a better if more subtle understanding of Redding et al.'s [14] "reduce[d] overall cost of ownership" by conditioning this with the overall objective of achieving highest overall value from the IPS^2 and thus the TES.

The work of Cambridge, Cranfield and others towards the BAE Systems/EPSRC research programme "Support Service Solutions: Strategy and Transition" (S4T) [10] also focussed on the challenges for manufacturers developing product-based services requiring "value in use" to be "co-created" between consumer and service provider. Both this work and Meier and Roy [6] argue for the need for an overarching framework to describe IPS^2 that could then be the basis for scoping TES: to provide as complete context and understanding for TES as is possible so that we can scope through-life engineering services as being that integrated set of technical services provided by an enterprise to allow the user of an asset to maximise his value in use at minimum cost.

However, the overarching framework for "a competitive IPS^2 "proposed by Meier and Roy [6] is, arguably, quite complicated and does not appear to lend itself to easy description but the key concepts discussed align well with Ng et al.'s "Common Integrating Framework" (CIF) for complex services [11, pp. 1–19, 439–454] which identified the three main "transformations" driving the co-creation of value in a PSS: creating value by the enterprise transforming (changing the state of) material such as physical assets, information and people or human assets, their behaviours and the way in which they are organised. Figure 21.1 is adapted from Ng et al. to provide a scoping context for through-life engineering services deployed in support of an overall "complex service" or PSS. Although this may not help us very much in defining TES in a way that will have broad and easy traction (with potential users) it does perhaps provide an high level framework that demonstrates that we should consider TES within the overall context of service provision, focusing on the co-creation of value in use for the physical assets. Roy et al.'s individual "technical services" [15] are then focused on the "transform materials" sector covering maintenance, repair, overhaul, update, replacement and retire for the physical assets.

Shaw et al. [17] take a singular approach to TES: "Through-life Engineering Services is a novel dimension in engineering services rendered during the lifecycle

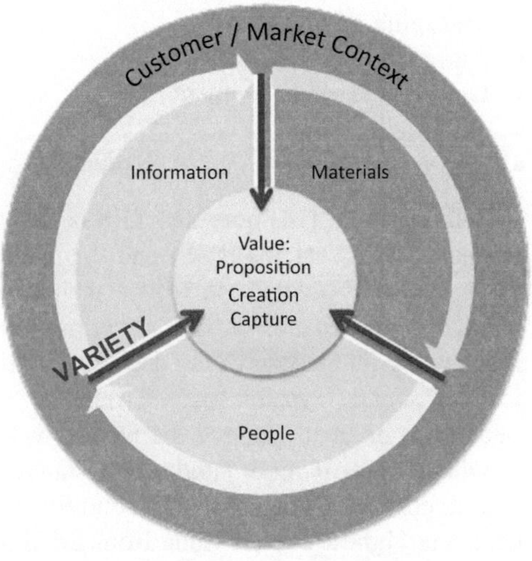

Fig. 21.1 Interpretation of Ng et al.'s common integrating framework for complex services [8]

of a complex engineering system" whereas the earlier work by Pliska et al.'s [13] perhaps more realistically but less helpfully described TES as "a broad concept that embraces different aspects and processes that can differ from sector to sector" suggesting Roy et al.'s examples of TES [15] as a small sub-set of topics within this "broad concept".

A wider definition of TES as a service (rather than an integrated system of individually identifiable of services) is provided by taking the principles advanced by Ng et al. represented in Fig. 21.1 to expand on the narrower aspects of Roy et al. and Shaw et al.'s definitions, unpacking "optimum whole life cost" and providing a focus on value: "a Through-life Engineering Service (TES) is the collaborative provision of a holistic customer capability (the ways and means of capturing value which will vary over time) based on the assured readiness and availability of complex engineering assets and products. The system boundary is set to ensure the service delivery is most effective and risk is appropriately distributed across the delivery network".

It is suggested that this overall definition of TES puts the individual engineering services in an appropriate context of an overall service, with all that implies about the need to transform people and organisations, as well as information and materials focusing on value in use as an effective service outcome: a new paradigm for design and manufacture [4].

Examples of a through-life engineering service comprising various individual and integrated engineering services may be:

- A military "availability service" such as that provided by BAE Systems and Rolls-Royce for fixed wing fast jets [7]
- Rolls-Royce's "power-by-the-hour" and TotalCare and Boeing's GoldCare (all of which are trademarked services)
- Fleet maintenance services found in the rail sector such as that for the Voyager fleet by Bombardier Transport
- Engineering services deployed to maintain a manufacturing plant and its component assets operational as and when required
- Engineering services deployed by car manufacturers and retail outlets to ensure cars are maintained effectively—with a variety of financing deals available.

21.3 The Role of Standards in Business Model Innovation

There is strong evidence that the development and adoption of appropriate standards has historically been critical in enabling the commercialization of new innovations [1, 18, 19]. One advantage in engaging with the standards making process at an early stage is that of attaining 'first mover advantage' ahead of the competition. These standards can add value for many reasons, including establishment of common vocabularies in multidisciplinary industries; enhanced confidence of investors and customers; and reduction of barriers to trade. Additionally there is evidence that suggests that an absence of standardization at an early stage can lead to unnecessary and prohibitive economic inefficiencies and reduced effectiveness.

BSI's Standards Development Briefing [2] says: "Standards play a vital role in bringing new ideas to market faster because they provide a formal process for knowledge transfer and consensus between the research community, investors and the future supply chain". They help to codify new and emergent knowledge to enable exploitation. They can promote the diffusion of knowledge, accelerate speed to market and reduce risk allowing the attraction of affordable capital, and can promote competitive advantage—"who sets the standard makes the market". Standards do not need to disclose intellectual property (IP) to promote innovation in these ways although market leaders will invest their IP in how standards are applied.

Innovations of different maturities often benefit from varying approaches to standardization. For example, new technologies with an immature and disparate stakeholder base require vocabularies and semantic standards to enable efficient communication of highly technical information. As the innovation matures and begins to reach the marketplace, characterization and testing standards become necessary to add confidence and ensure quality. As such products (or services) become established in the marketplace continued development of appropriate standards can promote ongoing innovation based on the compatibility of, and common specifications for foundation technologies, processes and behaviors.

In the case of an innovation such as engineering services, it is suggested that we must look to deploy a combination of relatively mature technology, already standardised, together with more emergent thinking related to service processes and organisational behaviour. As the key "enterprise-level" component of servitized manufacture, standardization will assist in integrating resources and capabilities across the eco-system (or the service supply network) requiring new framework standards and the development or tailoring of technical and process standards. Whilst some practice has addressed multi-actor complex engineering services [7], and new research into multi-actor eco-systems is emerging, development of a framework of standards must be an essential pre-requisite to enable practical development and innovation across these wider service delivery networks, whilst also assisting smaller enterprises to make the first steps towards servitized manufacturing with an assurance that has not been evident to date.

21.4 Standards Applicable to TES and Tailoring Guidelines

There are currently a great many standards and standards development activities that clearly relate to the broad concept of TES: extant and developing standards aimed at assisting the efficient life-cycle management of engineering assets (using the term "asset" to refer to any manufactured or built product or system which needs to be maintained in operational service over a long lifetime in order to support wider business objectives) with PAS-55/ISO-55000 being perhaps the prime example together with the work of BSI's Dependability Committee (DS-1, including for example IEC-60300-3-3 and BS EN 62402) and TDW/4/7 working on BS8887. It is however clear that with the possible exception of PAS-55, the wide variety of existing standards, and current development activity, is driven "bottom-up" as has been practice for standards development to date.

Shaw et al. [17] reported that some 83 standards, from 11 sources, had been identified as relevant to TES whereas a broader 2014 database survey by BSI identified two orders of magnitude more world-wide (of the order 2–3000). If we use Ng et al.'s Common Integrating Framework (Fig. 23.1) to illustrate the areas which standards might need to address in order to cover the important aspects of TES it seems likely that there are:

- Many extant technical and process standards concerned with "*material transformation*": ensuring the maintenance, repair, overhaul and upgrade of material or physical assets. Indeed there are many standards covering maintenance with little clarity over which may represent "best practice" for any particular application although unsurprisingly there are undoubtedly potential gaps in areas of emergent knowledge (from current research) such as "autonomous maintenance" and self-repair, degradation and the management of "no faults found" identified by Roy et al. [15]. Nevertheless many of the existing technical standards are directly relevant to TES and ought to require little tailoring.
- Many extant technical and process standards concerned with "*information transformation*": acquiring, sharing and transforming data and information, and codifying knowledge across the enterprise to inform action to enable the organisations responsible for assuring assets' performance through-life are able to provide the support service valued by the user [5]. Development of standards in this area tends to be dominated by aerospace and defence although automotive and rail have a rapidly developing interest. It is suggested that the main gap is in the ability to understand what information is needed for effective TES in light of use and customer variety, and in agile information and knowledge management.
- There are probably few current process standards and fewer framework standards that help with how individuals and organisations within the enterprise perform to ensure that the through-life engineering services deliver value to users and to each collaborating stakeholder. PAS-55 appears to help with the overall management of assets through life and the development of ISO-9000 to address services may help the understanding of quality in service (which is by definition intangible and dependant on customer context) [6].

PAS-55 is concerned with asset management and in its 2008 form (PAS-55:2008) focuses on managing the physical asset whilst considering the interface with other asset types (Fig. 21.2) each of which will have an influence on the technology, technical processes and behavioural standards of technology-based human resources needed for delivery of an effective product service system. PAS-55 provides a management framework and codifies practice for the specification, implementation and operational management of physical assets in support of business needs. It is

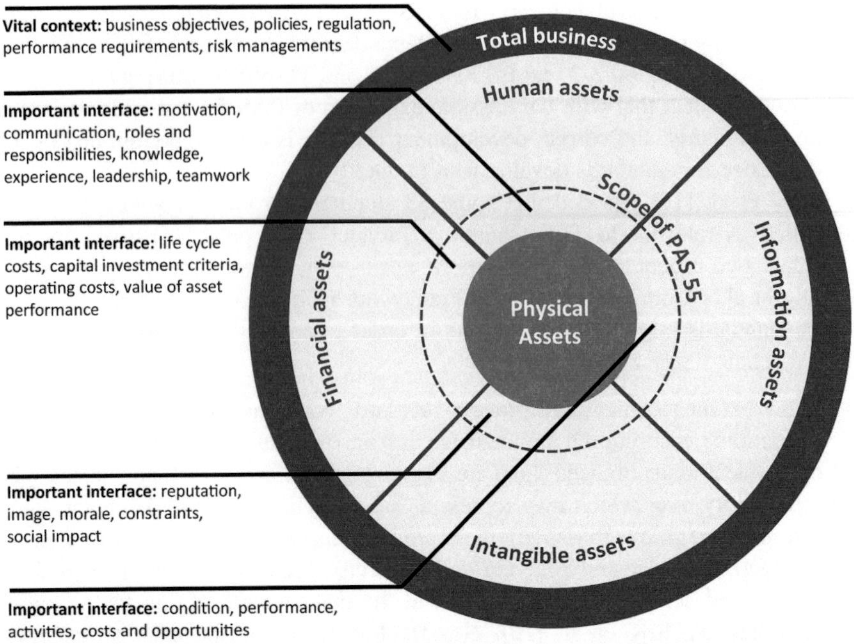

Fig. 21.2 Illustration of scope of PAS-55, before conversion to ISO-55000

predisposed to considering assets such as manufacturing plant but should be equally applicable to other types of asset (but appears to have little traction with potential users outside those interested in manufacturing plant management). However, the similarities between the PAS-55 "types of asset" (Fig. 21.2) and the CIF's (Fig. 21.1) three "transformations" that drive service value, are striking: PAS-55 must be, subject to further consideration, a prime candidate as a good first attempt at an overarching management standard for TES if the issues of branding could be overcome (asset management, TES, degradation, maintenance all appear to be largely overlapping terms which have protagonists prepared to fight their corner). In the form of ISO-55000 however, asset management is generalised to encompass assets of any type arguably resulting in a loss of focus from the engineering services activity specific to the management of physical assets.

21.5 Value-Driving Activities and Processes—A Framework Standard for TES

Shaw et al.'s "value map" at Fig. 21.3 [16] provides a view of those activities or attributes of engineering services that most contribute to realization of the overall value proposition based on the need to improve asset availability at minimum cost, although does not preclude other values. Although on-going development and

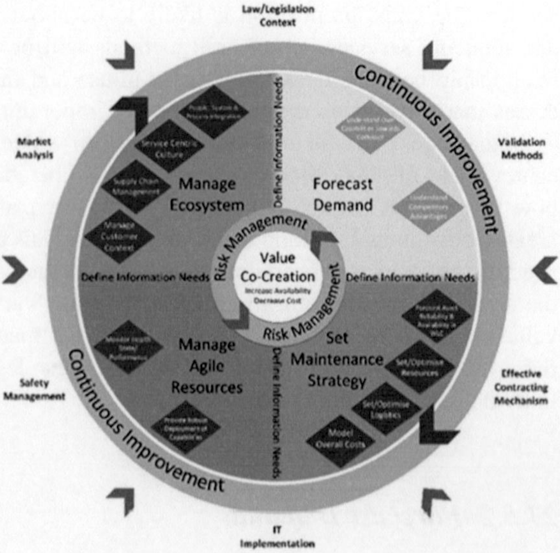

Fig. 21.3 Possible value map for an engineering service—taking an enterprise or eco-system perspective for each actor

validation would undoubtedly be desirable, the value map has had some qualitative cross-sector endorsement. Arguably, it may be considered a good starting point to comprehensively address those activities and process areas on which engineering services most critically depend, and as a basis for understanding where tailored or new standards, particularly new framework standards, can support innovation in TES.

The value map (Fig. 21.3) covers all the important activities—the application of tools, processes and resources, and demonstrates the need to understand the values, principles and behaviours across the overall organization (the enterprise or eco-system) it is proposed that the value map provides a good basis for a framework standard for engineering services. This would span a wide variety of manufacturing sectors with users being those manufacturing organizations with an interest in "servitizing" or that have "servitized" and now offer a service or services based on the availability and readiness for use of complex engineering or physical assets.

As a framework standard the intent would be that the document would codify the behaviors and overall approaches to be taken to ensure that the embedded engineering services can integrate effectively to deliver the outcomes required. The new framework standard will call on relevant extant process and technical standards, providing guidance on how these may be tailored for engineering services and, or specific application.

21.5.1 *Value and Risk*

At the center the framework focuses on value co-creation and determination of the worth [12] of the overall service and the resulting value proposition, how this is co-created by the eco-system and captured by individual actors. For engineering services the value proposition is likely to be about increasing service availability and reducing service cost. Overall methods will be described for establishing the relationship between delivery of value-in-use and the principle activities and processes that drive value creation, and other important enabling activities.

Management of the relationship between value co-creation and the risk to achievement of the desired outcomes will also be addressed in order to understand how this can be effectively managed across each of the key process and activity areas. Conventional risk and environmental analysis may be used, but attention will need to be paid to uncertainties and the impact of change and the ability of the service system to absorb customer and use variety [3]. Consideration of the value proposition, capture mechanism and risk management needs will establish the overall contracting framework between the legal entities within which the individual actors work.

21.5.2 Forecast Demand

As well as forecasting demand for the assets that support the service provision—how, when and in what environment they are to be used, demand for the overall service needs to be forecast in the context of the market and customer need. This means understanding the customers' overall operating environment in relation to the capabilities of service providers and the advantages that potential competitors may have. Consideration needs to be given to eliciting objective, subjective and implied requirements, such as ethos [8].

Establishing and maintaining a forecast of demand requires continuous and effective communication between the customer and service supplier, and all the actors within the eco-system establishing the principle requirements for information flows across the eco-system delivering the information needed to set and adapt the maintenance strategy.

21.5.3 Set Maintenance Strategy

Setting, adapting (during operation) and delivering the maintenance strategy for the physical assets that underpin the service—allowing the customer to enjoy value in use—is the "core" engineering activity. In the context of demand, setting the maintenance strategy requires a forecast of asset reliability, availability and readiness across the lifecycle, and the resources and logistics needed to deliver asset availability. Whilst the focus will be on physical assets, consideration needs to be given to the reliability, availability and readiness of non-physical assets including organizations and people.

Because the overall cost of providing a "complex service" or "product-service system" will tend to be driven by the cost of providing, maintaining and disposing of physical assets, the ability to model costs is an important feature of iterating to an optimum maintenance strategy. However, again the costs associated with non-physical assets, and with change and uncertainties must also be considered.

21.5.4 Manage Resources

Resources of all types within teams and across the eco-system must be deployed and adapted with agility to support the forecast demand and a maintenance strategy responsive to variety in use. Based on monitoring the health state and performance of physical assets and undertaking appropriate analysis, non-physical assets (people, processes, tools, finance, information etc) need to be configured and re-configured for optimum deployment. Processes will need to be integrated, adapted to reflect customer and use variety, and human resources re-configured to maintain an appropriate alignment of behaviour whilst ensuring capabilities are complimentary across the eco-system.

21.5.5 Manage the Eco-system

This section of the standard will address the approach to overall management of the eco-system based on information gained (and fed back to) demand forecasting, maintenance management and the deployment of resources. The activity will ensure the culture, behaviors, and capabilities remain aligned across the overall eco-system and supported by appropriate interface arrangement, business models and the basis for effective contracts focused on jointly managing risk and co-creating value in use. It will include what would be conventionally described as supply chain management (which will still be appropriate for some areas of activity) whilst enabling the network of collaborators to understand and manage the customer context. The need for the collective to manage customer context—to work together to change what the customer does—is predicated on the need for the customer to change in order to enjoy improved value and the continuous improvement of this.

21.5.6 Continuous Improvement and Enabling Activities

A mechanism of continuous improvement is proposed that will allow integration and ongoing performance improvement of the value driving activities, supported by the more "conventional" business and engineering support activities such as market analysis, legal and regulatory compliance, validation methods (including new methods for level of service validation) commercial and contracts management, information technology and safety management.

Of these enabling activities, safety management is the most challenging for which new techniques related to the service risk management approach will need to address the potential conflict between the need for clear lines of accountability and collaborative endeavor focused on co-creation of value in use.

21.5.7 Guidelines on Tailoring the Value Map

We have seen that there are very many technical standards applicable to design for life and the maintenance of physical assets, not least of which being those addressing dependability, together with some process standards. Most if not all these extant standards apply to conventional manufacturing and maintenance, repair and overhaul activity and are likely to need various degrees of tailoring to ensure they can be applied to service and engineering services in particular. This is driven by the collaborative nature of service and the need to co-create value in use realized at the time of use, rather than the normal paradigm of value in exchange derived from a transactional or sequential, rather than concurrent activity.

It should be intended therefore that this section of the standard will provide some general guidelines for tailoring extant standards for application to engineering services.

References

1. Allen R, Sriram R (2000) The role of standards in innovation. Technol Forecast Soc Chang 64 (2):171–181
2. BSI (2014) Standards outlook 2014, BSI
3. Erkoyuncu JA, Roy R, Datta P, Wardle P, Murphy F (2011) Service uncertainty and cost for product service systems. Complex engineering service systems. Springer, London, pp 129–147
4. Griffiths B (2012) Manufacturing paradigms: the role of standards in the past, the present and the future paradigm of sustainable manufacturing. Proc IMechE B: J Eng Manuf
5. Mason H (2011) The challenges industry faces with data integration in through life support of capability. In: UKCeB conference on through life engineering
6. Meier H, Roy R (2010) Industrial product-service systems—IPS^2. CIRP Ann Manuf Technol 59(2):607–627 (Elsevier)
7. Mills J, Parry G, Purchase V (2011) Enterprise imaging: visualising the scope and dependencies of complex service enterprises. Complex engineering service systems. Springer, London, pp 49–65
8. Mills J, Parry G, Purchase V (2011) Towards understanding the value of the client's aspirations and fears in complex, long-term service contracts. Complex engineering service systems. Springer, London, pp 87–103
9. Neely A (2008) Exploring the financial consequences of the servitization of manufacturing. Oper Manag Res 1(2):103–118
10. Ng I, Parry G, Wilde P, McFarlane D, Tasker PH (eds) (2012) Complex service systems, strategy and transition. Springer, London

11. Ng I, Parry G, Maull R, McFarlane D (2011) Complex engineering service systems: a grand challenge. Complex engineering service systems. Springer, London, pp 1–19, 439–454
12. Parry G, Tasker PH (2014) Value and servitization; creating complex deployed responsive services. Strategic change 23:303–315
13. Pliska A, Pandey A, Rigatti C, Chuku G, Matar F, Alajmi N, Qian R, Tang W (2012) Final report on through-life engineering services standards development (revised 10 Aug 2012). Cranfield University MSc project 2012
14. Redding L et al (2012) Challenges and opportunities in through-life engineering services—a survey of manufacturing organisations. In: 10th international conference on manufacturing research
15. Roy R et al (2013) Through-life engineering services. Manage Control 46(6)
16. Shaw A et al (2014) Establishing key performance drivers for through life engineering services. Procedia CIRP
17. Shaw AP, Pliska A, Pandey A, Rigatti C, Chuku G, Matar F, Alajmi N, Qian R, Tang W, Roy R, Sheridan B (2012) Through-life engineering standards development. In: Proceedings of the 1st international conference on through-life engineering services, Cranfield University
18. Swann GMP (2010) The economics of standardization: an update. Report for the UK Department of Business, Innovation and Skills. London
19. Van Merkerk RO, Robinson DKR (2000) Characterizing the emergence of a technological field: expectations, agendas and networks in lab-on-a-chip technologies. Technol Anal Strateg Manage 18:411–428
20. Wood LA, Tasker PH (2011) Service thinking in design of complex sustainment solutions. Engineering service systems. Springer, London, pp 397–416

Part VI
Autonomous Maintenance

Chapter 22
Building Dependable Electronic Systems for Autonomous Maintenance

Richard McWilliam, Philipp Schiefer and Alan Purvis

Abstract Maintenance repair and overhaul (MRO) of high value systems is expensive, time consuming and relies heavily upon back-to-base repair and overhaul activity. Autonomous maintenance of repairable systems is a rapidly developing area in through-life engineering services that specifically aims to reduce both mean time to repair and frequency of preventative maintenance. Modern engineering systems must perform reliably in the event of random upset events that threaten to induce malfunction or unpredictable behavior. These requirements are fuelling the integration of fault-tolerant and self-repairing techniques into electronic systems at design time. This chapter investigates emerging techniques being utilised in electronics that bring new self-repair capability to high-value applications such as aviation, land vehicles, renewable energy and space exploration. The cost/benefit trade-off of self-repair strategies is analysed in terms of redundant resource allocation and key performance metrics. The potential for future uptake is discussed in the context of current and next-generation platforms.

22.1 Introduction

The emergence of built-in self-test (BIST), self-reconfiguration and mechatronic assist technologies is driving new research and development of systems that will be capable of self-repair. Within the electronics domain these concepts promise to create the ability to maintain in-service operation in the presence of faults by either masking the effects of the fault or else performing self-reconfiguration to remove the faulty logic. The former is useful for handling random, non-permanent fault events while the latter re-establishes a fault-free system by deactivating logic that has suffered permanent faults. The primary benefit is that such systems become

---

R. McWilliam (✉) · P. Schiefer · A. Purvis
Centre for Electronic Systems, School of Engineering and Computing Sciences, Science Laboratories, Durham University, South Road, Durham DH1 3LE, UK
e-mail: r.p.mcwilliam@durham.ac.uk

L. Redding and R. Roy (eds.), *Through-life Engineering Services*,
Decision Engineering, DOI 10.1007/978-3-319-12111-6_22

better able to look after themselves by performing self-maintenance tasks, thus leading to increased availability.

A common feature present in all self-repair strategies this is that of additional redundant resources, implemented either at the fine-grained level (e.g., transistor or gate element) or modular sub-component level (e.g., chip or board level). Redundancy has been deployed to great effect in increasing the yield of electronics manufacturing processes for many years, especially for high density components such as memory chips. In recent years however there has been significant rejuvenation of established fault masking techniques for increased fault tolerance following the emergence of next generation nano-electronic fabrication techniques that will suffer lower yield than current semiconductor processes. Aside from manufacturing yield, there is significant interest in supporting through-life maintenance through the provision of self-repairing capability by incorporating self-repair by autonomous reconfiguration within both COTS FPGAs and custom ASICs.

The basic concepts of fault masking and self-reconfiguration suitable for electronics design are discussed in Sect. 22.2. Section 22.3 investigates the different ways in which redundant resources may be deployed and identifies key performance metrics. State of the art strategies that utilize COTS FPGA chips are discussed in Sect. 22.4. Finally, a brief commentary on the future of self-repair and key challenges is provided in Sect. 22.5 along with a qualitative summary of key performance metrics.

22.2 Basic Concepts and Motivation

The incorporation of fault masking, built-in test (BIT) and built-in self-repair (BISR) functionality brings attractive potential benefits: (1) reduction of the cost associated with MRO by assisting with efficient maintenance scheduling; (2) increased availability of electronic components and systems; and (3) extension of the predictable operational life time by better regulation of wear out. The fundamental challenge of implementing test and repair strategies stems from the fact that redundant resources are expensive and must therefore be applied sparingly or else the design quickly becomes unyielding and expensive. Efficient deployment of redundant resources requires design compromise in which the expected cost/benefit trade-off must be understood clearly in order to create an effective mitigation strategy. There is as yet no single optimal strategy for fault-masking or self-repair [1, 2]. Evidence of this is seen in electronic systems that incorporate boundary scan test interfaces, which are in common for production test and repair. This type of interface provides an effective trade-off between resource and benefit for production yield enhancement, however its scope is somewhat limited for in-service duties because the circuitry is not self-contained and relies upon external hardware in order to provide full BIST capability. This special interface is not generally optimised for low power, but rather high speed due to the fact that in the production environment each second of test time adds significant overhead[1] to product cost [3]. Thus, implementations that enable in-service BIT, fault-masking or BISR tend to rely upon a multitude of strategies that are constrained by minimal power and resource overheads, but which must also provide effective fault recovery and reporting. In the case of non-repairable faults being detected in the field, human interaction is often necessary in order to confirm faulty hardware to repair or replace the offending component or sub-component and such procedures are routinely carried in the repair shops. BIT, BISR and fault masking offers the potential for in situ diagnosis, fault monitoring and repair.

A fundamental difference between fault masking and self-repair is that the latter requires active management of redundant resources. This is related to practice of dynamic redundancy, in which spare components are available but held in a standby state. When a fault condition is detected a fault detection and reconfiguration unit initiates replacement of the faulty component. This type of strategy can be combined with other fault masking methods, for example triple active redundant hydraulic power systems for aircraft that also incorporate an emergency standby system.

BISR design relies upon an improved understanding of the characteristics of component degradation and random upset events and a suitable strategy for deployment of redundant resources. This challenge has been identified in the context of avionics [4] but is seen in a wide variety of high value systems. A detailed analysis of the trends in reliability of high density SRAM was carried out by White [5] in order to better understand actual failure characteristics for SRAM chips. The result was an apparent departure from the conventional "bath tub" curve (Fig. 22.1), influenced in part by improved understanding of counter-wear out

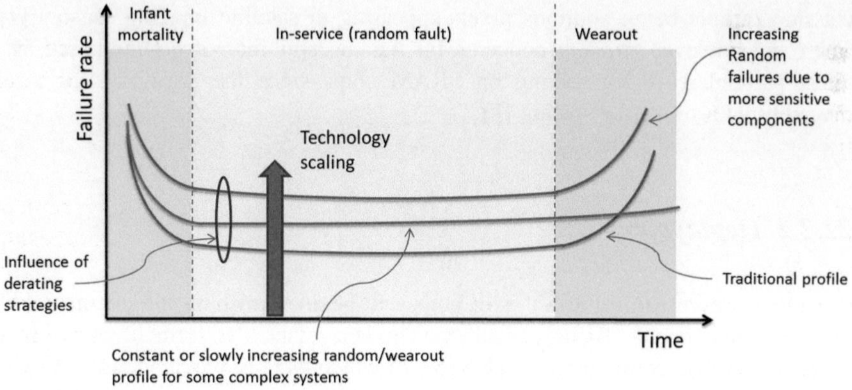

Fig. 22.1 Depiction of potential failure rate versus time profiles expected for modern electronic systems

---

[1] An often-quoted rule of thumb is that each hour of test time contributes 5 cent to product cost.

measures such as derating, and partly by careful modeling of fault models for production faults, wear out faults and random upset events. In some cases it becomes difficult to distinguish between particular wear out and random failures that assume nominally constant rate during in-service operation. This is a very important feature in the deployment of self-repair strategies because the selection of optimal mitigation strategy requires intimate knowledge of the underlying fault characteristic.

It is also important to keep in mind the robustness of wholly integrated systems that require sensors and actuators in addition to core circuitry. A holistic view should be formed whereby fault tolerance of sensor data and output data is included into the redundancy scheme whenever possible. A traditional approach here is to incorporate concurrent error detection using information redundancy, such as error correction codes (EDC) that seek to filter incoming sensor data and remove faulty code streams.

22.2.1 Sources of Errors

Electronic systems are bombarded by a variety of error-inducing events. Random errors may be caused by environmental factors such as hostile electromagnetic interference (EMI), high energy radiation particles and thermal cycling. Mitigation against EMI is provisioned at design time by applying appropriate grounding, shielding and transient suppression. Random events such as electromagnetic pulses (EMP) cause errors that are difficult to predict. High energy particles interact with semiconductor junctions and are also fundamentally random by nature [6]. Their influence can become of great concern for space and high altitude applications, a classic example being spurious pixels appearing in satellite imaging sensors [7]. There are however growing concerns for the susceptibility of ground-based systems, particularly those relying on SRAM chips since the error rate increases inversely with transistor scaling [8].

22.2.2 Deployment

A natural progression of BIST is of course to begin furnishing the circuitry with resources that enable BISR capability. This may take the form of either fault masking or self-reconfiguration techniques of which there is a great variety. In order to understand their relative merits, key performance factors need to be understood. There are many examples where redundant resources are added, often in modular fashion, for the purpose of fault masking where the goal is to preserve the correct functionality in the presence of faulty logic [2, 9–11]. While fault masking strategies may not be considered as being truly self-repair, it can be argued that they draw upon redundant resources in order to reassert the correct internal signal and thus represent a step towards self-repair of *information* rather than *logic*. On the contrary, self-reconfiguration strategies seek to eliminate faulty logic and re-establish a fault-free circuit and are usually more complex due to their need for both test and repair.

22.2.3 Key Performance Metrics

As with any new capability incorporation of self-maintenance requires an evaluation of cost/benefit trade-offs. The key features of self-maintenance have been identified for FPGA and custom ASIC architectures, an example of which is shown in Table 22.1. This classification, adopted from [12], originates from self-repair strategies implemented using FPGAs but the concepts are common to general self-repair strategies.

We now discuss in detail each of the metrics mentioned in Table 22.1.

Physical resources In order to incorporate self-repair capabilities, special resources must be added that may be of different design to that of existing logic design, hence requiring different testing and verification methods. In contrast, implementations for FPGAs commandeer a fraction of the available resources, which are then no longer available for application specific tasks. An example is online fault tolerance in FPGAs that require spare PLBs for built-in test and repair

Table 22.1 Evaluation metrics for self-repair

Metric	Description
Physical resources	Additional resources involved in implementing fault handling capability
Critical components	Additional components that must be assumed fault-free
Fault coverage	The type of fault manifestation to be handled and hardware involved e.g., permanent, transient faults that occur in logic blocks, interconnect, memory, IO resources
Fault capacity	Resources necessary to handle a second additional fault event. Used to quantify efficiency of redundant resources
Detection latency	Time incurred to detect presence and location of fault
Recovery time	Time during which system is not available or else executes at reduced speed while repair is carried out
Performance impact	Reduction of overall system performance as a consequence of including repair-mechanism
Recovery granularity	Smallest constituent part of system that can be repaired
Fault exploitation	Ability to reuse defective resources

that could otherwise have been utilised to increase the performance of the main task.

Critical components In addition to the correctly functioning circuitry, any fault repair or making strategy must reply upon additional fault-free resources that may be called upon in response to a fault. This includes redundant resources and any logic required to carry out repair. Examples include the majority voter in a spatial redundancy scheme and built-in reconfiguration units. Identification of critical components is especially important for reconfigurable chips that are provisioned with a finite set of resources.

Fault coverage Fault coverage refers to the type of hardware protected by the scheme and may include logic, interconnect, IO logic or memory. It may also relate to the essential characteristic of the fault event itself, be it transient or permanent, singular or persistent in nature. The nature of the design dictates the class of fault that can be tolerated, and for each case the most effective strategy is selected. It has been observed that interconnect failures are somewhat neglected in comparison to logic failures [12].

Fault capacity Once furnished with suitable resources the design is able to recover from certain fault events. Fine-grained fault masking strategies are designed to tolerate single random fault events at any location within a logic sub-circuit, but do not address multiple persistent fault events that accumulate over time. Thus their fault coverage is usually limited to random SEU or single permanent fault events. By contrast, self-repair strategies target cumulative errors by continuously circumventing faulty logic at the expense of consuming resources. As such their fault capacity is often quantified according to the necessary elemental redundant resource needed to address an additional fault. This resource may be allocated at design time (spare configurations) or else enlisted at runtime (spare configurable logic blocks).

Detection latency Many repair strategies need dependable fault detection circuitry in order to generate efficient repair configurations. Detection latency is an important metric defined as the period of time elapsed between the occurrence of a fault (during which the system is potentially untrustworthy) and completion of the fault repair operation. Fault detection must become an intrinsic design element and be optimised for the particular repair strategy. Also it is not necessarily the case that lengthy detection time will result in poor overall repair performance: roving self-testing areas (STARs) [13] incur high detection latency however faults are automatically quarantined once detected, enabling normal execution to continue whilst repairs are carried out.

Recovery time The speed of repair is critical in most applications, especially if when the system clock is frozen during repair. This results in a period after the fault event during which the outputs are inoperative or untrustworthy. An attempt is made by Parris to compare the recovery time for different FPGA-based strategies [12]. However, establishing a clear distinction between detection latency and recovery time can be challenging in some cases.

Performance impact Fault masking often achieves little or no performance reduction in the event of faults but at the expense of significant redundancy overhead. Self-repair strategies attempt to use resources in a more strategic manner, although performance degradation is still common. Normal operation may be halted whilst the repair is carried out or by conversion of existing resources into redundant resources.

Recovery granularity The smallest constituent part that is repairable using redundant resources is specified by the recovery granularity. For example, TMR is often applied at the component level where each module comprises a complex electronic sub-system such as a flight control computer. In this case repairs proceed by masking or replacing a faulty module and further investigative work is then needed to identify the precise cause of fault. Fine-grained approaches operate at the gate or even transistor level and offer a more diverse variety of repair mechanisms.

Fault exploitation Operates at the logic block, gate or transistor level, where it is possible to detect that a stuck-at or bridge fault has occurred. In this case it is sometimes possible to reuse the offending fault signal within the existing design, e.g. when a stuck-at high level does not affect the output behavior. Self-repair strategies for FGPA platforms are able to achieve fault exploitation, particularly when based upon evolutionary algorithms.

22.3 Deployment of Redundant Resources

Modern maintenance procedures rely upon BIT mechanisms coupled with human maintenance actions in order to find faults and carry out repair operations. In the production environment test and repair is often automated since the conditions are carefully controlled. An example of this is the test and repair of electronic memory chips containing defective transistors that would otherwise be discarded. The procedure involves detection and bypassing of defective cells using spare redundant cells. In order to achieve test and repair capability special circuitry specifically designed for BIST is inserted into the chip design that enables execution of test patterns within the chip and results to be relayed to external test equipment. The results provide an indication of defective cells that are subsequently deactivated and replaced by spare cells, often by laser-activation of fuses for a permanent reconfiguration. This process has many variations and boundary scan methods are applied in many production test situations.

Fault masking strategies seek to restore reliable operation in the presence of faulty transistors, gates and cells and, while common in some designs, are not considered by themselves as being self-repairing. They could instead be viewed as initiating restoration of logic signals propagating though the network using redundant elements incorporated at design-time. Redundancy is deployed at many different levels of design. Some examples are depicted in Fig. 22.2, where replication of transistor, gates, design cells (logic units) and modules are applied within a design. The three fundamental types of redundancy are: spatial, temporal and information, each of which may be applied in isolation or in combination.

Aside of fault masking, redundant resources are required by repair strategies that seek to restore a fault-free circuit i.e., to remove the presence of faulty logic. While

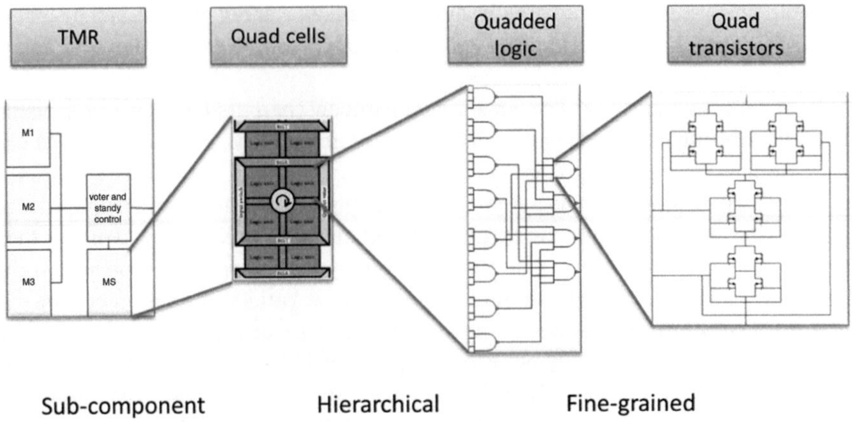

Fig. 22.2 Examples of redundancy deployed at various design levels including transistor, gate, cell and sub-component levels

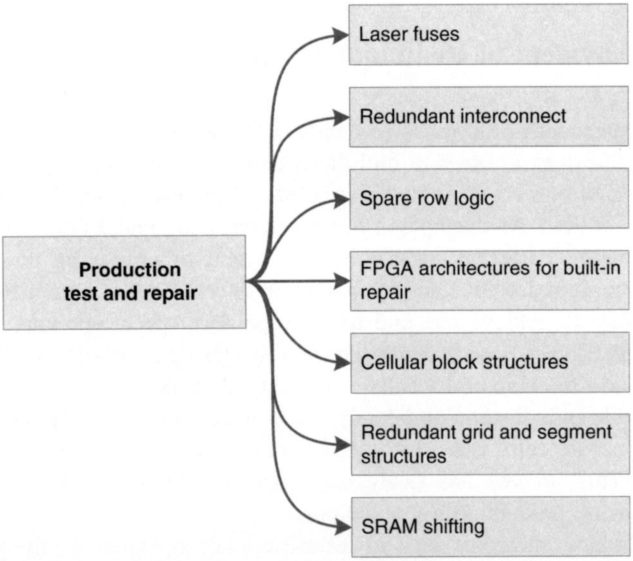

Fig. 22.3 Common repair methods used for yield enhancement (adapted from [1])

fault masking offers the capacity to 'hide' a limited sequence of simultaneous transient or successive permanent fault events, self-repair draws upon additional built-in reconfiguration capabilities to continually restore a fully correct network that is able to sustain repeated faults. A combination of both approaches is clearly desirable in some situations, especially where the operating conditions result in the occurrence of both SEU events and accumulation of permanent faults.

22.3.1 Spatial Redundancy

Spatial redundancy involves direct replication of physical resources that are used to mask faults or to replace faulty logic. Several well-known spatial redundancy schemes exist including triple modular redundancy (TMR), quadded logic and quad transistors. They are particularly attractive for SEU prone conditions since recovery is virtually instantaneous, however their fault capacity is generally low in the presence of cumulative permanent faults. Yield enhancement methods invariably rely upon spatial redundancy in order to provide a fixed set of resources available during production test and repair (Fig. 22.3). Here efforts are directed towards eliminating defects occurring during fabrication rather than random failures or wear out. This leads to a prediction of the defect tolerance of the given circuit that depends on the component failure rate and the effective deployment of redundant resources.

Considering once more random failures that occur in-service, a comparison between the reliability of common spatial strategies can be carried out using simple reliability analysis. A simple example is shown in Fig. 22.4, which compares the redundant strategies of TMR, two spares and quad design. Such comparisons are often analysed in the context of production yield since the probability of component failure, p, can be determined by the process quality. However we may also extrapolate this information into the corresponding in-service domain provided we assume that SEU events are random (i.e., statistically independent) and that the resulting failure condition is analogous to the condition of a manufacturing defect.

In Fig. 22.4 The line labeled "single part" shows the predicted reliability, R, of a non-redundant design. The line labeled "TMR" shows the probability that 2 out of 3

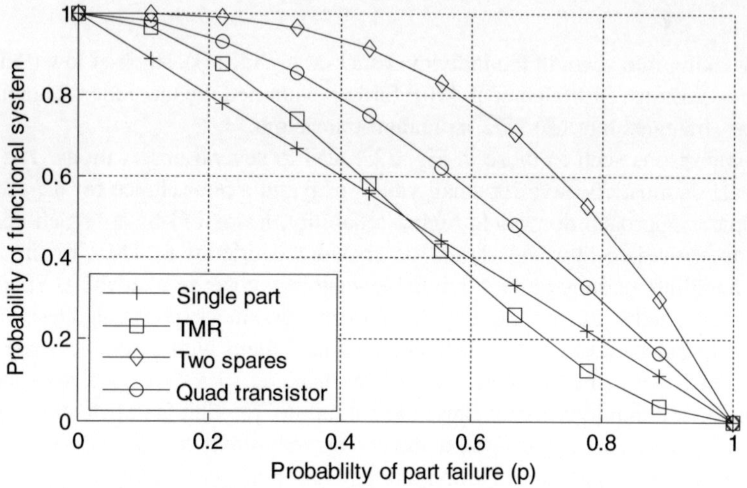

Fig. 22.4 Reliability characteristics for common fault tolerant design strategies, depending on probability of component failure, p

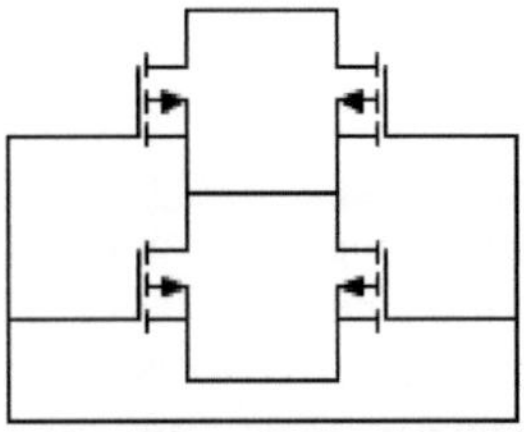

Fig. 22.5 Quad transistor circuit topology

parts are functional and does not take into account the reliability of the majority vote logic. In this case:

$$R = (1 - p)^2 + 3p(1 - p)^2. \tag{22.1}$$

A further method labeled "Two spares" is also shown, which requires at least 1 out of 3 functional components:

$$R = (1 - p)^2 + 3p(1 - p)^2 + 3p^2(1 - p)^2. \tag{22.2}$$

Equations 22.1 and 22.2 are evaluated by the well-known Binomial distribution [14] applied to system level reliability analysis. Finally, a method specific to transistor level fault tolerance labeled "Quad transistor" is calculated (see also Fig. 22.5) that that requires 3 out of 4 transistors to be functional. In this final case, the reliability is derived as:

$$R = 1 - \frac{3}{2}p^2 + \frac{1}{2}p^3. \tag{22.3}$$

By taking into account the individual suck-at high (SaH), stuck-at low (SaL) and bridging failures, El-Maleh et al. [10]. further examined the theoretical reliability of similar strategies that use N^2 redundant transistors.

Comparisons such as those in Fig. 22.4 lead to several observations. The TMR approach is most effective for small values of p and a poor choice for p > 0.5. The two spares approach does yield higher reliability, however this approach does not provide error detection and should be considered inferior to TMR for in-service fault handling (assuming that a reliable majority voter is available). Spares are commonly used in the production environment wherein a sophisticated test machine is able to detect faulty components and replace them with available spares. The quad transistor method exploits the failure characteristics of a quad network such that stuck-at conditions do not cause overall failure. Explicit fault detection does not occur however. This comparison demonstrates the importance of understanding the properties of the applied mitigation strategy and application requirements.

Fault masking will become especially relevant for next generation nanoscale technologies, where the basic resources of transistors and interconnections will be fabricated with high density but subject to lower low yield than achieved by current manufacturing processes. This has led to considerable activity in the area of online fault tolerant methods, which use massive redundancy to bring improvements to overall system availability [14, 15]. These strategies could be viewed as achieving the creation of reliable circuits in the presence of many faults as per Von Neumann's early work on the principle of building reliable computational networks. The primary benefits are twofold: built-in fault tolerance at the point of manufacture, and SEU fault masking. Note however that the allocation of redundant resource is allocated at design-time and hence there is a fixed resource for both manufacturing test and repair and in-service fault tolerance.

Fine-grained approaches In the case of fine-grained strategies such as the N^2 transistor design circuit, fault rate modelling should take into account behavior at both underlying transistor structure and gate level. In [10] it was demonstrated that fine-grained redundant methods offer favorable failure rates in comparison to gate and cell based redundancy for certain designs. An alternative detailed gate-level fault model presented in [16] accounts for transistor and interconnect level faults specific to CMOS NOR gates. This was used to build an accurate fault injection model for CMOS fault rate analysis. The properties of a gate-level redundancy scheme were investigated in [17] by adopting a simple fault injection model and insertion of various open and short circuit conditions. This enabled evaluation of detectable and undetectable fault conditions and potential repair mechanisms.

22.3.2 Temporal Redundancy

Circuits that incorporate temporal redundancy generate majority signals with minimal hardware by repeated use of logic units to calculate several results over time. If no faults have occurred during the time frame, then a fault-free network is assumed. This leads to reuse of physical hardware but requires a minimum time period before a valid majority vote signal becomes available. Hence detection latency can be considerable. Temporal and spatial redundancy can be combined in order to provide more flexible self-checking capabilities. For example, repeated operations across multiple identical hardware cells provides additional integrity checking beyond purely TMR implementations.

22.3.3 Information Redundancy

Concurrent error detection relies upon information redundancy in the form of additional information added to data patterns stored in memory. This enables recovery from corrupted data bit exploiting additional information redundancy placed into the data pattern. The additional information is referred to error detection and correction (EDC) codes. Commonly implemented for communication channels, this approach has also been implemented to protect embedded hardware design. In distributed cellular architectures, where special codes analogous to DNA sequences are stored locally in every cell, reconstruction of the correct data occurs continuously even in the presence of multiple upsets [18].

For digital logic whose functional behavior captured in the form of a state transition table, EDC can be applied to protect the data (and hence the functional behaviour). Finite state machines (FSMs) may be protected in this way mapping their state table to a suitable look-up table (LUT) stored in regular memory, which can in turn be protected using EDC codes [19]. When stored in read-only memory (ROM), high-speed compact hardware implementations result although additional (and vulnerable) execution logic is needed to implement the memory address look up and input/output interfacing. ROM-based FSMs have become prevalent in ASIC design [20] but also in FPGA implementations due to their speed and compactness [21]. Even so, the additional logic necessary to run the state machine is not negligible and must also be protected from faults. An elegant solution here is state mapping, where the original state codes themselves are modified to include single error correction (SEC) redundancy codes. This implementation, illustrated in Fig. 22.6, is able to tolerate SEUs present in the LUT and in the next state logic but cannot directly protect the EDC logic used to remove single errors present in the next state. Rochet observed [22] that the vast majority of random errors occurring in the EDC logic are fixed by the single error correction (SEC codes), however the resource overhead associated with additional LUT entries and the error decode logic leads to ever more complex designs [19]. This can ultimately lead to poor overall resource deployment in contrast to simpler strategies such as TMR, therefore careful consideration of the repair method is once again important.

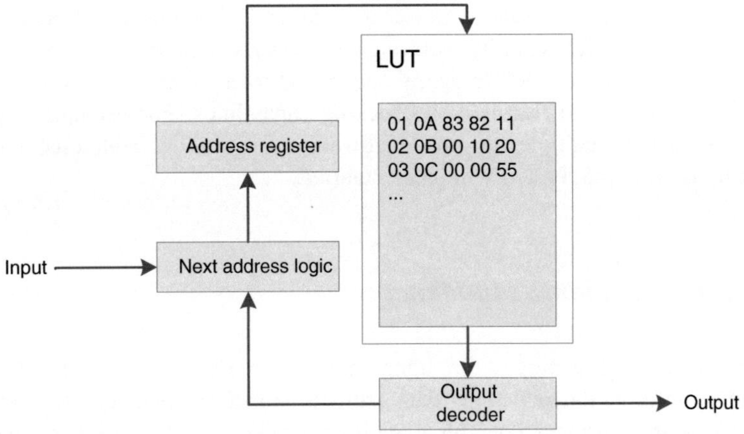

Fig. 22.6 Block diagram of ROM based FSM with LUT contents protected by EDC codes. The FSM requires minimal execution logic for next address generation and input/output interfacing

22.4 Platforms for Self-repair

Before examining the features specific to self-repair strategies, is it useful to briefly consider recent developments in compatible hardware platforms. The most common platform is the field programmable gate array (FPGA), which is essentially an ASIC chip furnished with a large pool of sophisticated reconfigurable logic resources, memory and interconnect routing. These chips are popular due to their in situ reconfigurability, wherein a bitstream file is loaded into chip in order to organise its resources at runtime. The resulting die is extremely densely populated with logic, SRAM and routing resources necessitating extensive production test and repair in order to enhance manufacturing yield. Because of this device-level (DL) fault tolerant methods are used that are transparent to the end user. To overcome this imitation, a large variety of configuration-level (CL) methods have been devised that seek to reuse the FPGA's resources for dynamic in-service repair [1, 2, 12]. Reconfigurations are carried by alteration of the configuration bitstream. Alternative full custom chip designs have been also proposed that are furnished with new redundant resources specifically tailored for fault mitigation rather than production test and repair. Strategies include fine-grained [23] and cell-based [24] redundancy. Another class of ASIC is the fully customised chip having a non-reconfigurable design that is constrained according to speed, efficiency and size restrictions. These designs may be equipped with fault tolerant resources at design-time however their target application is more likely to be performance sensitive hence highly optimised implementations are needed.

22.4.1 Key Strategies

Self-repair requires a combination of BIT and self-reconfiguration in order to detect and eliminate faulty logic using redundant resources. The principle of self-repair as autonomous design has been discussed at length [25], and principally involves the actions of detect-divert or detect-replace in order to circumvent fault logic. Another potential feature is self-preservation, which attempts to inhibit future degradation. Efficient maintenance scheduling relies on having a detailed knowledge of expected fault events and actions to be taken. An important feature of self-repair strategies is the provision of built-in logging (and possibly classification) of both random fault events. Accurate records of SEU and permanent errors are thought to be extremely valuable in gauging the remaining fault capacity and for refinement of overhaul work scheduling.

Considering strategies that are specific to FPGAs, numerous comparisons of resource-performance trade-offs have been reported, in most cases by estimating the key metrics described in Sect. 22.2.3 (primarily relating to detection latency, recovery time and resource overhead) [1]. A comparison of the relative benefits has also been presented in the form of a number of "sustainability metrics" that include

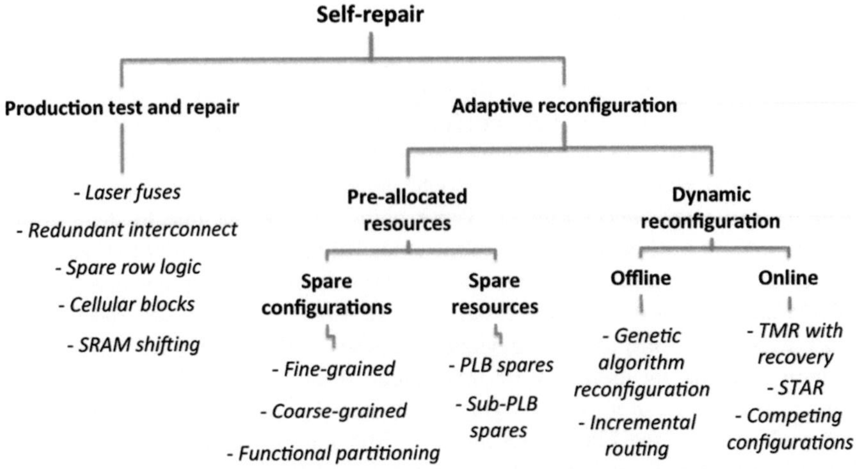

Fig. 22.7 Strategies for self-repair at point of manufacture and for in-service reconfiguration (see [1, 12, 27]). *PLB* Programmable logic block; *SRAM* Static random access memory; *STAR* Self-test area

fault capacity, granularity, interconnect fault handling and critical fault resources [12]. A further survey has considered fault detection as being an essential part of the self-repair process [26]. Findings from these studies have suggested that no single method is optimal and that the application priorities must be carefully evaluated in order to select an effective fault tolerant approach. This is true not only FPGAs but even more so for custom ASIC designs that must be optimised according to power and speed constraints. Figure 22.7 provides a summary of prominent methods reported in the literature, classified according to strategy. Pre-allocated and dynamic reconfiguration methods feature heavily in FPGA strategies, where their ability to alter the configuration bitstream is exploited to circumvent faulty logic. Pre-allocation utilises alternative configurations defined at design time in the form of spare configurations or spare logic. Spare configurations are allocated at design time and comprise a collection of alternative bitstream configurations that can be loaded in response to a faulty condition. These schemes depend upon effective BIT in order to determine the most appropriate replacement configuration that is most likely to circumvent the faulty logic. Similarly, spare resources are made available at design time and are activated upon detection of a faulty logic block. This results in a direct substitution of the block with a nearby spare block. Spare blocks are arranged at a finer granularity than spare configurations but their distribution must be allocated carefully in order to provide sufficient fault capacity.

By contrast, dynamic methods are able to generate new configurations in the field as a direct response to faults. Offline methods seek to derive a new configuration using an algorithm that attempts to process fault information, assess available resources and generates a new configuration. An FPGA's normal operation must be halted in order to reconfigure the device and in some cases to provide additional on-chip resources for calculation of the new configuration. Genetic algorithms have also been demonstrated to be capable of achieving impressive fault capacity with efficient resource usage and even fault exploitation. However, their recovery time tends to become unbounded due to the nature of the algorithms involved. Online methods seek to preserve active operation while BIT and reconfiguration operation are carried out. These approaches are somewhat complex due to their need for real-time bitstream manipulation. That said, the TMR with recovery method simplifies the complexity at the expense of performance reduction due to the need for high spatial redundancy within the active configuration. Application performance is also compromised by the STAR method approach. A detailed examination of this can be found in [13], where the performance penalty for a various design using STARs is quantified in terms of maximum clock speed and spare resource overhead. For the sample designs considered, the maximum design clock speed was reduced by some 2.5–15.1 % when 20 % of the chip was reserved for STARs. However, this penalty is countered by important benefits including the ability to deal with of dormant faults by adaptive re-usage of resources and incremental fault tolerance that escalates the provision of redundant resources as faults become more aggressive and frequent.

22.5 Potential Impact and Uptake of Autonomous-Maintenance

To date, no single fault masking or self-repair strategy has proved all encompassing; each particular design option must be evaluated in order to determine the best strategy. A wider understanding of cost/benefit trade-offs at different levels design level granularity of would be of great benefit in this area. A key issue is that accurate evaluations of fault rate behavior generally lead to complex and protracted analysis, resulting in lengthy design verification. This may be alleviated by developing useful optimisation metrics capable of exploring complex combinations of redundant strategies and application constraints. This process should begin by evaluating key metrics against the top-level strategies of fault masking and self-repair, then classifying the benefits brought about through each specific approach.

Table 22.2 provides a starting point for this process, however a common resource allocation model would be needed in order to generate different combinations of redundancy strategies.

Whether the characteristics of a particular self-repair strategy are considered acceptable depends on the application and the cost/trade-offs evident in Table 22.1. By example, the STAR strategy exerts a long detection latency amounting to several seconds typical for complex roving BIT mechanisms. However since test and repair coverage is 100 % and utilises only non-active logic at any single point in time, there is no actual down-time incurred for the self-checking process as far as the active logic is concerned. Should a fault strike then the worst-case effective

Table 22.2 Comparison between key metrics of fault masking and self-repair strategies

Metric	Fault masking	Self-repair
Mitigation strategy	Faulty logic is tolerated without error in output behavior	Faulty logic is circumvented or reused
Fault capacity	Limited capacity for SEU errors. Resources may be able to handle second additional fault—but only in limited locations	Resources designed to generate significant fault capacity
Detection latency	No explicit detection provided, but majority vote errors could be used	Varies from 100's μs to many seconds
Recovery time	Virtually instantaneous	Between 10's of milliseconds and several seconds (or even unbounded)
Fault reporting	Not inherent, but majority vote offers limited reporting	Event data available, but most solutions not sufficiently developed for reporting or logging

detection and recovery time could be unacceptable long, and redundant fault-masking resources may be needed at a more finely-grained design level. Conversely TMR fault-masking strategy offers negligible detection and recovery time however requires expensive triplicate structures and reliable majority voting logic in comparison to around 10–20 % spare resource for STAR implementations.

Another factor is the necessary complexity of the underlying fault-tolerant resources: FPGAs commonly employed for self-repair strategies are composed of blocks containing complex arrangements of logic and interconnect (albeit with a well-understood and regular structure). Fault masking strategies tend to be implemented using very simple logic that is no more complex than the non-redundant design. N-modular redundancy principles have been reported at the extremely fine-grained transistor and gate level [17]. The TMR principle is sufficiently flexible for FPGA design that use multiple configurations when combined with standby configurations [28]. Indeed, the replication of entire components or sub-components is commonly adopted in mission critical situations such as navigation and mission control computers, although there is great interest in exploring less resource-expensive strategies that achieve equivalent reliability through self-repair using COTS components.

22.5.1 Test and Verification

There are three key aspects testing of self-repair methods. First, verification of the core functionality is necessary in order to ensure that the fault tolerant mechanism operates as intended. Second, an evaluation of key metrics (Sect. 22.2.3) will lead to better understanding of resource/performance trade-offs. Third, test and verification of the complete sub-component is needed in order to ensure that its behavior is predictable. Self-repair systems must be able to adapt and continue in the presence of faults, as well as maintain predictable behavior at all times. For mission-critical applications autonomy raises concerns regarding predictability in addition to serviceability and certification, requiring that new standards be developed that set out suitable qualification procedures. The concept of design for *full fault coverage* using built in test is relevant here [29]. In production test, BIT logic must be able to cover every critical test combination in order to provide full repair of defect logic. Similarly, complete coverage is necessary when testing fault masking or reconfiguration strategies.

A useful technique for testing performance and compliance of autonomous self-repair is fault injection. This takes the form of asserting fault conditions within the circuit in a random fashion in order build up a statistical picture of its fault behavior. For example, the fault injection algorithm proposed in [30] involves temporarily setting fault conditions in a sequential manner where it is assumed that the DUT design remains unchanged during the test procedure. Chip production and in-field diagnostics exploit boundary scan logic [31] which can be used for testing of self-repair mechanisms. However the test at system level can become complex since appropriate resources must allocated for built-in test and control lines [32]. Production yield test and repair machines utilise highly optimised test-repair algorithms, allowing them keep track of permanently faulty logic resulting from manufacturing defects. Although the test requirements appear similar in nature, online reconfiguration strategies seek to alter the configuration of the DUT in response to multiple fault events, some of which may be permanent by nature. Hence a suitable fault injection tool must create a combination of transient and permanent fault conditions, and keep track of the complex sequence of faults during test. Due to the complexity of design for self-repair, proxy hardware is often employed during development that simulates the design before committing to ASIC manufacture. This is generally achieved using FPGA platforms that are configured to emulate the design. Whether used for emulation or test bench interfacing, the underlying proxy hardware must be sufficiently reliable and bug-free so that it may be assumed error-free.

22.5.2 Potential Impact and Uptake

Within the context of assisted-maintenance the problem addressed by self-repair strategies is that of increasing the robustness of sub-components that would otherwise perform unpredictably or else fail entirely in the presence of random and permanent failures. In the latter case, the hope is to achieve a graceful degradation that prolongs operational lifetime and provides valuable monitoring of remaining fault capacity as an indication of impending end of life at the sub-component level. Strategies for self-repair draw partly upon techniques developed for production yield enhancement. This means that the end customer is never aware of production-related repairs. However self-repair strategies must operate without access to external test hardware equipment and thus maintaining customer transparency is more challenging. In some instances the repair strategy relies on customer-level tools such as vendor-supplied synthesis and layout for FPGAs. It could be argued however that a sustainable model (in the context of predictability, serviceability and certification) is one in which fault tolerance operates at a guaranteed "self-repair hardware level" that is transparent to the customer and hence may be regarded as autonomic. Such concepts are especially conducive to custom ASIC designs.

Ongoing developments in autonomous maintenance aim to achieve in-service "on the wing" active repair during operation or perhaps during predicable periods of downtime. They would no longer be limited to scheduled maintenance periods. Aside from those online reconfiguration methods discussed (Sect. 22.4.1), there are other examples of in-service models. For example a self-tuning analogue RF circuit has been demonstrated capable of online self-correction actions and thus is able to sustaining optimal performance without minimal degradation [33]. In this case an in situ BIT mechanism is incorporated that constantly monitors and performs self-correction of parameters that would otherwise drift out of specification. This example might be considered as being self-optimising rather than self-repairing, however the concept is similar since both rely on self-contained detect-restore actions.

Briefly revisiting the classic model of component lifetime discussed in Sect. 22.2, for some systems random and wear out failures are difficult to differentiate and hence great care is needed when allocating self-repair strategies such that they target the correct failure mechanism (i.e., presumably prioritising random rather than wear out failures). With this mind, we must ask the question: how to we select the most appropriate mitigation strategy and prove its efficacy? Clear and conclusive assessment of the metrics discussed above combined with effective test and verification are essential.

The next question is then: who should carry out this task? This is a complex systems design problem since it is not obvious as to what level of autonomy and transparency is most beneficial. Abstraction from application specific development tools may seem appropriate in terms of resource allocation, however acceptable limits on impact to performance, power and timing are very much application dependent. One final design aspect is that self-maintaining systems will be expected record and report fault history and thus assist planning of MRO scheduling. Fault event logging is already present in many systems, for example in automotive Engine control units (ECUs). However, additional information is needed such as remaining fault capacity and should be reported in a sensible format that is easily understood. An example would be when a sub-component that is experiencing aggressive fault conditions, necessitating an escalation of resource allocation in order to maintain fault-free operation. Reporting of remaining capacity (and indeed trending of this information) provides valuable information to both maintenance scheduling and decision-making about when to replace depleted resources. An key benefit is be the ability distinguish between exhaustion of fault capacity and general wear out, which would better inform decisions involving whether to replace complete sub-component or to whether to carry out overhaul to replace depleted boards and modules contained therein.

Our demand for ever-increasing performance and efficiency of engineering system is driving the adoption of COTS components even for mission critical applications. This is fuelling the uptake of concepts in self-repair and autonomy that for example achieves similar design robustness to that achieved through component screening. Self-repair concepts are also generating significant interest in the area of next-generation nano electronics where transistor and interconnect reliability is expected to be much lower than of current technology. In the widest context, self-repair seeks to increase the robustness of cost-sensitive, high value systems and therefore bring about cheaper maintenance through autonomous strategies. However a better understanding the cost/benefit trade-offs and effective design methodology is essential for its uptake.

References

1. Cheatham JA, Emmert JM, Baumgart S (2006) A survey of fault tolerant methodologies for FPGAs. ACM Trans Des Autom Electron Syst. 11(2):501–533. doi:10.1145/1142155.1142167
2. Morgan KS, McMurtrey DL, Pratt BH, Wirthlin MJ (2007) A Comparison of TMR With Alternative Fault-Tolerant Design Techniques for FPGAs. IEEE Trans Nucl Sci 54 (6):2065–2072. doi:10.1109/TNS.2007.910871
3. Evans AC (1999) Applications of semiconductor test economics, and multisite testing to lower cost of test. In: IEEE proceedings of international test conference. doi:10.1109/TEST.1999.805620
4. Walter JD (2003) Methods to account for accelerated semi-conductor device wear out in longlife aerospace applications. DTIC document, Virginia
5. White M (2010) Scaled cmos technology reliability users guide. http://hdl.handle.net/2014/414912010
6. Sexton FW (2003) Destructive single-event effects in semiconductor devices and ICs. IEEE Trans Nucl Sci 50(3):603–621. doi:10.1109/TNS.2003.813137
7. (2007) Radiation Effects on Embedded Systems, 1 edn. Springer, Berlin
8. Nicolaidis M (2010) Soft errors in modern electronic systems. Springer, Berlin
9. Von Neumann J (1956) Probabilistic logics and the synthesis of reliable organisms from unreliable components. Automata studies 34:43–98
10. El-Maleh AH, Al-Hashimi BM, Melouki A, Khan F (2009) Defect-tolerant n2-transistor structure for reliable nanoelectronic designs. IET Comput Digital Tech 3(6):570–580. doi:10.1049/iet-cdt.2008.0133
11. Lyons RE, Vanderkulk W (1962) The use of triple-modular redundancy to improve computer reliability. IBM J Res Dev 6(2):200–209. doi:10.1147/rd.62.0200
12. Parris MG, Sharma CA, Demara RF (2011) Progress in autonomous fault recovery of field programmable gate arrays. ACM Comput Surv 43(4):31. doi:10.1145/1978802.1978810
13. Emmert JM, Stroud CE, Abramovici M (2007) Online fault tolerance for FPGA logic blocks. IEEE Trans Very Large Scale Integr (VLSI) Syst 15(2):216–226. doi:10.1109/TVLSI.2007.891102
14. Breuer MA, Gupta SK, Mak TM (2004) Defect and error tolerance in the presence of massive numbers of defects. IEEE Des Test Comput 21(3):216–227. doi:10.1109/MDT.2004.8
15. Han J, Gao J, Jonker P, Qi Y, Fortes JAB (2005) Toward hardware-redundant, fault-tolerant logic for nanoelectronics. IEEE Des Test Comput 22(4):328–339. doi:10.1109/MDT.2005.97

16. Wadsack RL (1978) Fault modeling and logic simulation of CMOS and M08 integrated circuits. Bell Syst Tech J 57(5):1449–1474
17. Kothe R, Vierhaus HT, Coym T, Vermeiren W, Straube B (2006) Embedded self repair by transistor and gate level reconfiguration. In: Design and diagnostics of electronic circuits and systems, IEEE. doi:10.1109/DDECS.2006.1649613
18. Jones D, McWilliam R, Purvis A (2008) Designing convergent cellular automata. Biosystems 96:80–85
19. Wendling X, Rochet R, Leveugle R (1996) ROM-based synthesis of fault-tolerant controllers. In: IEEE international symposium on defect and fault tolerance in VLSI systems
20. Gerbaux L, Saucier G (1992) Automatic synthesis of large Moore sequencers. Integr VLSI J 13(3):259–281. doi:10.1016/0167-9260(92)90031-S
21. Moreno G, Civit-Balcells A, Guerra-Gutierrez P (2007) ROM-based finite state machine implementation in low cost FPGAs. In: IEEE international symposium on industrial electronics
22. Rochet R, Leveugle R, Saucier G (1993) Analysis and comparison of fault tolerant FSM architecture based on SEC codes. In: IEEE international workshop on defect and fault tolerance in VLSI systems
23. Walker JA, Trefzer MA, Bale SJ, Tyrrell AM (2013) PAnDA: a reconfigurable architecture that adapts to physical substrate variations. IEEE Trans Comput 62(8):1584–1596. doi:10.1109/TC.2013.59
24. Basha C, Pillement S, Lagadec L New reconfigurable fault tolerant FPGA architecture: a design for mission critical applications. 8th HiPEAC workshop on reconfigurable computing, Vienna, Austria, Jan 20–22
25. Frei R, McWilliam R, Derrick B, Purvis A, Tiwari A, Serugendo GDM (2013) Self-healing and self-repairing technologies. Int J Adv Manufact Technol 69:1033–1061. doi:10.1007/s00170-013-5070-2
26. Campregher N, Cheung PYK, Constantinides GA, Vasilko M (2006) Reconfiguration and fine-grained redundancy for fault tolerance in FPGAs. In: International conference on field programmable logic and applications (FPL '06)
27. Stott E, Sedcole P, Cheung P (2008) Fault tolerant methods for reliability in FPGAs. In: International conference on field programmable logic and applications (FPL '08)
28. Zhang K, Bedette G, DeMara RF (2006) Triple modular redundancy with standby (TMRSB) supporting dynamic resource reconfiguration. In: IEEE autotestcon
29. McCluskey EJ, Bozorgui-Nesbat S (1981) Design for autonomous test. IEEE Trans Circuits Syst 28(11):1070–1079. doi:10.1109/TCS.1981.1084930
30. Quinn HM, Black DA, Robinson WH, Buchner SP (2013) Fault simulation and emulation tools to augment radiation-hardness assurance testing. IEEE Trans Nucl Sci 60(3):2119–2142. doi:10.1109/TNS.2013.2259503
31. Sedmark RM (1994) Boundary-scan: beyond production test. In: 12th IEEE VLSI test symposium
32. Arlat J, Aguera M, Amat L, Crouzet Y, Fabre JC, Laprie JC et al (1990) Fault injection for dependability validation: a methodology and some applications. IEEE Trans Software Eng 16(2):166–182. doi:10.1109/32.44380
33. Goyal A, Swaminathan M, Chatterjee A, Howard DC, Cressler JD (2012) A new self-healing methodology for rf amplifier circuits based on oscillation principles. IEEE Trans VLSI Syst 20(10):1835–1848. doi:10.1109/Tvlsi.2011.2163953

Chapter 23
Autonomous Maintenance for Through-Life Engineering

M. Farnsworth, C. Bell, S. Khan and T. Tomiyama

Abstract This chapter looks at the overall theme of automating maintenance practices with a particular focus upon the application of robotics within this field. Covering the current state of the art in automating maintenance processes this chapter also looks at the current challenges to moving beyond simple inspection and diagnosis to the design and construction of fully automated platforms for undertaking maintenance. This includes methodologies for capturing and classifying maintenance task processes so that they can be automated in some way and how to link this task classification with some level of automation. The chapter ends with a discussion on how the design process can be adapted to aid automated maintenance, self-healing and no fault found applications.

23.1 Introduction

The business model for the provision of a wide variety of high-value capital assets such as aero engines, trains and medical scanners is undergoing a fundamental shift [1]. There is now a growing value in maintaining the life of a manufactured product throughout its lifecycle, and a number of services have grown to meet this need.

Through-life Engineering Services (TES) is a field born out of the need to offer support for guaranteeing the performance and function of high value assets or products over their operational life cycle [2]. Today TES accounts for over 55 % of revenue for high-value manufacturing companies within the aerospace and defense sectors [2]. One such service is maintenance, often borne by the end user; this is

M. Farnsworth (✉) · S. Khan · T. Tomiyama
EPSRC Centre for Through-Life Engineering Services, Cranfield University, Cranfield, Bedfordshire MK43 0AL, UK
e-mail: m.j.farnsworth@cranfield.ac.uk

C. Bell
Department of Mechanical Engineering and Mathematical Sciences, Oxford Brookes University Wheatley Campus, Wheatley, Oxford OX33 1HX, UK

L. Redding and R. Roy (eds.), *Through-life Engineering Services*,
Decision Engineering, DOI 10.1007/978-3-319-12111-6_23

now evolving into a service-based model in which a maintenance provider takes over the responsibility for the operability and maintenance of the asset [3].

This is seen on UK railways where the time-limited franchise for operating a particular service is given to a train operating company who then lease the rolling stock from its owner and contract with a maintenance company under a service level agreement to ensure availability of the asset. An example of this is Bombardier providing the maintenance to East Midlands trains for the class 222 train sets. Such a business model provides motivation for improving the maintenance function—in order to reduce through-life costs and maximize profits on those contracts.

Recent decades have seen an increasing use of robots within manufacturing processes. The speed, power, availability, productivity, and improved accuracy of robots have had significant impact in reducing manufacturing costs while improving production quality. It therefore seems logical to explore the possibilities of automation and autonomous systems being used within the maintenance function to provide the same benefits as are found by their use in the manufacturing process.

Autonomous maintenance can be divided into two main themes, the first involving the concept of 'self' within the context of the high value system, looking at incorporating 'self-healing', 'self-monitoring', 'self-aware', 'self-configure' and 'self-protect' technologies or characteristics. The second focuses upon automating maintenance practices within organizations, with a particular focus upon the use of autonomous robotics to aid, guide or take over current maintenance tasks undertaken by human engineers/workers.

This chapter explores the latter, beginning with a brief delve into the background of maintenance, repair and overhaul (MRO) in Sect. 23.2 and followed up with a look at the current engagement of automation within industrial maintenance in Sect. 23.3. Next follows a section on classification of maintenance tasks and the presentation of a novel methodology designed to aid in this process. Section 23.4 covers integration of automation into maintenance practices, discussing how and at what level of complexity this should entail. Finally the chapter ends with a section on design strategies aimed at not only autonomous maintenance but also closely related fields of self-healing and no-fault found.

23.2 Maintenance, Repair and Overhaul

Maintenance, repair and overhaul (MRO) encompasses the organizations within a company tasked with the responsibility for the provision of a fully serviceable asset when required by the end user at an affordable and reasonable cost at reasonable quality [4]. Within the aerospace industry there is fierce competition, both nationally and internationally that brings with it the need to reduce costs, in particular within the service arm of MRO [5, 6]. Therefore the goal of MRO organizations is to reduce maintenance cost expenditure and turn-around time so as to maximize revenue for the owner.

Maintenance can be defined as the process of ensuring that a system continually performs its intended function at its designed-in level of reliability and safety. In the railway industry for example train overhaul maintenance facilities (TOMFs) are utilized to keep capital assets in good order. The operation of such a TOMF or depot is based upon either time-based maintenance in which a train's performance is checked periodically or condition-based maintenance when a train malfunctions, either mechanically or functionally [7]. Often time-based maintenance is treated as a repair system with the performance of the train inspected periodically and a number of known maintenance tasks undertaken to maintain both safety and reliability of the train [7]. These approaches can vary between countries as can be seen throughout Japan, Germany and France and their high speed railway systems, with different granularity or degrees of maintenance split among a number of levels. These levels cover simpler and more routine maintenance such as inspection and cleaning needed after daily operation to much more complex and intensive overhaul and repair (body update, paint, seat replacement) [8]. The integration of some form of automation into either of these maintenance levels may be a driver for further reduction in costs and also lead to improved through life performance of the asset.

23.3 Automation Within Industrial Maintenance

Automation has been a key driver for successive growth within manufacturing throughout the century with the automobile industry for example relying heavily on the assembly line.

The traditional image of maintenance is one consisting of 'Dull/Dangerous or Dirty' work. There is also a similarly negative connotation that maintenance is a 'non-productive' element to manufacturing which is only now changing. The introduction of automation into maintenance practices would help alleviate the need for humans to work in such dangerous or dirty environments but also improve its image into one with productive and cost saving elements requiring the need for highly skilled, tech-savvy engineers.

Where automation within manufacturing is often regimented, uniform, deterministic and standardized, which gives rise to easier integration, maintenance holds a different set of characteristics which make the task more difficult. Maintenance is often non-deterministic because one maintenance sub-operation (removing bolt) may differ depending on the age of the asset (i.e. rusted) which makes any pre-programmed autonomous behavior difficult. It can be non-uniform, with different actions required depending on the type of asset present, regardless of maintenance sub-operation. Therefore removing and fitting one type of disk brake pad may vary depending on the maker of the asset (i.e. train undercarriage). Finally it is often irregular, with varying levels of planned and unplanned maintenance which makes determining the right autonomous systems are in place at a given time difficult, particularly if such autonomous maintenance systems can only perform a specific function (inspection, manipulation) and are unable to multitask, something a human is often able to do to meet an increasing workload.

This is why the current focus of automation towards some level of maintenance within industry has fallen upon the field of non-destructive testing (NDT). Inspection of assets often requires little physical or applied force interaction and can often be integrated into smaller autonomous systems. Current examples can be found in industrial maintenance of pipe infrastructure from a wide variety of industries (chemical, oil, gas). Often the first level of application is in pipe inspection gadgets (PIGs), and because the environment of these pipelines means wireless communication cannot be used inside them, either teleoperation or some level of autonomy is required [9].

Other applications include the detection of cracks along pipe infrastructure, with a number of robotic solutions considered as an attractive alternative to manual inspection. These can be non-autonomous robots controlled by a human through some kind of teleoperation and tethered to a cable, semi-autonomous robots which contain functions that enable it to perform some task on its own (navigation) but require a human to undertake more complex tasks (manipulation, decision making) [10], and finally fully autonomous robots which carries with it all the necessary capabilities needed to perform its assigned task autonomously [11, 12].

The use of magnetic flux leakage (MFL) techniques are common [13, 14], and eddy current arrays (ECAs) have also been used in an autonomous NDT approach within the aerospace industry for routine inspection of fatigue crack detection of aircraft oil delivery tubes [15].

Mobility is an issue when looking to automate such NDT methods, particularly when access to dangerous environments is required, such as those found within the nuclear industry, or in large industrial sites [16, 17]. Pipe climbing robots of numerous types have been developed for inspection and NDT, some utilizing electromagnets to stabilize and grip to surfaces prior to performing inspection and analysis [18, 19].

Climbing robots have also been developed for the maintenance of vertical structures, with applications to welding of ship-hulls to the inspection of steal bridges or nuclear power plants [20, 21]. A survey of current state of the art climbing and inspection robots can be found in [22], covering over 100 hundred examples.

As discussed maintenance can also consist of dull or rudimentary tasks, which take away time from engineers to complete for more complex and needed maintenance. In large scale industrial infrastructures there are afforded many opportunities for the application of autonomous systems to take over dull or menial jobs otherwise done by humans. In [23] a multi-robot system for building maintenance and surveillance was demonstrated, with advantages of increased efficiency and reducing manpower. Here mobile robots are able to roam throughout the building and collect information through the use of RFID tags on monitored equipment and feed this information to a central management centre.

The need to remove humans from dangerous situations was the motivation for the creation of autonomous robots to perform maintenance tasks on power distribution lines [24]. Here a robot is able to receive instructions such as 'Insert the bolt' or 'clench the nut' from an operator and then autonomously generate the motion trajectory required by the arm to perform the task. These small actions all form part of a much larger maintenance task of exchanging a switch gear on a power distribution line. Additional intelligence features are also required to perform this task, such as image recognition to aid 'gripping the bolt' by allowing the robot to locate and plan future motion trajectories. In this example a number of edge extraction algorithms were used and tested against in environmental illumination conditions. This example highlights one of the interesting challenges associated with trying to automate maintenance, that of trying to capture the necessary actions and decisions that are required to perform the maintenance task successfully.

The next section looks to present a novel methodology on how to capture and classify the necessary information associated with a maintenance task and decompose it in such a way that it can be used to design and build automated or autonomous solutions for this maintenance task.

23.4 Maintenance Task Classification

Industrial Engineering (IE) as defined by the Institute of Industrial Engineers; "is concerned with the design, improvement and installation of integrated systems of people, materials, information, equipment and energy." As a consequence of such a wide ranging definition there are many processes and practices which can be classified as 'Industrial Engineering'. These include Taylors scientific management theory [25], Gantt's eponymous chart for planning and scheduling work activities [26] or Taguchi's methods for improving quality in industrial facilities [27].

IE techniques fall into two basic categories; they are either used for analysis of existing processes or improvement of those processes. Reported uses of IE within maintenance activities tend to describe how techniques are used for the latter, either by increasing their efficiency or improving their planning. So the Single Minute Exchange of Dies (SMED) techniques are used to reduce setup times [28] and scheduling rules have been used to improve repair system performance within the system's existing constraints (e.g. labour hours, repair facility availability) [29].

Because of the characteristics of maintenance make it difficult to automate, the vast majority is undertaken by humans. In order to automate maintenance tasks it is therefore important to capture information on what are the most important requirements needed to complete the task. The first is the necessary motions and functions performed by the human and secondly the decisions and intelligence that guide them over the timeframe of the maintenance process. This information also needs to be able to be easily transferrable to a form of automation, be it machine or robotic, so that either platform can then begin the task of performing automated maintenance.

In order to capture information involved on the complex motions undertaken we look towards the time and motion study techniques developed by Frank and Lillian

Table 23.1 Therbligs

Therbligs				
Search	Hold	Assemble	Pre-Position	Plan
Find	Transport Loaded	Disassemble	Release Load	Rest
Select	Transport Empty	Use	Unavoidable Delay	
Grasp	Position	Inspect	Avoidable Delay	

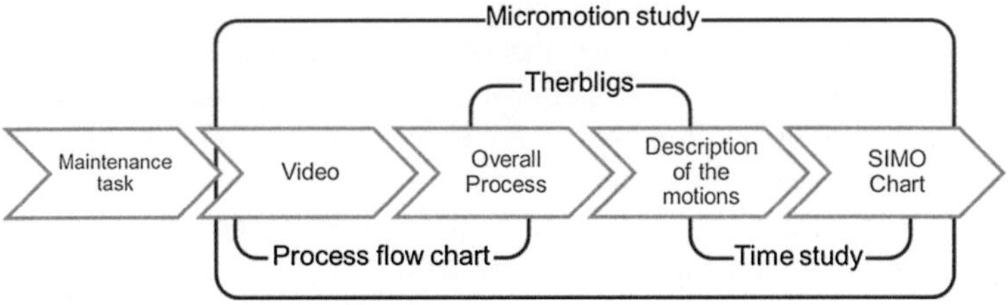

Fig. 23.1 Task classification methodology

Gilbreth [30]. Here 18 elemental motions called 'Therbligs' related to motions performed by humans undertaking manufacturing processes have been identified and are listed in Table 23.1. Other motion time study methodologies include, 'methods-time management' (MTM) and the 'Maynard operation sequence technique' (MOST). An extension of the therblig methodology is to include information on recorded movements for both right and left hands of the human performing the task and the time taken to achieve each motion. This is achieved through the use of a Simultaneous Motion (SIMO) chart approach [31]. A focus of this research is therefore on IE techniques which allow for the capture and analysis of both the motions and decision making used throughout the maintenance task, with these techniques combined into a 'Task Classification Methodology' [32].

A two stage process is therefore outlined, with video recording of the maintenance event; this is transcribed into a process flow, and then followed with more detailed analysis of the video to capture motion information through Therblig and SIMO classification. Figure 23.1 captures this process for maintenance task classification which was applied to a number of train maintenance tasks of which the details, results and discussion are outlined and given below.

23.4.1 Class 222 Diesel Engine Train Undercarriage Maintenance

The class 222 is a diesel-electric multiple unit high speed train. Underneath each carriage on the train is a diesel engine which drives an electrical generator which then drives an electric traction motor. As well as driving the train, the traction motor provides rheostatic braking to reduce brake pad wear and in order to shield it from the elements a series of side skirts is employed. In order to perform maintenance it is necessary to lower and raise the side skirts to gain access to the traction unit and any ancillary equipment.

A breakdown of the generic steps required to perform this maintenance is shown in Table 23.2. The electric traction motor in the undercarriage is also connected to a pair of bogey wheels through a final drive gearbox. One of the regular maintenance tasks on the gearbox involves checking and analyzing of the oil filler. A breakdown of the generic steps required to perform this maintenance is also shown in Table 23.3, and the oil filler is shown in Fig. 23.2.

The task classification methodology leads to the creation of a number of IE process and motion information of the maintenance task involved. For each maintenance example this includes video information, process flow charts, Therbligs and SIMO chart motion information. For the class 222 diesel engine train undercarriage example a number of results are shown.

The aluminum oil filler cap incorporates a magnetised steel rod with knurled markings that denote the maximum and minimum oil levels. It is retained in the cast iron gearbox casing by a bayonet fitting. Once the cap has been removed, the magnetic end of the rod is inspected for metal deposits before it is wiped and placed back into the gearbox casing and allowed to rest unfastened before being removed

Table 23.2 Lowering and raising side skirt breakdown

Step	Description
1	Unscrew left and right hand securing bolt
2	Open latch and lower skirt to horizontal position
3	Remove spring clips on retaining lanyard from both sides
4	Lower skirt to vertical position
5	Raise skirt to horizontal position and re-attach spring clips
6	Raise skirt and close hatch
7	Screw in left and right hand securing bolts

Table 23.3 Checking oil filler level breakdown

Step	Description
1	Assemble sampling device
2	Remove filler cap
3	Visually inspect oil on spigot for contamination
4	Wipe spigot
5	Replace filler cap
6	Remove filler cap and check oil level point on spigot
7	Take oil sample
8	Replace filler cap

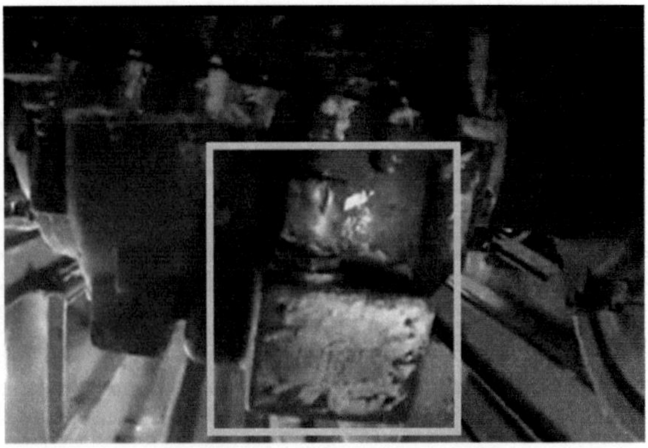

Fig. 23.2 Gearbox oil filler

again and the oil level noted. It is then removed a second time to allow an oil specimen for debris analysis to be taken before being replaced once more and fastened into place. A process flow chart for this maintenance is shown in Fig. 23.3.

A number of issues were identified during the process of oil filler maintenance. An oil-tight seal is produced because the elongated inside of the aluminum cap has a close fit to the accurately-bored hole in the casing. Consequently the cap can be quite challenging to remove—usually requiring the shaft of a long screwdriver to lever it out. Bronze particles produced by wear of the brushes in the gearbox will not be attracted to the magnetised steel rod; however they will cause the oil to have an orange tint which can be identified by an experienced maintenance fitter. Similarly water contamination can be identified by sight without having to wait for the results of the laboratory analysis of the oil sample. A number of lessons were also learnt from the skirt maintenance. Lowering the skirts is a two-stage process. Firstly the securing bolts at each end are unscrewed before the central latch is released which causes the skirt to drop. Its descent is arrested at the 90° position by wire lanyards at each end attached to the skirt by spring clips. Once these are disconnected the door can be fully lowered. Most maintenance is carried out with the skirt remaining attached to the unit but should it require completely removing, the hinge design is such that this can be achieved by simply lifting the skirt some 30 mm.

Releasing and refastening the spring clips can be challenging and the lanyard is easily trapped when the skirt is raised. While using power tools to screw and unscrew the securing bolts may appear to save time, it can result in difficulty in unscrewing when the unit next comes in for maintenance.

The latch on the heavy skirt is the same design as that on the lighter skirt. Consequently over time the latch on the heavier door is inclined to wear and ultimately fail. It is therefore imperative to support the weight of the skirt when unscrewing the second securing bolt in case the latch is incapable of keeping the skirt raised. An air filter assembly forms part of the skirt and should remain in place at all times. However with wear and tear, when the skirt is lowered to its final position it is not uncommon for this to slide out and drop onto the floor. Should this happen, the assembly will need replacing and holding in place during the subsequent raising of

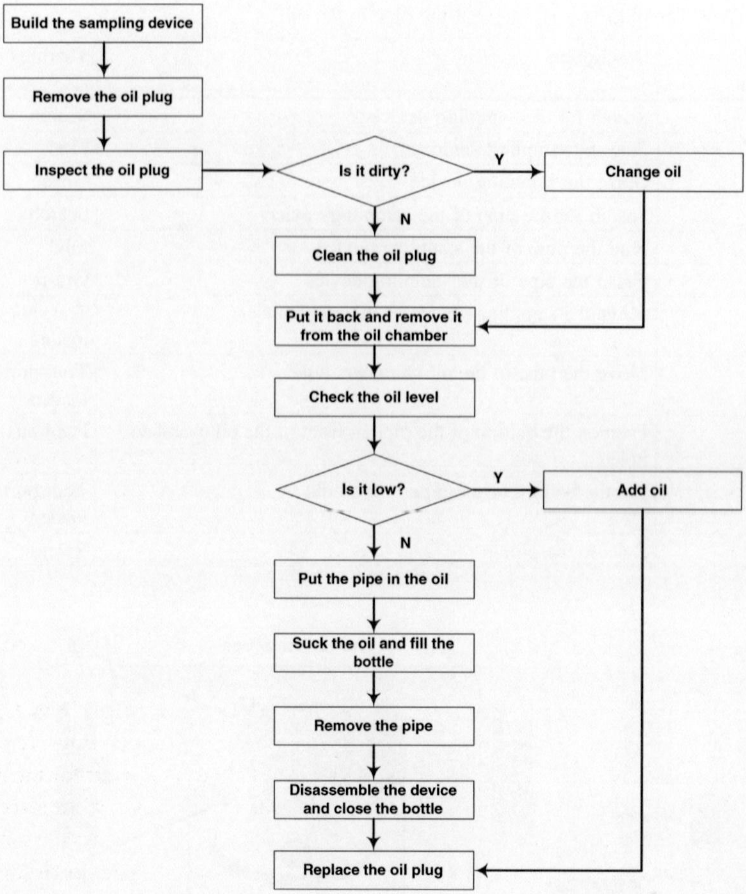

Fig. 23.3 Process flow chart for sampling oil filler

the skirt. The weight of this skirt is such that when it is being raised the hinges are inclined to separate unless a second person presses them back into place.

A detailed breakdown of the motions required to perform a specific step of the oil filler maintenance task is shown in Table 23.4. Here the methodology captures ten motions for the task 'Put the Pipe in the Oil', highlighting the often large number of motions required to perform what would seem to be a simple task. Finally it is also possible to gather information on the time taken and importantly the left or right hand motions required for each step of the maintenance task as seen in Fig. 23.4. This is important as future automated solutions may include multiple robotic arms or find the task decomposed into a number of automated or robotic solutions, each one equipped to perform one motion or step of the maintenance task in a collaborative way.

Table 23.4 Therbligs step seven—Put the pipe in the oil

No of motion	Description	Therbligs
1	Search for the sampling device	Search
2	Find the sampling device	Find
3	Grasp the sampling device	Grasp
4	Search for the pipe of the sampling device	Search
5	Find the pipe of the sampling device	Find
6	Grasp the pipe of the sampling device	Grasp
7	Move the sampling device to the oil chamber	Transport loaded
8	Move the pipe to the oil chambers hole	Transport loaded
9	Position the bottom of the pipe in front of the oil chambers hole	Position
10	Put the bottom of the pipe in the hole	Transport loaded
11	Release the pipe	Release

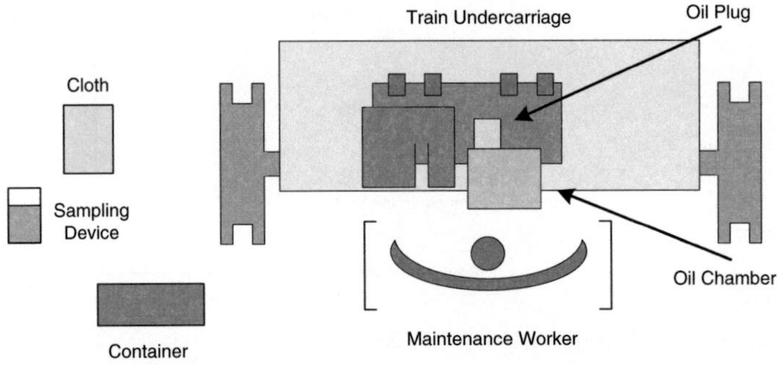

Left Hand Description	Symbol	Time (s)	Time (s)	Symbol	Right Hand Description
Search for the sampling device	SH	0.1	1	R	Rest
Find the sampling device	F	0.1			
Grasp the sampling device	G	0.5	0.1	SH	Search for the pipe of the sampling device
Move the sampling device to the oil chamber	TL	2.5	0.1	F	Find the pipe of the sampling device
			0.5	G	Grasp the pipe of the sampling device
			0.5	R	Rest
			1.5	TL	Move the pipe to the oil chamber's hole
Hold the sampling device	H	3			
			1	P	Position the bottom of the pipe
			1	TL	Put the bottom of the pipe in the hole
			0.5	RL	Release the pipe

Fig. 23.4 SIMO chart seven—put the pipe in the oil

23.4.2 *Task Classification for Maintenance: Outcomes and Discussion*

The use of the chosen IE techniques can help to describe the process and motions of maintenance by decomposing the tasks into a number of 'unit tasks'. However there is often not enough detail to ascertain whether this task can be accomplished by a robot. Consider the task 'Put the pipe in the oil' shown in Table 22.4. The Therbligs used for its description and breakdown consist of Search, Find, Grasp, Transport Loaded, Position and Release. These Therbligs describe the process quite correctly; however some details are missing from this picture, particularly when you consider the need to automate this process. For example step 3, 'Grasp the sampling device' is quite easy to perform for a human being; however several issues may occur if an autonomous robot has to carry out this operation. Firstly how to approach the object, what level of orientation or angle it should take to reach it? When it has reached the object how much force should it put into grasping it? And what happens if it fails, or it is unable to grasp it in the correct orientation? Once again if we look at the Therblig for 'Transport loaded' there is a lack of information on where the device is to be moved with regards to 'XYZ' coordinates, or relative position with regards to the oil chamber hole.

It also appears that some Therbligs are too ambiguous. This is the case of "Use" for example. This is a general term to describe the motion to be done. In fact, if the operator is using a screwdriver, the way of performing should be more precise: should the rotation be clockwise or anticlockwise? What design of screw head is being used? Cross-head, torx, slot, etc.? Or, considering another example, what should be understood when speaking about "using" a tissue to clean a surface? So while Therbligs are helpful in visualising the process, their focus on providing a structured methodology for calculating the time an activity should take to perform means they are not specific enough to describe exactly what is happening from a physical perspective. Therbligs such as "Transport Loaded", "Grasp" or "Position" do not give enough information about the required physical attributes to lift or move an object. They do not consider the questions "what is the required force to lift an object?", "what pressure must be applied to grab an object?", "what is the speed, velocity or acceleration required to move an object?"

Similarly there are no Therbligs to take account of decision-making activities. During the maintenance on the train a coolant leak was observed. Therbligs would be needed to deal with the questions "What is to be done when a leak is found?" "Is the leak significant enough to require attention?", "where is the source of the leak?"

The IE techniques outlined in this research can describe and analyse processes but something is missing, especially if these techniques are used to convert activities performed by humans into activities to be undertaken by robots.

For a better description and analysis of a complex activity, more Therbligs are needed. For example the Therblig "Use" is too generic and some additional Therbligs should be created depending on the tool being deployed. For example if a screwdriver is used, Therbligs like "turn clockwise" or "turn anticlockwise" could be more appropriate.

In addition, more information to support the Therbligs are required, such as physical or geometrical parameters. For example:

- If a hammer is used the Therblig "Use" should be integrated with information about the angle to hit the part.
- If a fragile object is grasped, information about the pressure and the force to apply is needed.
- When a part is being lifted, if it is heavy, more force will be necessary.
- The distance the object is being moved—is it within the physical constraints of the robot being considered for the activity?

Furthermore, Therbligs for decision making are needed to better answer questions like "what to do in this particular situation?", "is the task done properly?", "if something happens, what should be done?"

Referring to the train maintenance tasks performed, it is important to evaluate different situations using all the different senses. For example, it is crucial to recognise an oil leak from the smell or the sight of drops or spots. Other senses involved are hearing, to notice suspect noises, or touch to detect heat or smoothness of surfaces.

It was asserted in Sect. 23.3 that key characteristics of maintenance are that it is irregular, non-deterministic and non-standardised. This was demonstrated during the experimentation on the train. The results of the oil analysis from the final drive could require that unplanned maintenance be undertaken, removing the filler plug from the final drive often requires extra assistance from a large screwdriver and the work required to lower and raise a skirt varies depending on whether it is heavy or light.

23.5 Integrating Automation and Robotics into Maintenance

The ability to classify and decompose the main steps or actions involved in undertaking a particular maintenance task is an important first step. In order to develop fully capable automated systems we need to develop methodologies which can link each of the decomposed 'unit tasks' with some form of automation, be it physical or cognitive.

Thankfully engineering design research has resulted in the formulation of a number of systematic methodologies over the past few decades [33, 34]. Here we look to the idea of conceptual design, where a list of requirements is abstracted to identify the essential problems of the system [33], and solutions to each of these requirements are chosen to meet these requirements. In this instance each of the decomposed 'unit tasks' created in the maintenance task classification step essentially fulfill the overall functionality of that required system.

One method able to exploit such a construction is the morphological table. A powerful combinatorial tool, it utilizes the principles of morphological thinking to guide the creation of conceptual designs through a systematic combination of all possible combinations of fragments that together can constitute a system [34]. Here a visual exploration of the design space is undertaken with the functions represented as our maintenance 'unit tasks' are listed against the means or solutions to achieve each of those functions in a two dimensional matrix [34].

In order to aid the process of conceptual design for automated maintenance the standard morphological table has been adapted [35]. As discussed previously the task classification process creates a series of individual 'unit tasks' that capture the main processes of the maintenance task undertaken.

In order to build a system that can perform the entire maintenance task in an automated fashion it is important to go beyond simple hardware and look into more detail the motions and information required to complete a particular 'unit task'. Here we adapt the functions of the morphological table to include information necessary for automation, for example how much force or torque is required to unscrew a bolt, or Cartesian coordinates when looking to transport an object or tool. Also we use knowledge at the granularity of physical phenomena associated with a particular function as shown in Table 23.5.

Successfully completing the maintenance task outlined previously requires the ability to undertake certain cognitive functions [36]. In the case of the oil filler example, the ability to search for and find the sampling device, a task which requires suitable object recognition algorithms along with the means to gather the necessary information to act upon. Finally the decomposition of the maintenance task allows for the possibility of multiple systems accomplishing specific tasks, in particular for example the application of collaborative or swarm robotics. This it brings with it a whole new challenge as communication and coordination is needed between these individual systems if they are to achieve each maintenance step successfully.

The adaptation of the proposed morphological table is designed to link up with the previous maintenance task methodology so as to provide a 'joined-up' process for capturing, classification and concept design of automated maintenance solutions. An overview of how the methodology is able to capture and classify

Table 23.5 Physical, cognitive and information phenomena

	Means			
Function 1 'Grasp'	Force + Friction	Suction	Magnetic Attraction	Electrostatic Force
Function 2 'Find'	Object Recognition	Classifier Based	Case Based Lookup Table	Pattern Recognition
Function 3 'Signal'	Infra Red	Sound	Ultrasonic	Laser

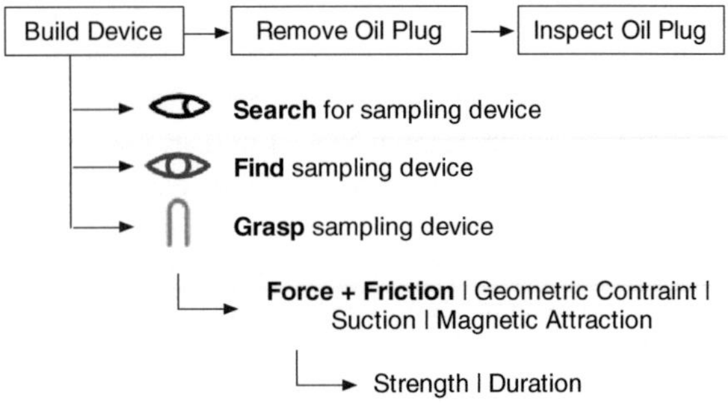

Fig. 23.5 Concept generation breakdown

maintenance tasks and link these with conceptual design is shown in Fig. 22.5. Here a small section of the main processes for the oil filler maintenance example are expanded to show their associated motions and in the case of 'Grasp', the associated solutions to perform this particular motion. Included are also some characteristics required to perform the motion in an automated fashion, with the ideal strength and duration noted.

The application of fully autonomous systems that remove the need for human workers is an ideal final goal when looking to automate industrial maintenance tasks. However there are a number of alternative levels of integration or approaches to improving maintenance processes through some level of automation.

Automation through autonomous robotics: The fully integrated approach, which boils down into some form of autonomous asset which acts as a delivery system for some specific need (inspection, physical interaction etc.). This in itself could be one single unit/robot, or consist of several individual robots each specified to a particular role, and in conjunction with each other, perhaps in some collaborative way look to complete the maintenance task. Such an approach may prove difficult, with tasks that involve a significant amount of manipulation, dexterity and physical power, designing (mobile) robotic platforms that go beyond simple inspection will require more advanced power systems, that can support stronger manipulators needed to perform tasks such as removing a damaged bolt/nut. The removal of humans from the task also requires thought into how decision making and communication/cooperation are to be undertaken by the autonomous system employed. Though advancements in decision making, object recognition and task planning are making strides in the robotics research community, it may still prove difficult to provide the necessary cognitive functions/abilities needed to perform complex tasks in a non-deterministic/irregular maintenance environment.

Automatic mechanisms: Maintenance tasks consist of multiple steps, involving some level of action or decision to be taken by the human worker. An alternative solution to the inclusion of autonomous robots is to actually look to remove the need for either a robot or human to perform the step altogether. In the case of the class 222 diesel engine train maintenance we may for example look to design a carriage to operate on its own, with integrated sensors which can read oil quality of the engine, or mechanisms which can open and close the train skirt automatically.

Semi-automation with human-robot collaboration: Given the current state of robotics, the removal of humans from complex tasks such as maintenance is difficult. Therefore strategies designed around 'human-robot collaboration' which try and get benefits out of both worlds could prove to be the solution to incorporating some level of automation in industrial maintenance tasks. Here we can utilize the decision making and cognitive abilities of experienced maintenance engineers to help guide and aid robotic solutions. Conversely we can look to robots to perform more dangerous physical tasks, such as lifting and transporting heavy objects, either through individual robotic platforms or some form of exoskeleton expander that can amplify human power and support external loads. Robots may also be employed to take over some of the more mundane and dull jobs of maintenance, with the task of gathering information through inspection and relaying this to an engineer to decide whether to perform maintenance or not.

23.6 Design Strategies for Autonomous Maintenance, Self-healing and No Fault Found

Looking towards the future through life engineering services of capital assets is the important consideration of how their design can be guided to improve their maintainability, reliability and supportability through their entire life span. It is important now more than ever that engineers focus on designing maintainable equipment that can be repaired quickly and easily, that enhances reliability through the concepts of robustness and low failure rate, and finally enhances supportability by providing an increased clarity of system faults and failures should they occur. This section looks to outline guidelines for each of these three key areas, with a discussion firstly on how to aid autonomous maintenance through design, secondly how to incorporate self-healing into products to improve reliability and resilience to damage, and finally how systems and assets can be designed to reduce the risk of no fault found.

23.6.1 Autonomous Maintenance

Design can impact the field of maintenance within TES in a number of ways. This can include methods for aiding the design of maintenance depots so as to aid overall system performance as shown in a case study on a TOMF [7], or more applicably the product itself. The reliability of a mechanical system depends on its design, the quality of its components, and its maintenance. However, maintenance is typically considered during product design in an ad hoc, ineffective manner, leading to unnecessary life-cycle costs [37].

Many of the challenges faced by autonomous robotics applied to perform some kind of maintenance task often stem from the design of the product or asset they are working on. Because of the limitations in what is achievable in today's current state of robotics it is very difficult to overcome these challenges without some form of design change.

Sensor integration: A large number of maintenance tasks simply involve some level of inspection and analysis and if required some form of action to be taken. When considering autonomous maintenance the first step is to think on how this degree of inspection and analysis can either be removed or done in situ within the asset itself. The integration of sensors, for example as discussed previously those able to monitor oil conditions within the undercarriage of a 222 diesel engine train remove the need for building platforms able to performing the complex motions and decision making needed to perform the inspection.

Reaction force reduction or removal: The motions and actions involved in a majority of maintenance tasks involve some degree of reactive pressure or force to be applied, be it through the removal of a nut or bolt, or the lifting and guiding of a heavy object such as a train skirt. The difficulty faced when looking to apply some form of automation or autonomous system is that this necessary force can often not be applied when needed. If we look towards mobile robotic solutions for example, such a force is likely to tip over or move the robot itself rather than perform the necessary action required, particularly if the object in question is degraded, for example if the object consists of a rusted (stuck) bolt. It is better to look towards a fixed automated solution, such as guided machines/tools on rails which can be maneuvered towards the target area, this way there is a backbone of rigidity to work with when applying force. Another solution is to simply remove the necessity for force or pressure altogether, by removing objects that may need to be acted upon during the maintenance task, such as bolts or heavy skirt panels, and as discussed above regarding inspection, simply remove the need to access the area in question.

Decision making: Throughout a maintenance task a number of decisions or analyses will have to be made, either through direct interaction with the asset or object in question, i.e. how much torque should I apply? Can I smell an oil leak? Where should place this tool? Or through some kind of assessment, i.e. is this surface clean enough? Can I detect signs of corrosion in this oil sample?

The implementation of intelligence and cognitive ability into autonomous systems is a difficult task and there are many challenges associated with it in relation to application towards industrial maintenance. Given the non-deterministic nature of maintenance, the situation and environment is likely to change often, and therefore how does an autonomous system adapt to this, how does it recognizes objects it is supposed to work on, which may change over time due to degradation. Building a knowledge base of objects and shapes for which a robot is likely to act upon, and then using these where necessary during the design process may help reduce some of the cognitive burden required to perform a particular task. Hopefully this can alleviate the possibility of large number of decision making scenarios if the choices are restricted through design.

Another challenge is the development of autonomous systems which are capable of gathering the necessary information required to perform reasoning for fault diagnosis, prognosis, and maintenance planning. Once again if this information could be provided by the asset itself that is undergoing maintenance then this removes some of the need for information gathering, and leaves it to focus upon reasoning and decision making based upon accurate and up to date information.

Module design: Disassembly is a difficult task, particularly for an autonomous system. The modularization of large and complex assets such as train carriages could allow for the reduction in downtime through removing the need to perform maintenance there at the scene, and replace it with a process where an autonomous system simply removes and replaces whole modularized components. The damaged or in need of service components can now be maintained or repaired at a later date at separate and more ideal facilities. Incorporating such an ideal of modularization into design could be an alternative route to building ever more complex autonomous robots.

Assisted maintenance: Not all maintenance has to be completely autonomous as discussed previously. Looking towards a form of human-robot collaboration, it would be beneficial for designers to incorporate methods for providing users with data about their internal state, as well as information about necessary maintenance operations and specific instructions about how to execute them. This form of 'assisted maintenance' [38] requires that designers identify which kind of failures are likely to occur and under which conditions, and then incorporate methods for aiding technicians to execute maintenance and repair operations should they be required.

Self-maintenance: The concept of 'self' incorporates a varied and large number of applications and fields. This includes 'self-healing', 'self-repair', 'self-maintenance', 'self-reconfiguration', 'self-assembly', 'self-monitoring' etc. In the case of self-healing and repair we look towards systems that can overcome damage sustained over the life-cycle of the device, through either a level of redundancy in which to maintain some degree of functionality, or the ability to regain functionality after damage has occurred. An approach which takes this philosophy and investigates the approach for 'self-maintenance' is not new and has been successfully undertaken previously by the authors [39, 40]. Here a design methodology for a self maintenance machine (SMM), in this instance a photocopier which integrated levels of functional redundancy which looked to use the potential functions of existing parts within the system as a whole.

23.6.2 *Self-healing*

We are currently undergoing an era of rapid technological development driven by significant advances in computational power and global collaboration. These developments have led to exponentially more complex systems, which inherently tend to be less reliable. There are a myriad of reasons behind this but perhaps the first and foremost is the increasing demand for the latest technology, which isn't always fully tested and developed.

Whilst improvements in maintenance regimes and through-life services can enhance the reliability of systems, it is almost inevitable that failure will occur at some point during their operation. One reason for this is that despite significant recent improvements in the modelling and prediction of systems, most machines still fail due to unexpected events [41]. As systems become increasingly more complex, the operational uncertainty increases, and one of the long-standing challenges of creating a reliable system is achieving robust performance under uncertainty [42].

What is briefly proposed here is a method of designing machines and systems that are resilient to both expected and unexpected failure. This novel 'design for self-healing' concept is perhaps still in its early stages, however a number of technologies are currently being developed that could ultimately lead to a fully self-aware system that is able to monitor its own status and perform any action required to maintain its intended function. By adopting this paradigm, it is not only expected failure that is mitigated—unexpected failure modes are also inherently compensated for, extending the potential life of a system and reducing the need for through-life servicing [43].

Efficient and effective design is a crucial component of a product lifecycle, and numerous strategies have been proposed to improve this. One such strategy is 'Design for X', which is a design philosophy intended to focus a design around a particular parameter [44]. In the case of 'Design for Self-healing' the parameter inherently covers a number of other issues, such as 'Design for reliability' or even 'Design for Maintenance' in that the system must be designed to be self-aware and hence maintenance can shift from being reactive to preventative.

To be successful self-healing techniques must be designed to compensate for a wide-variety of failure modes, thus overcoming some of the problems associated with uncertainty. Although specific solutions are not suggested, proposed strategies for developing a self-healing system should not focus on a finite number of underlying causes. Instead the focus should be on how these causes manifest, how they can be detected and ultimately how they can be corrected autonomously. To achieve a self-repairing system, it is clear that the system must have an element of self-awareness. Amor-Segan et al. [45] state that the ultimate aim is to develop a system with "the ability to autonomously predict or detect and diagnose failure conditions, confirm any given diagnosis, and perform appropriate corrective intervention(s)". A general approach is thus proposed that can be applied to any systems:

0. Cause of fault
1. Detection of fault
2. Diagnosis
3. Correction action

Cause of Fault: The 0th step "Cause of Fault" is numbered as such because in an ideal 'self-healing' system, the underlying cause is irrelevant. The system should be designed such that any and all underlying causes are compensated for by instead focusing on how these causes manifest. For example in a simple mechanism, a particular rod or element's dimension maybe altered due to shock loading, manufacture defect, thermal expansion etc. all of which lead to a change in the behavior of the system. Likewise in an electrical system, a component may change its behavior due to radio waves, voltage peaks, poor soldering or simply because it was initially defective. Ultimately however, the system must simply recognize it as a damaged component. The next steps are therefore of more importance in a self-healing system.

Detection: Any fault within a system will almost invariably lead to a fundamental change in the behavior or output. This could perhaps be most easily interpreted as a deviation from the prescribed behavior, utilizing either internal or external telemetric data. Currently technology typically allows for a number of externalized sensors, and a number of high-end industries, such as Formula 1 have already implemented this. Using these sensors the operators are kept constantly aware of the health and status of their system. This data could theoretically be used by a master-controller to continuously watch for any deviation in expected behavior then used to diagnose and correct any fault that is found.

Diagnosis: Whilst the detection of a fault is within the realms of current technology, the accurate diagnosis of this fault is perhaps a more difficult proposition. One of the reasons for this is the difficulty in validating large, complex system models that can exhibit a vast number of possible states [45]. Furthermore, another issue arises, which is the confidence in diagnosis, i.e. how much certainty must be present to initiate a corrective action? Because of this, an additional step could be required in which the diagnosis must first be confirmed, to avoid undesirable events such as 'good' components being unnecessarily removed or routed around. Current methods for diagnosis include:

- Model-based: Abductive reasoning: compare observation with predicted observation: I expect 'X' but get 'Y', therefore I must correct 'Y' to get it to match.
- Bayesian belief networks: probabilistic graphical model (or statistical model) that represents a set of random variables and their conditional dependencies: If 'X' and 'Y' happen, it's likely a failure with 'Z'
- Case-based reasoning methods: anecdotal evidence, if 'X' happens, do 'Y'. (Simplest, but only accounts for expected failure)

Corrective Action: Perhaps the most significant aspect of a proposed self-healing system is the corrective action. The nature of the corrective action will likely be application specific; however a number of possible approaches are available. These can be broadly broken down into either self-repair, where the system has the ability to partially or fully fix a given fault to continue operation, or self-healing, where the system is able to physically bring itself back to its initial state of operation after a fault has occurred [43]. Self-repair corrective strategies are more within the confines of existing technology. A simple example of this is having adaptable redundant components that are able to alter their function to stand-in for which component is currently diagnosed as at fault. This concept of 'self-repair through self-reconfiguration' does not necessarily require additional, redundant materials; instead performance could be sacrificed to ensure continued functionality utilizing only the currently available resources. This approach would use degenerate modules that have the ability to perform the same function or yield the same output even if they are structurally different [46]. Conversely a true self-healing system would typically require a fully adaptable system, such as 'smart dust' [47], where there is a finer level of granularity and near infinite possibilities for reconfiguration. Alternatively the system could mimic biological organisms and possess the ability to recreate parts that fail. With the significant recent advances in 3D printing and similar technologies, this concept is perhaps not as far-fetched as it first appears.

23.6.3 No Fault Found

An important philosophy of maintenance is *to increase the operational availability of a system*. It outlines the importance of rectifying any failures before they could have any effect on the operational performance, calling quick and correct fault finding strategies [48]. *But what happens when these faults or failures cannot be located?* Recently, increasing system complexities and interactions have seen a rise in the number of *unknown failures* that are being reported during service. Components tagged as 'No Fault Found' (NFF) are evidence that a serviceable component was removed, and attempts to troubleshoot for any root causes have been unsuccessful [49]. In essence, the existence of the NFF phenomenon has had a definitive negative impact upon critical system stakeholder requirements, which at the top level, often includes systems safety, dependability and life-cycle costs [50]. For that reason, it is essential to prevent NFF events or (at the very least) reduce the impact they can inflict on the business operation.

NFF is the result of an inadequate diagnostic task during the maintenance of complex engineering systems and is common to all. However, the term 'No Fault Found' is somewhat ambiguous, and could be misconstrued as meaning that 'no fault exists' and that the equipment under maintenance is rather healthy. Even though the problem (to some extent) does include this descriptor, the more general class of NFF problems refer to 'the inability to identify a real and existing fault within the equipment', or in other words, be regarded as *failure to diagnose*. Investigations in the past have shown that there are a large variety of potential drivers that can lead up to a NFF such as incorrect diagnostic techniques, poor training, inadequate processes and procedure, operational pressures and poor system design—all of which merit exhaustive discussions in order to develop the principles of NFF, and to guide practitioners in the development and deployment of effective solutions to their NFF problems.

So what is the No Fault Found phenomenon? A much simplified maintenance process within an organization can be observed in Fig. 23.6, which separates the rectification process into three key levels within the organisation. Here, it is

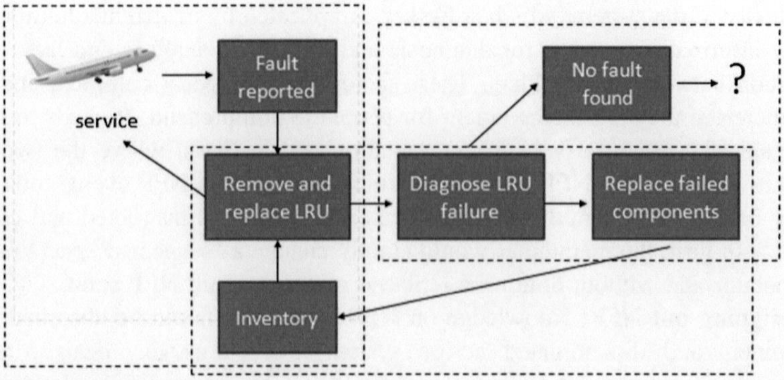

Fig. 23.6 Maintenance sub processes

important to understand the concept of how NFF instances can manifest themselves at various levels. When an operator records a system error, maintenance personnel are notified, who will attempt to investigate the reason for the system malfunction. For the most part, faults are diagnosed, isolated and rectified. But what happens when bench tests do not reveal any faults? What happens when removed components work perfectly under unit testing?

It is tagged as an 'NFF'.

There may be various reasons that contribute to this overall process and does raise some concerning questions. Perhaps the operator (or maintainer) lacks knowledge of the system? Or is the test equipment is incapable of detecting a fault? Or maybe the technicians received insufficient support to carry out their fault diagnosis? There can be many other reasons including having minimal understanding of the manuals, lack of equipment or operational pressures, etc. In any case, it should be noted that factors such as delays, inventory supply, operational pressure, system availability, cost implications, contracts, etc. will influence the decisions taken in each of the three levels. The aerospace industry has reported the majority share of electronic NFF faults, primarily within aircraft avionics. Although, some studies suggest that NFF events generally occur after an initial warning alarm has been triggered, indicating a system fault. This alarm does not provide any other direct diagnostic information, it simply triggers maintenance activities that 'repair' the faulty unit as it is removed for testing. During the testing phase, the situation arises where the same symptoms cannot be detected (or reproduced) with the standardized test equipment and procedures, or the exact nature (or location) of the fault is unable to be determined; as a result the unit is labeled NFF.

Categorization of NFF: NFF causes can be resulted from gaps in design in OEMs, diagnosis in dealers, and testing in suppliers. It shows problems with both the software and components at different system levels plus diagnosis and testing methods and equipment, and in addition, involves user's misuse, mistakes in packaging and damages during delivery. In general, it can be driven by the level of complexity of the system, which is further complicated by inadequate training, low quality information support for diagnosis and root cause analysis and lack of collaboration between stakeholders. These factors are obviously complex, and intricate, therefore they are inconvenient for people to comprehend. It needs to follow an organised, structured way for presentation and analysis, where the two main branches are technical NFF and non-technical NFF. Some NFF events took place simply because an operator did not operate the system as instructed and claimed "faulty". In turn, the maintainer would simply change a "suspected" part to satisfy the operator and without doubt the replaced part will result NFF soon.

Designing out NFF: Knowledge on NFF resolution should be absorbed by all maintainers and disseminated across system lifecycle stages, design, testing, manufacture, support, and across stakeholders. This can assist designers in improving the reliability of the equipment. However, such practical onsite knowledge is mostly kept by individual engineers rather than being documented and shared within the organisation and extended to the suppliers. At the core of the challenge for better troubleshooting is this difference between *expected failures* captured within the design and the *real-world failures* that occur during service. With an increase in equipment complexity, designers usually recognise the potential failure modes and their effects on the system using a FMEA. With this information, it can be determined how best to employ on-board diagnostic (or Built in Test) technologies to detect failures. This can also be used to prepare troubleshooting procedures, in advance, for analyzing the functionality of the system in order to differentiate among the many possible root causes of these expected failures.

However, when equipment enters service, it is exposed to the 'practical or real world', as shown in Fig. 23.7. Here, some faults that were expected during design will actually happen; but some never do. When a fraction of the theoretically possible failure modes occur, the weaknesses in a piece of equipment will become evident during the operation. It can then be extrapolated that equipment which fail on one aircraft, are more likely to fail on other aircraft of the same design, operated in similar conditions. But most importantly, many real-world faults are not

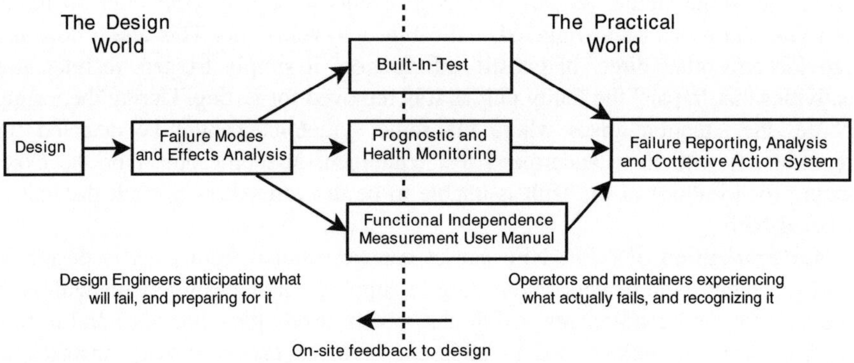

Fig. 23.7 Troubleshooting: expected versus real-world faults [52]

anticipated by the design engineers, and therefore the traditional diagnostic systems do not resolve them. In those cases, human experience and ingenuity may help solve the problem, but where is this *innovative* knowledge stored after its creation? It does seem that some of this knowledge makes does make its way back into troubleshooting manual updates, and may be fed back to engineering designers to modified their current designs for much more reliable parts. But more importantly, most of this knowledge only resides within the heads of a few key experts, or in personalised organisational databases which usually are consulted only after a problem has resisted several attempts at resolution. Therefore, on-site experience must be blended with other diagnostic and prognostic tools and techniques. The use of 'Failure Reporting, Analysis, and Corrective Action System' (FRACAS)' procedures can help in this process and provide valuable insights to designers.

Tracking spare parts: The capability to correctly identify suspected NFF units is the most importance factor in mitigating the overall implications of NFF events. The key here is put into place the necessary procedures to track NFF suspected units by serial number, showing installation/removal dates, the platform on which the unit was commissioned, number of operating hours/cycles, number of hours since its last maintained and a concrete reason for the generated removal codes. In addition to this, the history of the operating platform must be recorded with an easy to use retrieval system [51]. The significance of such historical information is to help in finding out the effects the failure has on the overall system and whether if the replacement unit offered any level of confidence in rectifying the original problem.

References

1. Baines TS, Lightfoot H, Steve E, Neely A, Greenough R, Peppard J, Roy R, Shehab E, Braganza A, Tiwari A, Alcock J, Angus J, Bastl M, Cousens A, Irving P, Johnson M, Kingston J, Lockett H, Martinez V, Michele P, Tranfield D, Walton I, Wilson H (2007) State-of-the-art in product-service-systems. J Eng Manuf 221:1543–1552
2. Roy R, Shaw A, Erkoyuncu JA, Redding L (2013) Through-life engineering services. Measur Control 46(6):172–175
3. Zhang Z, Chu X (2010) A new approach for conceptual design of product and maintenance. Int J Comput Integr Manuf 23:603–618
4. Ayeni P, Baines T, Lightfoot H, Ball P (2011) State-of-the-art of 'Lean' in the aviation maintenance repair overhaul industry. Proc Inst Mech Eng, Part B: J Eng Manuf. 225 (11):2108–2123
5. Borestein S (1992) The evolution of U.S. airline competition. J Econ Perspect 6(2):45–73
6. Andrew T, Byard P, Henry P (2008) Self maintenance works for repair firm. Eng Technol 3 (7):69–72
7. Um IS, Cheon HJ, Lee HC (2011) A simulation-based optimal design and analysis method for designing a train overhaul maintenance facility. Proc Inst Mech Eng, Part F: J Rail and Rapid Transit. 225:523. doi:10.1243/09544097JRRT309
8. Fu Z, Wang G, Gao F, Tian X, Li Y, Lu B (2012) Review of high-speed train maintenance. In: International conference on quality, reliability, risk, maintenance, and safety engineering (ICQR2MSE), Chengdu, June 2012, pp 418–422

9. Schempf H, Mutschler E, Goltsberg V, Crowley W (2003) Grislee: gas main repair and inspection system for live entry environments. Int J Robot Res 22(7–8):603–616
10. Kirkham R, Kearney P, Rogers K, Mashford J (2000) PIRAT—a system for quantitative sewer pipe assessment. Int J Robot Res 19(11):1033–1053
11. Hertzberg J, Kirchner F (1996) Landmark-based autonomous navigation in sewerage pipes. In: Proceedings of the first Euromicro workshop on advanced mobile robot, 9–11 Oct 1996, pp 68–73
12. Nassiraei AAF, Kawamura Y, Ahrary A, Mikuriya Y, Ishii K (2007) Concept and design of a fully autonomous sewer pipe inspection mobile robot: KANTARO. In: IEEE international conference on robotics and automation, 10–14 April 2007, pp 136–143
13. Dobmann G, Walle G, Höller P (1987) Magnetic leakage flux testing with probes: physical principles and restrictions for application. NDT Int 20(2):101–104
14. Davis RJ, Nestleroth JB (1997) The feasibility of using the MFL technique to detect and characterize mechanical damage in pipelines. Review of progress in quantitative nondestructive evaluation, vol 16. Plenum, New York
15. Bo L, Feilu L, Zhongqing J, Jiali L (2010) Eddy current array instrument and probe for crack detection of aircraft tubes. In: International conference on intelligent computation technology and automation
16. Urdaneta M, Garcia C, Poletti G, Ejarque G, Saltaren R, Aracil R (2012) Development of a novel autonomous robot for navigation and inspect in oil wells. CEAL 14(3):9–14
17. Hong D, Velinsky SA, Kazuo Y (1997) Tethered mobile robot for automating highway maintenance operations. Robot Comput Integr Manuf 13(4):297–907
18. Eich M, Vögele T (2011) Design and control of a lightweight magnetic climbing robot vessel inspection. In: IEEE mediterranean conference on control and automation, Corfu, Greece, pp 1200–1205
19. Yukawa T, Suzuki M, Satoh Y, Okano H (2006) Design of magnetic wheels in pipe inspection robot. IEEE international conference on systems, man, and cybernetics, Taipei, pp 235–240
20. Luk BL, Cooke DS, Galt S, Collie AA, Chen S (2005) Intelligent legged climbing service robot for remote maintenance applications in hazardous environments. Robot Auton Syst 53:142–152
21. Nagakubo A, Hirose S (1994) Walking and running on the quadruped wall-climbing robot. In: International conference on robotics and automation, ICRA, San Diego, pp 1005–1012
22. Schmidt D, Berns K (2013) Climbing robots for maintenance and inspections of vertical structures—a survey of design aspects and technologies. Robot Auton Syst 61:1288–1305
23. Lopez J, Perez D, Paz E, Santana A (2013) WatchBot: a building maintenance and surveillance system based on autonomous robots. Robot Auton Syst 61:1559–1571
24. Xianjing Y, Hida M, Suzuki Y, Emoto K, Tatsuno K (2013) Image recognition of a bolt in power distribution line maintenance tasks for an autonomous robot. In: International symposium on micro-nanomechatronics and human science (MHS), Nagoya, 10–13 Nov, pp 1–4
25. Taylor FW (1911) The principles of scientific management. Harper & Brothers, UK
26. Gantt HL (1903) A graphical daily balance in manufacture. Trans Am Soc Mech Eng XXIV:1322–1336
27. Taguchi G (1988) Introduction to Taguchi methods. Engineering (London) 228(1):i–ii
28. Cakmakci M (1995) Process improvement: performance analysis of the setup time reduction-SMED in the automobile industry. Int J Adv Manuf Technol 41:168–179
29. Dutta U, Maze TH (1987) Application of industrial engineering techniques in transit maintenance. ITE J 57:45–49
30. Gilbreth FL (1911) Motion study. Princeton NJ, Van Nostrand
31. Price B (1989) Frank and Lillian Gilbreth and the manufacture and marketing of motion study, 1908–1924. Bus Econo Hist 18:88–89 http://www.thebhc.org/publications/BEHprint/toc181989.html
32. Akrout H, Anson D, Bianchini G, Neveur A, Trinel C, Farnsworth M, Tomiyama T (2013) Maintenance task classification: towards automated robotic maintenance for industry. Procedia CIRP 11:367–372

33. Pahl G, Beitz W, Feldhusen J, Grote KH (2007) Engineering design—a systematic approach. Springer, London
34. George DJ (2012) Concept generation using morphological and options matrices. Thesis, Clemson University
35. Farnsworth M, Tomiyama T (2014) Capturing, classification and concept generation for automated maintenance tasks. CIRP Annals—Manuf Technol. http://dx.doi.org/10.1016/j.cirp.2014.03.093
36. Vongbunyong S, Kara S, Pagnucco M (2013) Application of cognitive robotics in disassembly of products. CIRP Annals—Manuf Technol 62(1):31–34 (ISSN 0007-8506)
37. Hernandez G, Seepersad CC, Mistree F (2002) Designing for maintenance: a game theoretic approach. Eng Optim 34:561–577
38. Frei R, McWilliam R, Derrick B, Purvis A, Tiwari A, Di Marzo Serugendo G (2013) Self-healing and self-repairing technologies. Int J Adv Manuf Technol (JAMT) 69:1033–1061. doi:10.1007/s00170-013-5070-2
39. Umeda Y, Tomiyama T, Yoshikawa H (1991) A design methodology for a self-maintenance machine. In: Stauffer LA (ed) Proceedings of design theory and methodology—DTM '91—, DE-vol 31, ASME, New York, pp 143–150
40. Shimomura Y, Tanigawa S, Umeda Y, Tomiyama T (1995) Development of self-maintenance photocopiers. AI Magazine 16(4):41–53
41. Bongard J, Zykov V, Lipson H (2006) Resilient machines through continuous self-modeling (sic.). Science 314(5802):1118–1121
42. Thrun S, Burgard W, Fox D (2005) Probabilistic robotics. MIT Press, Cambridge
43. Bell C, McWilliam R, Purvis A, Tiwari A (2013) Concepts of self-repairing systems. Meas Control 46(6):176–179
44. Desai A, Mital A (2013) Design simplification and innovation through adoption of a time-based design for maintenance methodology. In: Kumar S, Mital A, Pennathur A (eds) Human work productivity: a global perspective. CRC Press, Boca Raton
45. Amor-Segan ML, McMurran R, Dhadyalla G, Jones RP (2007) Towards the self-healing vehicle. Automotive electronics, In: 3rd institution of engineering and technology conference, IET, pp 1–7, June 2007
46. Edelman GM, Gally JA (2001) Degeneracy and complexity in biological systems. Proc Nat Acad Sci 98(24):13763–13768
47. Kahn JM, Katz, RH, Pister KS (1999) Next century challenges: mobile networking for 'Smart Dust'. In: Proceedings of the 5th annual ACM/IEEE international conference on mobile computing and networking, pp 271–278, August 1999
48. Mobley RK, Higgins LR, Wikoff DJ (2008) Maintenance engineering handbook
49. Khan S, Phillips P, Jennions I, Hockley C (2014) No fault found events in maintenance engineering Part 1: current trends. Implications Organ Practices, Reliab Eng & Sys Saf 123:183–195
50. James I, Lumbard D, Willis I, Goble J (2003) Investigating no fault found in the aero-space industry. In: Reliability and maintainability symposium, 2003. Annual (pp. 441–446). IEEE
51. Hockley C, Phillips P (2012) The impact of no fault found on through-life engineering services. J Qual Maintenance Eng 18(2):141–153
52. Khan S, Phillips P, Hockley C, Jennions I (2014) No fault found events in maintenance engineering Part 2: root causes. Tech Dev Future Res, Reliab Eng Syst Saf 123:196–208

Part VII
Future Challenges and Opportunities in TES

Chapter 24
New Approaches to Through-Life Asset Management in the Maritime Industry

Christian Norden, Karl Hribernik, Zied Ghrairi and Clemente Fuggini

Abstract European shipbuilders are facing a strong, worldwide competition. Innovation is often discussed as the key to increase the competitiveness of the European maritime industry. However, the market penetration of innovative technologies is difficult due to higher investment costs and uncertainty regarding functionality, reliability and reparability of the new technology resulting in an overall higher risk. Consequently, new approaches for through-life asset management have to be established to achieve a successful market introduction of innovative technologies. Therefore the European funded research project ThroughLife pursues the development of promising and innovative new technologies on the one hand and the identification and elaboration of new approaches for through-life asset management with the overall goal to optimise the lifecycle performance of vessels on the other hand. One of the most promising Throughlife asset management approaches is to transfer the concept of comprehensive after sales services, like a "worry free" service package established in the automotive and the aeronautic industry, to the maritime industry. It is expected that the implementation of this business model would lead to significant lifecycle cost savings for the ship operator and arises the opportunity to enhance the business area of repair or new building yards. Moreover the holistic lifecycle services would also foster the market introduction of new technologies by reducing the uncertainty about proper maintenance and repair of innovations. As a result, the business model implementation could end up in a win-win situation for all involved stakeholders. However, in the assessment of the full service business model implementation barriers like difficult cost calculation and the dependency of the lifecycle cost from the vessel treatment have

C. Norden (✉)
BALance Technology Consulting GmbH, Contrescarpe 33, 28203 Bremen, Germany
e-mail: Christian.Norden@bal.eu

K. Hribernik · Z. Ghrairi
Bremer Institut Für Produktion Und Logistik GmbH (BIBA), Hochschulring 20, 28359 Bremen, Germany

C. Fuggini
D'Appolonia S.p.A., Via San Nazaro 19, 16145 Genoa, Italy

L. Redding and R. Roy (eds.), *Through-life Engineering Services*,
Decision Engineering, DOI 10.1007/978-3-319-12111-6_24

been identified. The application of comprehensive condition monitoring offers the opportunity to control the vessel treatment and to gather relevant information about the vessel to optimise the expected lifecycle costs and the corresponding service fee. Moreover applying condition monitoring enables additional benefits like the potential to optimise the maintenance and repair scheduling based on the actual and predicted condition of the component/system on the vessel. The chapter introduces the mythology of how the combination of the full service business model with comprehensive condition monitoring adds value for all involved parties and leads to an overall improved lifecycle performance of vessels to strengthen the European maritime industry.

24.1 Full Service Business Model in the Maritime Industry

24.1.1 Introduction

Full service contracts are well known in the automotive or aeronautics industry, where a number of services are offered in service packages in return of a fixed, time based fee. Table 24.1 presents two exemplary services: “Full service leasing” in the automotive industry, which is offered by the car rental company SIXT [1] and Total Technical Support (TTS) [2] offer by Lufthans Technik in the aeronautic industry.

These two examples point out that holistic service approaches for means of transportation are realised in other industries and lead to benefits for both involved parties. The service provider gains a long-term customer relationship with continuous revenue flows on the one hand and the customers gain calculable costs, expands his liquidity and reduces the lifecycle cost of the vehicle due to economies of scale effects and lower management costs.

Table 24.1 Full service packages in the automotive and aeronautic industry

Automotive	Aeronautic
• Car leasing	• Line maintenance
• Maintenance and repair	• Customized maintenance planning
• Insurance	• Troubleshooting
• Taxes	• Engineering services
• Mobility insurance	• Repair and overhaul of aircraft
• Tire replacement	• Engines and components
• Restitution	• Spare-parts pooling
• Fuel consumption management	• Spare engine leasing
	• Painting, cabin modifications
	• Airline Support Teams (AST®)
	• Logistics
	• Training

However, full-service contracts for ships are not established in the maritime industry. In the EU funded project ThroughLife, we developed a transfer of full-service business models to the maritime industry and assessed the potential advantage for the new building yards, repair yards and ship operators.

24.2 Adaption to the Maritime Industry

A modern way to illustrate and describe business models is the concept of the business model canvas [3]. Within the ThroughLife project, the adapted full service model has been developed and described in a business model canvas (Table 24.2).

The centre of the business model canvas is the value proposition for the customer which is the ship owner/ship operator in this case. In the full-service business model the customer benefits from calculable costs for the contract duration. Moreover, one point of contact would reduce the management costs for coordinating maintenance and repair activities and accelerate maintenance and repair processes which reduce opportunity costs. In addition, the customer can concentrate on his core business which increases the efficiency in the supply chain of the customer.

On the right hand side the of the business model canvas, the marketing of the value proposition is described in terms of targeted customer segment, marketing channels and customer relationship. The full-service contract is applicable for every vessel segment, which could be reached by the new building yard in conjunction with the new vessel or by the repair yard in return of provision in case they are not the service provider to establish a long-term customer relationship.

Table 24.2 Business model canvas for full-service contracts

Key partners:	Key activities:	Value proposition:	Customer relationship:	Customer segment:
❑ Marine equipment manufacturer ❑ Ship owner /operator for detailed information	❑ Managing the offered services	❑ Calculable costs for the contract duration ❑ Reduction of management costs ❑ Concentrate on the core business	❑ Long-term contract	❑ Every vessel segment
	Key resources:		**Marketing Channels:**	
	❑ Key information about the vessel ❑ Skilled Personnel		❑ Distributed by the new building yard in addition to the new vessel ❑ Distributed by the repair yard	

Cost structure:	Revenue Streams:
❑ Calculable costs for scheduled services ❑ Costs for unscheduled services	❑ Continuous revenues: time based fee

The left hand side of the business model canvas represents the way of producing the value proposition. The key activity is the management of the service contract, which includes the coordination, execution and administration of all services included in the contract. Therefore, the service provider requires detailed information about the vessel in terms of incorporated systems, components, materials, etc., skilled personnel as the key resources to fulfil the contract and a close collaboration with marine equipment manufacturers as key partners.

In the lower part of the business model canvas the financial matters in terms of cost structure and revenues are addressed. The main benefit for the service provider is the continuous revenue flow for the service provider. The cost structure consists of calculable costs for scheduled services and uncertain costs for unscheduled services like accidents. In this regard, the service provider needs to have statistical information about failure probabilities in order to reflect the risk-level in the calculation of the time-based fee. The main driver for the business model is the potential to realise economies of scale by serving multiple contracts to offer the service package at an attractive price point for the customer.

24.3 Business Model Assessment

The implementation of the introduced full-service business model in the maritime industry offers a range of benefits for the service provider and the customer on the hand, although on the other hand several weaknesses need to be considered. Table 24.3 summarises the identified strength and benefits for the service provider and the customer.

Table 24.3 Strength and weaknesses of the full-service contract business model

	Assessment of the full-service business model	
	Strengths	Weaknesses
Service provider	• Continuous revenues	• Potential legal issues
	• Serving multiple contracts to diversify risk	• Costs depend on the reliability
	• Cost reduction due to economies of scale	• Uncertainty about vessel treatment
	• New business segment for yards or entrepreneurs	
Customer	• Calculable costs for the contract duration	• Higher dependency from business partners
	• High service quality	
	• One point of contact	
	• Reduced vessel management costs	
	• Applying new technologies with reduced risk	

The biggest strength of the introduced business model for the service provider are the continuous revenue flow throughout the contract duration. Moreover, the business model foresees that the service provider serves multiple contracts to diversify the risk and to realise economies of scale effects. The business model additionally opens up the opportunity to expand the business of yards.

The customer benefits from calculable costs which reduces the uncertainty and avoids short-term financing issues. It is expected that the service quality is excellent since the service provider has a high interest in performing the services as cost-effective as possible. In addition, the vessel management efforts can be reduced thanks to the one point of contact concept. An additional opportunity is the application of new technologies or materials at a reduced risk level since the service provider would take over maintenance and repair activities. Thereby the full-service contract can serve as an instrument to foster the market penetration of new and innovative technologies. On the other hand, the customer has to have the willingness to reduce his level of self-reliance in a long-term service contract.

The weaknesses for the service provider are potential legal issues and the corresponding costs for payment prolongation and law suits, which affects both parties in this business model. Despite the potential legal issues, the lifecycle costs of the vessel and the corresponding costs to perform the services depend on the reliability of the vessel which is affected by the vessel treatment of the customer. As a result, the service provider has to deal with uncertainty regarding the vessel treatment, which could result in an unprofitable business.

24.4 Condition Monitoring

24.4.1 The Concept of Condition Monitoring

Condition Monitoring is the process of monitoring a parameter of condition of machinery, component and/or a system such that a anomalies and developing failure are detected. Condition monitoring is largely applied at industrial level in different sectors (aerospace, automotive, machine related, etc.) and for different engineering applications. By implementing a condition monitoring system, the idea is to keep under control the operating conditions of the component/system under investigation and to detect anomalies with respected to the foreseen designed or calculated operating behaviour. In such a way, condition monitoring supports condition-based maintenance as well as predictive actions across the lifecycle of the system. When dealing with the implementation of a condition monitoring solution for a specific application, issues related to the technological implementation of condition monitoring, the compliance of the system to standards and the costs for designing and installing the system need to be analysed, quantified and decided.

In maritime sector the shipping industry faces a major challenge being able to ensure a consistently high standard of condition monitoring expertise on every ship with a workforce that tends to move frequently between ships, when it comes to running a condition monitoring program. Without the use of a sensor-based monitoring strategy combined with the conserved knowledge of inspection specialists, the diagnostic data collected using visual inspection may be difficult to interpret to carry out optimised maintenance, and too often this result in the anticipated gains failing to materialise. The best way to make sure that full benefits are reaped is to develop a system where the timing, sharing, storage, interpretation/analysis and use of the collected data are automated and optimised. This can be achieved by implementing an intelligent sensor-based approach. The Condition-Based Maintenance System should provide the optimum approach to safety of ship's critical compartments, such as a BWT. This can be only achieved by automatically collecting and processing of sensor data. This enables analysis or interpretation of data at any time, either on the ship or onshore by remote diagnosis. Not only this: collected data relating to critical ship compartments need to be transmitted via the ship's communication system to a remote diagnosis centre, where it is monitored and interpreted automatically.

24.4.2 From Condition Monitoring to Condition-Based Maintenance

Every condition monitoring system is constituted by sensors whose aim is to monitor the component, machine, system under investigation with the goal to detect anomalies. Generally sensors and sensing systems are essential elements in enabling the introduction of monitoring solutions in new and different areas than the ones where it is already implemented, fostering new applications and the link among monitoring and maintenance. This is more than needed in the marine sector where the need for the introduction of innovative concepts and systems is often discouraged by the absence of technologies that can assess, validate and verify the feasibility, reliability and impact on lifecycle performance of an innovative solution. Sensors and sensing technologies as part of a condition monitoring system can offer a solution to this issue since they can provide the tool to understand the behaviour of the innovation in operation, assessing the performance and providing feedbacks on their implementation. On the other hand when monitoring solutions are linked to maintenance tasks, they can offer the mean to allow for better maintenance services that are based on real conditions of a monitored element (e.g. condition-based) rather than on scheduled activities. As a consequence of this approach one can optimise the management of available resources including capital, since maintenance tasks are planned on demand. To this end it is clear that a systemic approach is needed to be conceived in implementing monitoring solutions for marine applications.

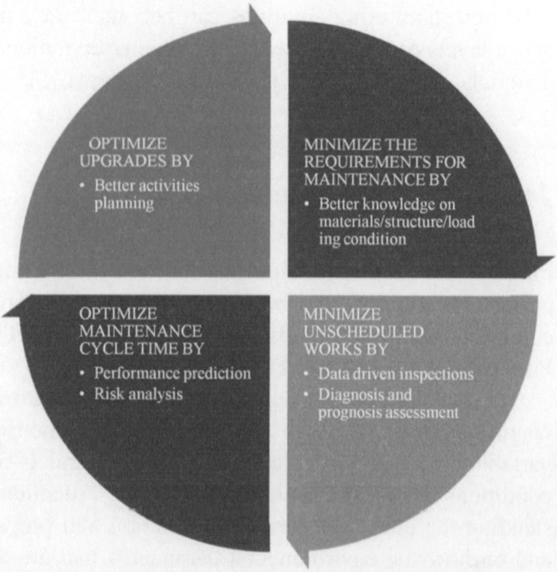

Fig. 24.1 Monitoring for reducing life cycle costs (LCC)

Within this framework, monitoring, maintenance and management activities are key aspects of any life cycle approach. The identification and the elaboration of feasible and cost effective solutions as part of a holistic lifecycle decision support structure represents a key issue the scientific community that industries need to face. Apart from the aerospace industry, this is especially needed in the maritime industry, where the different stakeholders involved in the value chain (e.g. ship owners/operators, yards) are demanding for new technologies to cope with future challenges of fuel cost increase and environmental legislation. Thereby the implementation of condition monitoring solutions can facilitate the day-to-day management by supporting operational and maintenance decisions, providing knowledge and triggering inspections autonomously, thus reducing maintenance costs throughout the system's lifecycle (Fig. 24.1).

This is more than true when monitoring solutions are conceived for problems that are unavoidable and when occur have a strong impact on maintenance tasks and management activities. Among the others corrosion induced effects that affect ship structures is one of the most oppressing issue that need to be faced, monitored and controlled. Indeed corrosion is a phenomenon that can reduce the lifecycle of a structural system. When corrosion occurs, an early intervention (maintenance) can reduce follow-up costs, sustain the availability and lead to overall lifecycle cost reduction. During long and fixed maintenance intervals, corrosion processes may be unnoticed and can reach a huge extend, which increases the repair efforts respectively the probability of failure. A continuous corrosion monitoring system with suitable sensors is able to prevent this by detecting the corrosion in early stages in difficult and non-accessible areas of structural elements. Moreover, the adoption of condition monitoring solutions can become a key factor by enhancing the confidence levels of the end users and reducing operational risks when introducing new materials, new designs or new repair concepts [5].

24.4.3 Technical Implementation

Proof of the implementation of condition monitoring strategies in the EU-funded project ThroughLife has been provided through a full-scale validation test of the condition monitoring of Ballast Water Tanks (BWT) structures on-board of a Ro-Pax ferry of Consortium Partner Balearia.

The aim here was to understand in which terms and up to which extent the conditions inside the tank modify the daily response of the structure subjected to variable imposed loads (e.g. cargo loads) and if a long-term prediction of the conditions inside the tank can be made. Attention was mostly paid on understanding the effects of corrosion initiation and propagation on the steel structures and on how the environmental parameters that are at the base of this propagation can be monitored and controlled through the use of temperature, humidity, pressure, pH, salinity and turbidity sensors. Indeed the idea behind this, it is to define condition-based maintenance tasks that are the results of accurate data analysis and processing through a combined use of corrosion, environmental and structural sensors. The latter are fiber optics sensors that measure strain variation on a steel plates induced by load cycles. If one can correlate their measurements to those of environmental sensors, thus one can create a predictive model that allow accounting for initiation and propagation of phenomenon like corrosion across the entire lifecycle of the monitored structures (Fig. 24.2).

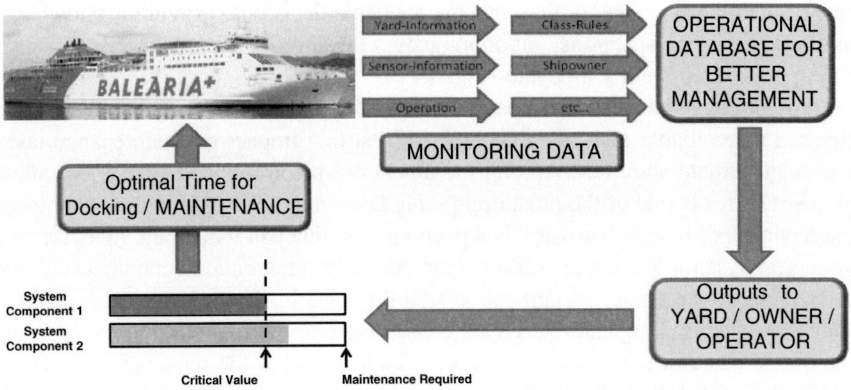

Fig. 24.2 Case study fort the overall implementation of a condition monitoring system within the EU-funded project through-life

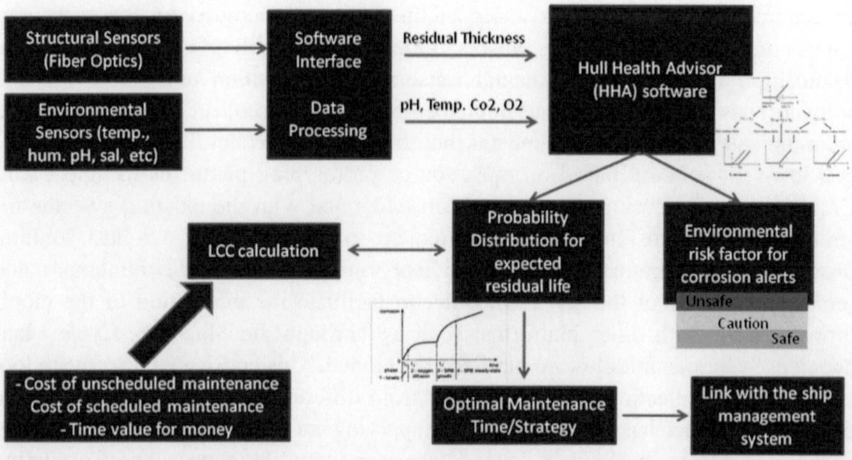

Fig. 24.3 Implementation of condition monitoring strategies in the case study

Figure 24.3 shows the logic flow that from the collection of raw data from structural and environmental sensors lead to the determination of optimal maintenance strategies that accounts for the estimation of the residual lifetime of the structure or the structural component, the environmental risk factor and life-cycle-costs (LCC) calculations. The result of this implementation is a so called "3Ms" approach: Monitoring for optimizing Maintenance towards better Management.

24.4.4 Results of the Case Study

The main outcome of the study has been the testing and validation of a condition monitoring system for corrosion detection and control in close areas such as ballast water tanks (BWT). This was the result of a deliberate methodology and a holistic approach in understanding how a combination of sensing technologies, data analysis and processing solutions can solve problems like corrosion and its effects on structural members, they face daily. This is why in implementing the solution, particular emphasis was put in the definition of the purposes of the system, what is the strategy behind it, with technical and market barriers need to be faced/overcome and what need to be done to move from the identification of a preferable monitoring system to its implementation and installation. Overall, the end-users perspectives, their needs and the capacity of implementing the identified solution from a technical and economical perspective have been taken as the key drivers.

The developed condition monitoring system, which consists of hardware components and software tools, enables the availability of a full-service business approach as described in Chap. 1, that can be adopted by ship owner/operators to trigger inspections and to perform condition-based maintenance tasks as a result of an accurate analysis and processing of the raw data recorded by the corrosion, environmental and structural sensors. One of the enabling components is the Acquisition Unit for environmental sensors. An acquisition unit (hardware and software) has been designed and implemented in order to collect, process, store and communicate ambient/environmental data from Ballast Water Tanks (BWT). This is a low-cost solution based on open-source prototyping platforms (Single Board Computer). The developed unit has been integrated with the existing systems for structural and corrosion sensor measurement to build together a robust solution providing ship operators/owners an efficient solution to monitor continuously and remotely the state of the BWT. In order to facilitate the interaction of the monitoring system with other platforms, such as ThroughLife Ship Life-Cycle Management System, a middleware has been developed. This consists of a software tool ensuring data collection and aggregation from different systems/data sources. The traceability aspect has been ensured by applying an efficient centralised and decentralised storage mechanism using database and text files approaches. Correlating the measurements from corrosion and structural sensors with the measurements of environmental sensors, it is possible to know at a fixed point in time what is occurring and if, how and when maintenance is needed based on the actual condition of the monitored component. Additionally, the monitoring system generates a predictive model, of what will happen in the future and what actions are foreseen to be done at a certain point in time in the future.

In particular, the condition monitoring system implemented in the ThroughLife project is designed to capture three distinct types of data related to corrosion in ballast water tanks. The types of data, the parameters collected and the aim of the data collection are shown in Table 24.4.

The three types of information generated and collected by the condition monitoring system can be analysed individually or put into relation with each other. Individual analysis can be carried out in order to achieve the aims described in Table 24.4. Put into correlation with each other, a more holistic understanding of the corrosion process can be achieved at both micro and macro levels. For example, Error! Reference source not found. shows timelines of salinity and pressure measurements on the left, and temperature and pH measurements on the right. The correlation of the different measurements over time can indicate and help shop operators interpret different phenomenon, situations and conditions in the ballast

Table 24.4 Types of data collected by the condition monitoring system

Type of Data	Parameters	Aim
1. Environmental (macro level)	Temperature, pH, humidity, salinity, pressure	Identify how corrosive the environment is at any given time and deduct the corrosion propagation rate
2. Corrosion (micro level)	First cracks and coating impairments	Indication of coating conditions and initial detection of corrosion occurrence
3. Structural	Strain measurement	Detection of impairments and structural damage

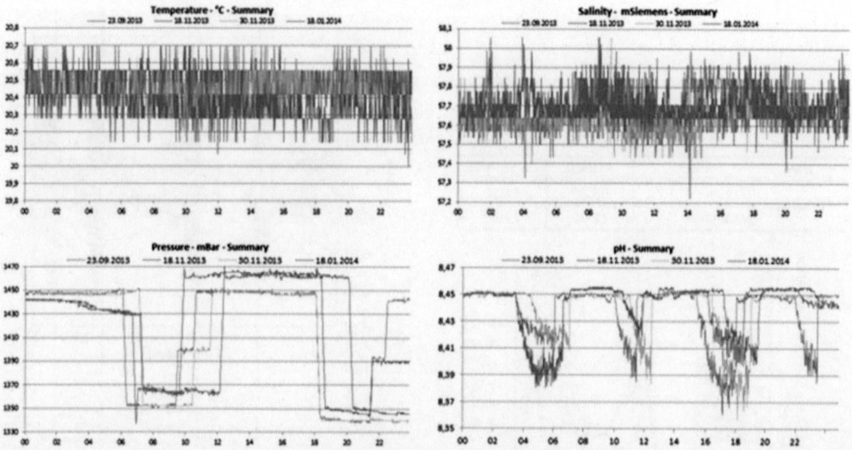

Fig. 24.4 Salinity and pressure (*left*). Temperature and pH (*right*)

water tanks. With information thus correlated, different analyses can be carried out to better understand the different phenomenon, situations and conditions and take action based on the improved understanding.

Coating conditions and other qualitative information can be displayed as a heat map with a spectrum of colours, for example, of green (good) to red (bad). The temporal distribution of sensor measurements can be very useful at investigating whether the occurrence of certain effects correlated with specific operations, such as looking at how loading periods relate to strain levels (Fig. 24.4).

Coating conditions and other qualitative information can be displayed as a heat map with a spectrum of colours, for example, of green (good) to red (bad). The temporal distribution of sensor measurements can be very useful at investigating whether the occurrence of certain effects correlated with specific operations, such as looking at how loading periods relate to strain levels.

A further use of the correlated data can be found in the definition of environmental risk indicators. Here, different levels of risk can be defined and visualised, e.g. using a spectrum of colours from safe conditions (green), conditions which call for caution (yellow) to unsafe conditions (red). Risk levels can be calculated through the elaboration of the environmental sensors data.

Another use can be found in the calculation of the residual life of structures subjected to corrosion and fatigue, where upper and lower values bounds on estimated residual life in weeks versus time.

With reference to the correlation among environmental and structural data, a correlation can be expected between pressure measured by sensors and structural strain sensors. On top of this, the following figure also shows a correlation among structural information and environmental information related to the pH level inside the BWT. This is a significant finding, because it means that ship owners can deduct a direct correlation between the "environmental safety" inside the ballast water tank

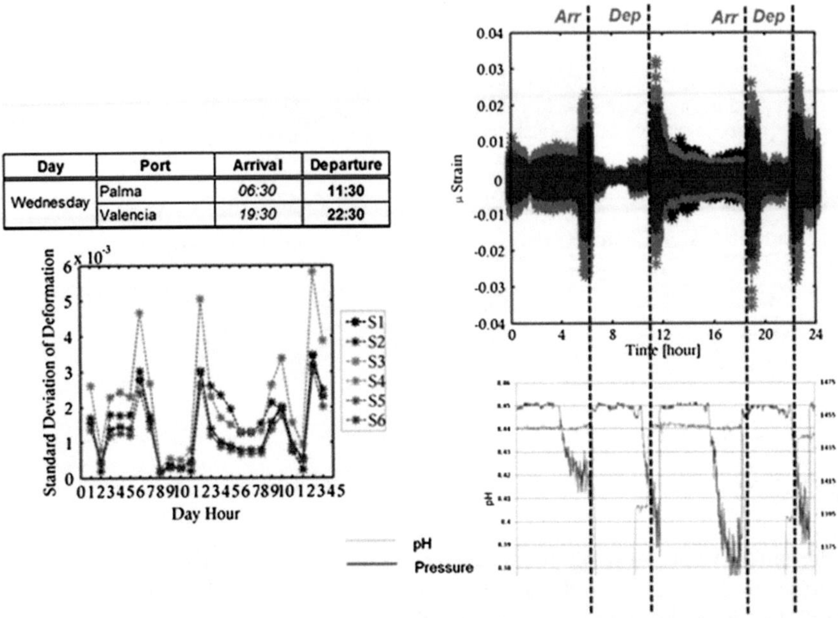

Fig. 24.5 Standard deviations of strain structural sensors (*left*); correlation among structural data and environmental data (pH, pressure) (*right*)

measured as a function of pH variation of water and the impact on the "structural safety" to the structures of the tank. If this information is gathered on the long perspective it can be used to allow the ship owner to take appropriate decisions on how, where and when to empty and fill the tanks, according to the most safe and valuable conditions. The information depicted in Fig. 24.5 can be easily correlated to the timeline of the ship shown in the left side of the picture, together with the standard variations of the structural data for one selected day.

With reference to the lifetime prediction of structural components such as steel plates in BWT structures, one interesting issue to be monitoring is the effect induced by stress-strain cycles that cycles can cause fatigue failure during the lifetime of the structure. A classical method for fatigue counting is to make use of an S–N curve for steel subjected to corrosion as can obtained from Det Norske Veritas recommendations for fatigue assessment of ship structures:

$$\log \mathrm{N} = \log a - \mathrm{m} \log \Delta\sigma \tag{24.1}$$

where

- N = predicted number of cycles to failure for stress range $\Delta\sigma$
- $\Delta\sigma$ = stress range (MPa)
- m = negative inverse slope of S–N curve = 3
- $\log a$ = intercept of log N-axis by S–N curve = 12.436

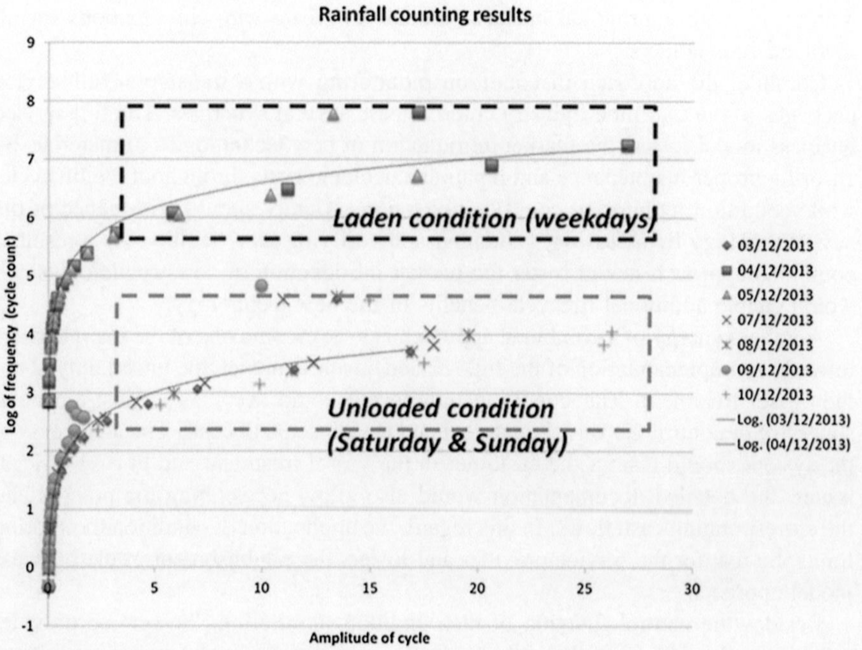

Fig. 24.6 S-N curves for fatigue analyses by using the data from the condition monitoring system installed on-board

The application of the algorithm leads to the calculation of the frequency and amplitude of cycles to which corresponds to a certain stress level. Figure 24.6 presents the results for two selected periods: weekdays and weekend. During the weekdays the ship is constantly under operation therefore continuous loading/unloading cycles happen. On the contrary during the weekend the ship is stopped in the port and no loading/unloading conditions happen. This is clearly depicted in the following figure where it can be noted the different stress level during the weekdays if compared during to that calculated during the weekend. If these measurements are extended for longer periods, it can be possible to determine of the basis of a fatigue analysis what would be the resultant decrease in percent in the lifetime of the selected component.

24.5 Synergies of a Combined Approach

The introduced business model and the application of condition monitoring offer opportunities to optimise the lifecycle performance on their own. The EU-funded research project ThroughLife (Grant agreement No. 265831) [4] pursues the synergies by combing the application of new technologies like the condition monitoring with appropriate business models to create win-win situations for all involved stakeholders.

Combing the approach of condition monitoring with a transfer of full-service packages to the maritime industry could release several synergies. The full-service business model fosters the market introduction of new technologies or materials by ensuring proper maintenance and repair at calculable costs throughout the lifecycle, while condition monitoring could reduce the uncertainty about performance of the new technology by assessing, validating and verifying its reliability. As a result, a combined approach would foster the market introduction of new technologies and would enable additional lifecycle benefits of the new technology.

Another synergy of a combined approach is to overcome one of the main barriers towards an implementation of the full-service business model, the uncertainty about the vessel treatment. The condition monitoring could serve as an instrument to supervise or control the vessel treatment and the subsequent costs. The awareness of the system should restrict the customer in the vessel treatment and in case of legal issues, the detailed documentation would also allow accelerating the process and the corresponding cash flows. In this regard, the application of condition monitoring limits the risk for the service provider and fosters the establishment of the business model approach.

Besides the control function of the condition monitoring, the service provider would also benefit from the opportunity to optimise the maintenance and repair schedule and the corresponding cost structure. As described in Chap. 3, these lifecycle cost savings ensure a profitable business for the service provider that could enable him to set the service fee at a competitive price point and create a win-win situation for the customer and the service provider [6].

The application of condition monitoring also utilise to gather operational data, which would serve perfectly as input for the service fee calculation and thereby optimise the income statement of the service provider. In case new materials or technologies are applied on the vessel, the condition monitoring would allow to track the behaviour of applied technologies and to document the operational performance, which could support the market introduction. Figure 24.7 summarises the synergy potential of combining condition monitoring with the full-service business model.

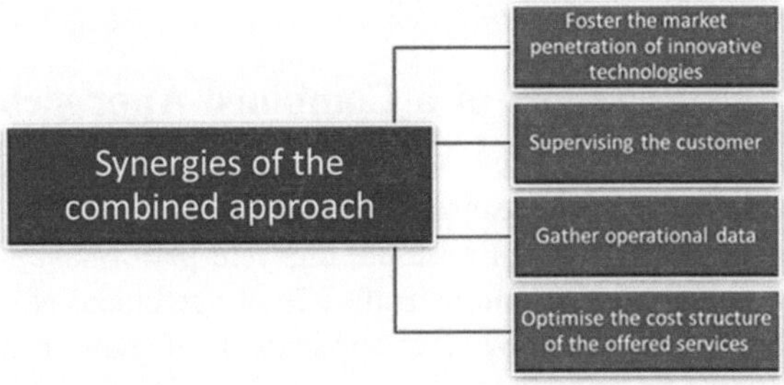

Fig. 24.7 Synergies of the ThroughLife approach

The condition monitoring sensors have a product lifetime of more than 30 years and could easily de-installed form on vessel and installed on another one. The service provider runs multiple contracts, which would result in a high demand for sensors and could end up in economies of scale. Moreover, the service provider would benefit from the synergies the most in the described business model. As a result, the additional costs for the investment and the assembly of the condition monitoring would have been defrayed by the service provider.

24.6 Conclusions

The transfer of full-service business model to the maritime industry and the technology of condition monitoring offer several benefits on their own. However, combining these two approaches could create several synergies and thereby boost the advantages for the involved stakeholder, especially in terms of market introduction of new technologies.

While the technical feasibility of the condition monitoring has been proven in the introduced case study in a ballast water tank within the ThroughLife research project, the data interpretation and the subsequent prediction of the system behaviour to utilise the full potential of the condition monitoring needs further development. Furthermore, there are some improvement potentials such as the development of wireless sensor nodes for such areas in order to avoid the complexity of cabling as well as profiting from the positioning flexibility of sensors into BWTs. On the other hand full-service business models are not established in the maritime industry, although feedback from various industry stakeholders during ThroughLife's final workshop indicated the potential of the approach, but also identifies the risk. In this regard, the introduced combined approach could foster the implementation of full-service business models in the maritime industry.

The next steps include the further development of data gathering using wireless networks, data interpretation and the predictive model as well as the assessment of the addressed synergies and their impact on the lifecycle performance of a vessel.

Acknowledgement The authors thank the partners and the European Commission for support in the context of the ThroughLife project, funded under contract number FP7-265831.

References

1. Sixt Leasing AG (2012) http://www.sixt-leasing.de/full-service-leasing/?c=g. Accessed 7 Feb 2012
2. Lufthansa Technik AG (2014) http://www.lufthansa-technik.com/tts. Accessed 5 Apr 2014
3. Osterwalder A, Pigneur Y (2010) Business model canvas. Wiley, Hoboken
4. ThroughLife project www.throughlife.eu

5. Fuggini C (2012) Towards a monitoring paradigm for the shipping industry: solutions and applications. In: Proceedings of the 5th European conference on structural control–EACS2012, Genoa, Italy, 18–20 Jun 2012
6. Norden C, Hribernik K, Ghrairi Z, Thoben KD and Fuggini C (2013) New approaches to through-life asset management in the maritime industry. In: Proceedings of the 2nd international through-life engineering services conference, Procedia CIRP 11, pp 219–224

Chapter 25
Future Challenges and Opportunities in Through-Life Engineering Services and Concluding Remarks

Louis Redding, Andy Shaw, Rajkumar Roy and Bill Bardo

Abstract This chapter presents the challenges and opportunities which were identified from a review of the literature and during several key events which brought academics, researchers and industrialists together to discuss the way forward. It presents opinions and insights taken from comments made by stakeholders and from event transcripts and of discussions and presentations. From the findings presented the chapter gives insight into the way forward and the benefits that can be obtained with the successful adoption of TES. Drivers and inhibitors are also identified.

25.1 Introduction

Much has been written in the literature relative to the ongoing evolution of manufacturing organisations. Today, one sees that organisations manufacturing complex engineering products are undergoing a paradigm shift in their mode of operations and the means by which they obtain revenue as they move from a position of that of pure design and manufacture with perhaps the supply of spares and a bolt on maintenance provision, to that of whole-life integrated manufacturer and service provider. The emergence of the 'manu-service' operation for want of a better description is being observed.

This has seen the emergence of '*availability contracting*' as the users of the product's function seek to only pay for it's availability for use. The oft cited Rolls Royce '*Power by the hour*' business model is a typical example and is perhaps the leading template for such a solution. In consideration of Rolls Royce it is said that….

> "For those who don't know Rolls-Royce, we have around about a £10 billion a year turnover, more than half of that comes from providing services for our clients and predominantly nowadays, the method of transacting we have on services is that on the day that

L. Redding (✉) · A. Shaw · R. Roy · B. Bardo
EPSRC Centre for Through-Life Engineering Services, Cranfield University,
Cranfield, Bedfordshire MK43 0AL, UK
e-mail: louis.redding@btinternet.com

L. Redding and R. Roy (eds.), *Through-life Engineering Services*,
Decision Engineering, DOI 10.1007/978-3-319-12111-6_25

> we sell an engine to a customer, we also sell a fixed price maintenance contract, and more and more today that is for a 25 year expected life products and it covers all eventualities. At the same time we guarantee the cost of that maintenance activity".
>
> (Source: Rolls Royce Speaker at recent I.Mech.E. event)

Essentially, with such contracting arrangements the risk to the revenue stream is of paramount importance and this has resulted in the emergence of Through-life Engineering Service (TES) as an important product support strategy in order to mitigate this risk. The previous contributions in this book are from leading academics, researchers and practitioners within the field. Each chapter has reported on various activities being conducted relative to TES. The remaining sections of this chapter present to the reader some of the challenges, opportunities and directions for further work within the field.

25.2 Sources Consulted for the Identification of Future Challenges and Opportunities Relating to TES

In seeking to identify both the challenges and opportunities that exist relative to the successful adoption of through-life engineering services which in turn serve to define the future directions that should be followed, *four* information sources were reviewed by the authors with the findings within the chapter contributions being left to speak for themselves.

They being:

- A review of the emerging literature relative to Through-life Engineering Services.
- The findings of an initial 'Think Tank' workshop held at the EPSRC Centre for Through-life Engineering Services held in 2012 and discussed in Chap. 1 of this book [1].
- A recent survey of UK practitioners which sought to give insight into "The adoption and use of through-life engineering services within UK manufacturing organisations" [2].
- A recent networking event held at the Institution of Mechanical Engineers (London—8th July) attended by leading academics and practitioners during which invited speakers presented their views as to potential future developments, challenges, opportunities, drivers, and inhibitors to the evolution and adoption of the concept. This was followed by an informal forum during which the above was discussed.

Attending the event in London one found speakers from various industrial sectors which included Aerospace, UK-MOD, Rail, Nuclear, Machine Tools, Power Generation, UK Based OEM supplying military hardware. This provided an opportunity to identify issues that cut across various sectors relative to TES from these sector speakers. This in turn was supported by informal transcription of key points identified from table discussions during the event.

25.3 Findings from the Review of Data Sources

This section of the chapter presents a summary of the findings from the resources identified above.

25.3.1 Findings from the Literature

As discussed in Chap. 2 of this book the literature relative to TES at the time of writing is only just emerging. Whilst there is much to be found within the literature which relates to the engineering problems relating to technical elements of a typical TES solution (sensors, system architectures, component degradation, knowledge management) they are not identified as being as such. Of those contributions which explicitly relate to TES they are mainly be found within three sources of the literature, namely:

- The Proceedings of the 1st and 2nd International Conferences of Through-life Engineering Services
- The CIRP Annals: Manufacturing Technology
- A limited number of peer reviewed journals which include (not exhaustive):
  - Proceedings of the Institution of Mechanical Engineers, Part B: Journal of Manufacture

Of the contributions identified the majority of subject areas addressed within the papers reviewed relate to specific elements of the concept (Table 25.2) with no contribution seeking to deal with or discuss TES holistically. It is seen in Table 25.1 that the contributions address specific problems within the engineering science and application area (sensors, systems, system architectures, non-destructive testing, self-healing and autonomous maintenance etc.). Whilst every contribution is equally valid and reports both the current and future work of the authors, the analysis is based on elements of the holistic solution and offers very little for the company wishing to ask the questions:

- So how do I implement a TES solution?
- Can my product offering accommodate such a solution?
- Is my organisation in the right position within the value added chain?
- How do I select and apply the technology?
- How do I inform my business and operating strategy to align with my desire to offer TES?
- How do I evaluate my organisations suitability to adot TES?
- What is the knowledge gap and how do I fill that gap?
- What are the essential building blocks for TES and how do I train personnel to acquire them?

Table 25.1 Elements of TES discussed within data sources

1st international conference on through-life engineering services	• Design for through-life service support • Through-life service delivery, operations and management • Service informatics, performance, and modelling • Repair technologies • Technologies of self-healing and self-repair
2nd international conference on through-life engineering services	• No fault found in through-life engineering solutions • Service and support engineering • Failure and diagnostics • Design and manufacturing for through-life engineering solutions • Maintenance and asset management • Maintenance, repair, and overhaul • Through-life engineering services feasibility studies • Wind turbine maintenance • Flow diagnosis • Life-cycle engineering and obsolescence management • Self-healing technologies and autonomous systems • Human influences in through-life engineering services • Service system modelling
CIRP annals	• NDE—thermography • Taxonomy • Knowledge management
I.Mech.E	• Level of adoption of TES

That said several authors do describe ongoing visions and progress made in the adoption of the TES philosophy in business operations (e.g. NEDTRAIN, Swedish Rail, Metro Rail, and Marine Operations).

Whilst gaps in the literature are far too numerous to discuss meaningfully within this book (perhaps the subject of another publication), the questions identified give a tantalising insight as to the many future challenges and therefore opportunities that exist from the perspective of the literature.

25.3.2 Findings from the Initial 'Think Tank' (Chap. 1)

In seeking to identify future directions for research into the development and application of TES which aligned to the needs of the stakeholders (academic, research, industrial) the UK-EPSRC Centre for Through-life Engineering Services hosted a 'Think Tank' and discussion forum attended by invited guests who had an interest in the concept development and application (Chap. 1). During the associated workshop several key themes emerged which were subsequently ranked by subjective assessment by all those attending. Whilst this process lacked academic rigor it did serve as a '*straw poll*' from experts and practitioners in the field and thus gave initial insight into the areas of concern. The findings were used to inform the

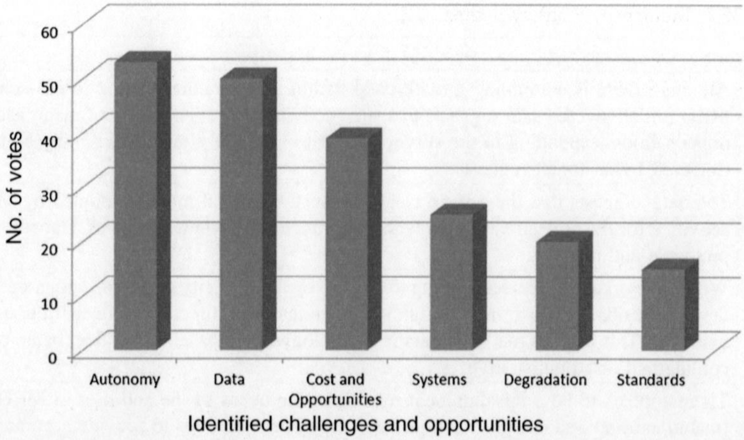

Fig. 25.1 Identification of challenges and opportunities (ranked) by stakeholders attending 2011 TES 'Think Tank' (adapted 1)

direction of future research by the team within the research community. In addition the findings gave insight by way of 'feeling the pulse' of those attending and gave increasing confidence that the work being planned (or currently undertaken) aligned to the needs of stakeholders. Finally, the findings were to become the rationale for the writing of this book and the identification of the section divisions within, with the contributing authors responding to a call for contributions (Fig. 25.1).

25.3.3 Survey of UK Based Manufacturers

In seeking to understand the levels of adoption of through-life engineering services (TES) within the UK manufacturing base *this section refers and reports the findings of a survey conducted in 2011* by the authors. In conducting this survey a stratified sample of the population of the UK based manufacturing sector was identified and studied, returning a sample of 404 UK based manufacturing organisations having a turnover of 10 million GBP or greater, offering B2B contracting and who offered, or had the potential to offer, TES solutions in support of their product offerings. Whilst the survey return rate was low the study did identify interesting insights relative to the opportunities, challenges, drivers and inhibitors as seen through the lens of the organisations responding. A summary of the findings identified by the survey is presented in Table 25.2.

Upon review of the data it is reported that the use of service as a means of achieving competitive advantage driven by a deliberate strategy [3] is at best only just emerging in those companies responding and their application is at best reactionary (Table 25.2—Finding 1). Of those organisations who did state that they actively pursued such a strategy (Rolls Royce, Xerox, Mann Trucks, GE, etc.) [4]

Table 25.2 Summary of survey findings [2]

Findings	
1.	Although there is continuing growth cited within the literature relating to pro-active MRO initiatives, the data suggests that such activities for the majority of manufacturing organisations responding to the survey appear to be mainly reactionary, only being triggered by customer request
2.	The data suggests that there is no clearly defined universal method adopted by MRO activities for the collection of data with such functions using a variety of data collection protocols and methods
3.	Within the organisations responding to the survey the majority of organisations state that they do feedback data to their design and manufacturing functions (either 'in house' or external). This implies that such data is technology/usage based rather than being purely commercial or administrative
4.	There appears to be a misalignment relating to the needs of the end user of MRO data (manufacturing and design functions) and the system abilities to provide accurate and timely responses (technical and usage data) and related information
5.	Challenges exist relating to the quality, quantity and completeness of data. Respondents to the survey suggest that there is too much data, it is fragmented and of poor quality in many cases and seldom complete
6.	Within the MRO field relating to TLES, there are no clear definitions, ontology, or taxonomy which can assist in the categorisation of data, information, and ultimately knowledge
7.	There are issues relating to incomplete understanding of product and component performance due to intermittent failures and No Fault Found (NNF) error signals in MRO systems. This skews the data and can affect decisions relating to MRO strategies
8.	Although organisations stated that they collected MRO data its content, structure, storage and retrieval appears to be ad'hoc in all but a few leading world class OEM's. The data is unduly skewed by the existence of NNF recording within the data, the effect of which is not clear or widely known
9.	Condition based monitoring/management and RFID technology are identified as being key opportunities within the field of TLES
10.	Interest in TLES continues to grow. The majority of organisations responding to the survey state that they have plans to move into TLES with a greater majority state that they have plans to extend the use of acquired MRO data and use this to inform the manufacturing and design functions within their organisation
11.	Whilst there are undoubtedly significant challenges and opportunities relating to the successful adoption of TLES the perceived threats to successful adoption within the practitioner base responding to the survey relate to data management issues arising from the lack of standards and supporting procedures

one finds that this tends to be the result of an emergent strategy [5] (*at least at the beginning of the transformation*) driven by their markets rather than of conscious strategic planning and defined initiatives.

The definition, execution, and successful implementation of such a strategy is also inhibited by the lack of a clearly defined (and standardised) method of acquiring, storing, trending, and reporting MRO and usage data with existing applications (where they exist) employing a variety of protocols and methodologies (Table 25.2—Finding 2).

When seeking to understand the use made of TES data by those companies who state that they harvest MRO/Service/Usage data one finds that organisations are seeking to employ the information gained to inform both design and manufacture (Table 25.2—Finding 3) although further subsequent discussions reveal that the effective holistic and closed loop service performance to design application is only just evolving and the benefits are limited. This could be due to a lack of alignment between the needs of the stakeholders (users, operators, designers, service engineers, and the manufacturing function) (Table 25.2—Finding 4).

Several challenges identified also included the quantity, quality, context, structure, and completeness of the data acquired and its alignment to the needs of the end user (Table 25.2—Finding 5). This is further hindered by the fact that at the time of conducting the survey, no clear definition, ontology, or taxonomy for TES existed (Table 25.2—Finding 6). This is in part addressed within this book (Chap. 2) with a an initial definition for TES being presented which addresses TES's *content, application*, and *purpose* whilst also proposing that the underlying foundation of any TES solution is the generation of knowledge from information obtained by the application of context to the data gathered. This is supported by a proposed illustration of a Taxonomy for TES (Fig. 2.4). Whilst the identity and Taxonomy proposed by the authors in this book is a first documented attempt to address these issues, they are not meant to be substantive and all inclusive, rather an attempt to trigger further much need discussion relative to these issues.

The findings also identified that the existence of 'No Fault Found' (NNF) error signals within systems were also a problem for those organisations who stated that they were using Condition Monitoring technologies in support of their TES solutions. This is well documented in Chap. 4 of this book. The existence of this phenomenon is identified and recorded within the survey (Table 25.2—Findings 7 and 8) and has a significant impact when considering the design of system architecture (Table 25.2—Finding 8).

The development of emerging technologies (i.e. RFID) is also seen as an enabler of successful TES applications as they offer significant part tracking opportunities (logistics) and well as weight reduction solutions when considering the wiring of extensive sensor systems in weight constrained assets (i.e. aircraft) (Table 25.2—Finding 9).

Finally the survey reveals that whilst TES has significant developmental challenges to overcome which are technological, structural, infra-structural, strategic, commercial, contextual, and cultural; organisations still see the adoption of TES as being strategically important to maintaining and increasing competitive positions

25.3.4 Findings from Network Event and Forum Held at the I.Mech.E (London—July 2014)

A recent networking event and forum held at the Institution of Mechanical Engineers in London and attended by invited leading academics, researchers and industrialists sought to further define future challenges and opportunities for TES. Of the invited speakers to the event one finds contributions from UK Government (UK All Party Group on Manufacturing), a leading aero-engine manufacturer, a leading global supplier of military hardware, the Manufacturing Technology Association, the energy sector (major power generator and infrastructure operator), an MRO contractor to high value military assets, and several other representatives of industrial organisations.

In the opening remarks to the event the attendees were introduced briefly to the growing importance to the economy of service in support of manufactured products and a brief insight into the scope of TES was also given (Fig. 25.2).

The importance and drivers of the increasing levels of development and adoption of TES was also illustrated by the increasing levels of Product Service Systems and the increasing levels of servitization being observed as one sees the increasing emergence of the 'manu-service' operation. It is well documented within the literature that manufacturing OEM's are increasingly competing through the application of availability contracting with the majority of contributions to the literature citing the aerospace sector as leading the way. However, the attendees were also informed of the growing importance of services to the UK Marine Sector with the following metrics being cited at the event:

The UK Maritime Sector....

- ...directly created 262,700 jobs
- ...contributed £13.8 Billion to UK GDP
- ...generated £2.7 Billion in UK tax revenues
- ...when including direct, indirect, and induced effects it ...
- ...supports 634,900 jobs
- ...contributes £31.7 Billion to UK GDP

(Source: Oxford Economics—The Impact of the UK Maritime Services Sector)

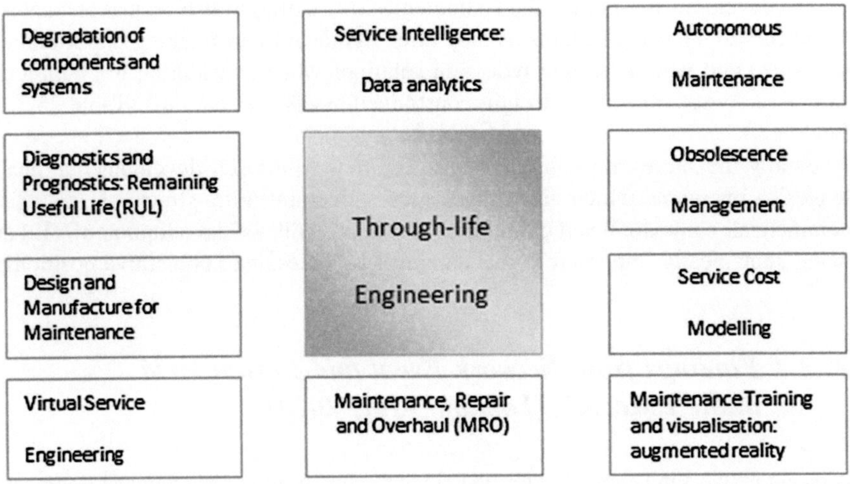

Fig. 25.2 Evolving scope of TES

Table 25.3 Summary of challenges and opportunities identified

	Opportunities	Challenges
Speaker 1	• Emergence of whole new sector	• Attracting youth into engineering
	• Job creation	• The skills gap and requirement to fill emerging positions
		• Working collaboratively across stakeholders
		• Using politicians to influence policy
Speaker 2	• No opportunities offered by this speaker but the editors view the challenges discussed as opportunities for those developing TES	• Getting the design right!
		• Design to a target cost which includes whole life support
		• Education and training
		• Increased knowledge of degradation from a through-life perspective
		• Knowledge management issues—how to capture, trend, apply etc.
		• Short term approaches limit thinking
Speaker 3	• Potential for the supporting technologies to benefit industry is immense	• Need to address and understand how to transition technological benefits into tangible and exploitable benefits for industry
		• Transition processes and times are to lengthy
		• Cultural issues between academia and industry—the need for publications versus the need to address the needs of industry (exploitable outcomes)
		• Co-locating industrial and academic research facilities
Speaker 4	• View is that TES is going to be large	• Asked for feedback relative to success of TES from association members "to a man and woman, they came back with nothing at all!"
	• EU environmental legislation is a driver for TES—EU mandate of ECO-DESIGN for machine tool sector	• TES is important to operations but not yet identified as a sector in its own right
	• Greater efficiencies within whole life delivery of function	• Policies and standards
	• Sustainability of solutions	
	• Increased levels of competitive service strategy	
	• Reduction in whole life cost	
	• Smart logistics and reduced inventory levels	

(continued)

Table 25.3 (continued)

	Opportunities	Challenges
Speaker 5	• Extend operational life of the product or asset	• Improved life-time management understanding and control systems
	• Increased reliability of product function	• Greater understanding' of 'in the field' or 'as installed' asset/product stock potential
	• Closer aligned strategies to whole-life function	• Improved cost models to understand the whole-life economic of running the system
	• Improved management of whole-life deployment	• Improved understanding of the needs of the supply chain and alignment of operating and support strategies
	• Improved whole life systematic approaches	• Challenges with knowledge transfer regimes—alignment and improvement
	• Greater understanding of degradation and obsolescence issues	
	• Improved investment agreements	
	• Greater levels and security of employment through-out the products life	
	• Increased and sustained revenue	
Speaker 6	• Increased competitive advantage	• Investment levels
		• Optimum MRO systems
Speaker 7	• Extension of operation life for assets	• Improved cost modelling and whole-life forecasting to work within target long term financial budgets
	• The need to inform standards through-out all levels of the concept	• Need for greater understanding of obsolescence issues
	• Improved information systems	• How to define and design through-life contracts which align to the reality of extended life function
	• Biggest opportunity and also challenge is this is all about understanding contracting…the move from short term to long term agreements	• Cross sector standard ontologies, taxonomies
		• Culture—conduct throughout a whole-life contract
		• The data rich but knowledge poor problem
		• Biggest issue—"how can I train people…."

(continued)

Table 25.3 (continued)

	Opportunities	Challenges
Speaker 8	• Through-life support systems create a through-life record of the company performance	• Complexity—business, sector, technology, products etc.
	• Systems are able to inform with greater accuracy where products/ assets are and arrival times	• Culture—view is in the short term and should move to whole-life delivery of function
	• Improved MRO decisions	• Cash—winners are new projects, losers are the holders of the support budgets especially when corporate figures have to be balanced
	• Information systems aid diagnostics and in some cases remove the need for MRO investigations	• Lacking skills within the business
	• Improved visual aids will help MRO engineers with augmented reality making repair process shorter.	
	• Improved information and knowledge management	

In opening the event the opening speaker stated that the objectives of the event were threefold, namely:

- To identify facts and business opportunities for the 'through-life support' business within the UK
- To identify the challenges faced by UK Manufacturing and Government organisations in through-life support
- To identify policy implications for the through-life support business growth.

(Source: Unpublished but circulated event documentation)

It can be seen when comparing the comments made (Table 25.3) by the invited speakers with the previous findings (Sects. 25.3.1–25.3.3) that there appears to be an emerging consensus relative to drivers, challenges (inhibitors) and opportunities for TES.

The drivers are seen to consistent. They being the need to mitigate risk to the organisation's revenue stream resulting from the loss of the product's design function when competing through the application of availability contracting. In addition, as the manufacturer seeks to compete in the global market they look at ways of protecting and improving their competitive position. Whole-life agreements can result in customer 'lock in' and the service support infrastructure becomes harder to reverse engineer thus further protecting revenue streams and commercial agreements. The concept is further driven by sustainability issues and obsolescence considerations.

In reviewing the inhibitors they are seen to fall into several groupings (i) Technical, (ii) cultural, (iii) commercial, (iv) financial, and (v) human resource. These inhibitors however also present significant opportunities and are not seen as a negative proposition. In seeking to overcome the aforementioned inhibitors (and therefore challenges) major benefits can be achieved in all of the categories identified. As one of the speakers stated, “the potential for the supporting technologies…[TES]… is immense”. The concept is seen by many as a ‘Cinderella’ application of technology but as it matures it offers the potential to enable a major shift in the way that manufacturing businesses generate revenue. This is seen as an essential eveolution of the business paradigm by such organisations as GE, Rolls Royce, Xerox, (to name but a few) who have all seen the majority of their revenue generated by the provision of through-life services.

25.4 Concluding Remarks

This compendium of contributions from leading exponents within the field of TES offers a structured reference to the reader. It has been written for multiple audiences which include postgraduate students, academics, researchers, practitioners and consultants. The sections of the book fulfill a need that was identified by conducting an initial workshop to discuss and identify the focus of further work to fulfill stakeholder needs.

This work offers the first definition for TES which addresses content, context, structure and scope. In addition an initial taxonomy is offered which it is hoped will initiate further development and discussion. The hypothesis that the underlying foundation to effective TES application is knowledge and the application of that knowledge in a through-life context is also defined.

The significance of data, diagnostics, prognostics and the impact of No Fault Found (NFF) on system design is also introduced in addition to system degradation and it’s relation to design. This is supported by the acknowledgement of the importance of sound knowledge of component degradation mechanisms (the root cause, mitigation, identification using novel NDE techniques, and their repair.

The need to conduct effective maintenance planning through the use of effective modelling techniques becomes ever-more important as organisations seek optimum solutions to the application of TES in support of manufactured products. In order to address this and offer insight into some of the techniques being developed, several modelling tools and methodologies are offered.

The authors suggest that the primary role of TES is to mitigate risk. It does this by increasing the whole-life product function availability for use. Key to this are considerations relative to cost of TES solutions and consideration of such risk and associated uncertainty though-out the life cycle (including obsolescence). An essential element identified is the complexity of such solutions (technical, cultural, structural, infrastructural, commercial etc.). Whilst not always explicit, the importance and relevance of the concept of risk management, obsolescence, and standardization has been present throughout each of the chapter contributions. These are introduced explicitly within section E of this book.

Autonomous maintenance and the use of self-healing has also been introduced. As ever-more complex products are designed and built it is inevitable that many will work in environments which are either not easily accessible, or totally inaccessible to man (e.g. Submarines, Space Craft, Satellites, Nuclear Power Stations (reactors), or within the human body applications (nanobots). With the emergence of such applications the need for self re-configuration/repair becomes increasing evident. The drive for more autonomy for machines in systems could influence through-life engineering services in two ways. Firstly, the recognition that systems, including unmanned vehicles, will feature longer periods of autonomous operation with less frequent recourse to human intervention; secondly, the results of research into the myriad aspects of autonomy will be applicable to the engineering life cycle of manned as well as unmanned systems whether it be as improved decision aids for humans or the means of identifying early signs of degradation or failure, or indications of cyber-attack.

The systems of interest are the partnerships between humans and machines with the division of tasks between them dependent on circumstances. They can be vehicles, military and civil, critical infrastructure, and many other forms in combination, and can involve networks of global extent. This drive to achieve greater autonomy is sometimes expressed as the need to replace humans when they are faced with dull, dirty or dangerous tasks. There are many cases that are military, and the civil examples include long flights, nuclear plant, driverless cars, and care of the elderly. Among the topics of research are: decision making by machines, interpretation of data from sensors (in particular replacing interpretation by humans of images), secure and resilient communications, defence against cyber-attack, planning algorithms that can re-plan as the situation changes, power and energy management, resilience through self-healing in electronics and in mechanical structures, diagnostics and prognostics. Systems issues are demanding: these include adaptive architectures, verification and validation, safety and certification.

Trust between the human and machine is a critical factor and an important consideration in deciding the structure of decision making. When a human does not have to be accommodated in a vehicle then much longer endurance can be contemplated for, say, an air vehicle to engage in surveillance operations. Several mechanisms then compete to limit the duration. Among these are fuel efficiency [hence the interest in new energy sources and energy scavenging] and reliability. This drive for higher availability stimulates research into means of replacing known sources of reliability problems such as cables and connectors. An example from the work of the UK programme Systems Engineering for Autonomous Systems Defence Technology Centre {SEAS DTC} is the use of surface waves instead of cables and connectors. A possible application which can be envisage is propagating electromagnetic signals across surfaces with electronics embedded in structural materials comprising the interior surface of a wing or fuselage.

New materials will enable the use of multiple small sensors to monitor the condition of equipment. Smart materials and miniaturisation in manufacture projects are important to TES and extensive continuous monitoring would contribute to solving the vexing problem of no fault found. Facilitated by increased computing power enabling the processing of large amounts of data from many small sensors. The drive for autonomy has stimulated attempts to construct computer architectures that borrow aspects of the brain's architecture. A recent example is IBM's TrueNorth with 5.4 B transistors arranged into 4,096 neurosynaptic cores with a power consumption of 70 mW. This features high fault tolerance in a distributed highly parallel event-driven architecture.

Finally, the problem of the lack of suitably trained engineers who have knowledge and understanding of the 'new world', that is the design and manufacture of complex products based upon a thorough understanding of the aforementioned principles and to which enable extended and guaranteed delivery of the design function throughout a product's extended life is sadly lacking (Table 25.3). This book and the compendium of contributions within offers a support text to the training of students and the provision of better understanding of some of the dimensions of Through-life Engineering Services for those at a later stage of their career.

References

1. Shaw A (2012) Through-life engineering services: think tank report (Presentation slide pack)
2. Redding LE, Tiwari A, Roy R, Phillips P, Shaw A (2014) The adoption and use of through-life engineering services within UK manufacturing organisations. Proc I Mech E, Part B, J Manuf (in press)
3. Mintzberg H (2000) The rise and fall of strategic planning. Prentice Hall, London
4. Baines TS, Lightfoot HW, Evans S, Neely A, Greenough R, Peppard J et al (2007) State-of-the-art in product-service systems. Proc Inst Mech Eng Part B-J Eng Manuf 221:1543–1552
5. Mintzberg H, Lampel J (1999) Reflecting on the strategy process. Sloan Manag Rev 40 (3):21–30

Appendix A
Cyber Security and the Internet of Things (IoT)

A.1 Introduction

Upon finishing this manuscript and during the preparation for the publication phase of this undertaking, the editors having discussed the content and format of the work, had, as is often the case, additional insight relative to future challenges and opportunities that emerged from both their own research focus and that of the contributing authors, supported by the continuing developments within the literature. In Chap. 2 of this book, Redding had proposed that the foundation upon which Through-life Engineering Services (TES) is built is related to the collection, storage, trending, retrieving, and application of knowledge. In this chapter he proposes an initial taxonomy for TES based upon six facets of knowledge in the hope that it will spur others to develop a much need development of this area.

Informed in part by this taxonomy Redding, supported by one of his industrial sponsors suggests an initial TES knowledge architecture which can be used at the organisational level across several stakeholder lenses (Fig. A.1).

The illustration shows the distinction between '*Data*', '*Information*', and '*Knowledge*'. In each column the reader can see the data sources; the tools, platforms and techniques used to generate information; and the stakeholders to the knowledge presented. Hopefully from the aquisition and application of such knowledge, stakeholder (and organisational) wisdom is achieved.

The figure shows that central to the architecture is a set of TES databases and libraries that could be knowledge based (referring to initial taxonomy), which in turn could interface with standard business operating systems (e.g. SAPP). Typically such a solution would not necessarily be solely organisation based. For example, the design function could be a sub-contract entity, the inspection function could be within a third party manufacturer of service provider, and the service engineer may communicate with the database or elements of the architecture remotely. It is easily seen that access to the architecture is not only within the OEM but also by way of the virtual (or remote) team working at all stages of the life-cycle. In addition to direct functional activates (design, manufacture, and service) there is also the need for support activities (logistics, inventory, and product

L. Redding and R. Roy (eds.), *Through-life Engineering Services*,
Decision Engineering, DOI 10.1007/978-3-319-12111-6

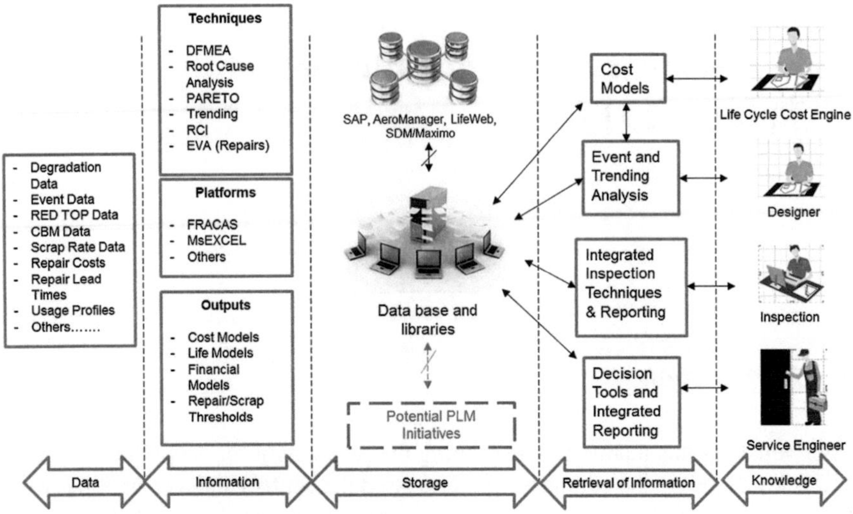

Fig. A.1 A potential future service knowledge based system architecture (*Source: L Redding 2014 – Unpublished Report*)

tracking) to be facilitated if such an architecture is to fully support the design function, the service delivery system, and the interaction between them.

In consideration of these developments the authors identify two related future research areas which are aligned to the development of TES. They being (i) Cyber Security and (ii) The Internet of Things (IoT).

A.2 Cyber Security

It does not require a very big leap in the imagination to appreciate that as organisations seek ever more complex system architectures in support of their products (Fig. A.1) which have the facility to acquire data from third party systems and portals which are external to the parent OEM architecture the issue of cyber security becomes evermore important. In addition, when reviewing the media it does not take long to discover recurring media articles relating to attacks on corporate systems from other overseas bodies (be they national or commercial bodies) [1, 2].

These vulnerabilities in system architecture has served to ensure that the development of cyber security initiatives to defend against hacking, and uploading of Malware, Trojans, Viruses and Tracking Algorithms for the purpose of State and/or Industrial sponsored espionage or just wanton vandalism is of paramount importance. Whilst not exhaustive or academically rigorous a basic list of the types of attack that can take place upon the system and associated network is offered by Wikipedia:

- Backdoors
- Denial of Service Attack
- Direct Access Attack
- Eavesdropping
- Exploits
- Indirect Attacks

[Source: Wikipedia] [3]

This book and its appendix does not seek to discuss these differing methods of cyber attack. It is enough for the editors to advise of the concern and direct the reader to the academic literature which deals with these different elements of cyber security. However, the existence of such computer 'warfare' is significant to the successful adoption and application of TES solutions if delivery of the design function to the product user is not to be impeded. The editors suggest that this is an area of TES that has not, at the time of publication of this book, received any research focus explicitly to TES solutions and serves as a rich vein of future opportunity, the results of which are essential to effective TES applications. It is of particular vital importance when considering the autonomous and remote configuration of complex engineering products.

Central to any future development are the increasing role of standards and Information Assurance (IA) procedures which are relatively young in their development. Typically one sees the following standards and governing bodies emerging:

- Standards
  - ISO/IEC: 27000:2005 *Information Technology—Security Techniques—Information Security Management Systems—Requirements*
  - ISO/IEC 21827:2008 *Systems Security Engineering Capability Model*
  - ISA-62443 Parts 1–4 General, Policies and Procedures, System, and Component Standards
- Bodies
  - NERC (North American Electric Reliability Corporation)
  - NIST (National Institute of Standards and Technology)
  - ISO (International Standards Organisation)
  - BSI (British Standards Institute)
  - ANSI (American National Standards Institute)

Future research into system architectures in support of TES should include consideration of the above. Currently there is very little by way of standards in this arena with the landscape resembling the 'wild west' frontier. Any initiatives and developments that exist appear to be in silos which are protected by research establishment and industrial IP agreements. This area potentially is a '*winner takes all*' with the '*first past post*' potentially dictating the future standards, domains, protocols, and architecture design standards. This is of particular relevance when we consider the '*Internet of Things*' (IoT).

A.3 The Internet of Things (IoT)

The Internet of Things (IoT) is an emerging area of research which seeks to develop the interconnection of existing and emerging products and entities which possess uniquely identifiable embedded computing ability (or sensors) to the existing internet (or web) architecture and the 'Cloud'. The aim of this evolutionary development is to improve the connectivity of machines, systems and architectures which goes far beyond existing dedicated machine to machine communications (M2M) [4]. The vision for the concept is a world where all sensor enabled products (or assets) are interconnected thus creating a smart network in which product usage profiles, forecasting and mitigation of degradation of design function, billing for use, and planned obsolescence strategies can be applied. Quite simply IoT holds the potential to facilitate the application of remote solutions to a plethora of issues, the magnitude of which we have yet to discover.

> *Today computers—and, therefore, the Internet—are almost wholly dependent on human beings for information. Nearly all of the roughly 50 petabytes (a petabyte is 1,024 terabytes) of data available on the Internet were first captured and created by human beings—by typing, pressing a record button, taking a digital picture, or scanning a bar code. Conventional diagrams of the Internet ... leave out the most numerous and important routers of all—people. The problem is, people have limited time, attention and accuracy—all of which means they are not very good at capturing data about things in the real world. And that's a big deal. We're physical, and so is our environment ... You can't eat bits, burn them to stay warm or put them in your gas tank. Ideas and information are important, but things matter much more. Yet today's information technology is so dependent on data originated by people that our computers know more about ideas than things. If we had computers that knew everything there was to know about things—using data they gathered without any help from us—we would be able to track and count everything, and greatly reduce waste, loss and cost. We would know when things needed replacing, repairing or recalling, and whether they were fresh or past their best. The Internet of Things has the potential to change the world, just as the Internet did. Maybe even more so* [4].

In this new world, every product is fitted with at least a sensor which communicates with the macro-network, through to an on-product CPU which could process data autonomously, only communicating with the network as required to do so (at usage intervals, or by exception—function degradation or failure). The challenges and opportunities which exist this frontier of technology are too numerous to even start to categorise. With the rise of IoT the research and practitioner community, not to mention society in general, are about to enter a voyage of discovery, the direction and destination of which holds no bounds. It really is a leap into the unknown with many challenges ahead which reside in all domains (scientific, technological, legislative, economic, political, environmental, procedural, etc).

This appendix has not sought to inform the reader of the academic content relative to the concepts identified, rather it served to documented by précis a passing discussion between the editors. TES is developing during the emergence of a 'new world'. The challenges and opportunities which exist are extensive. The editors hope that through the publication of this book they have given insight to the reader into what lies ahead and helped lay part of the foundation to address the new world.

References

1. NBC News (2013) Bank Cyber Attack: a major attack on finance industry. http://www.nbcnews.com/video/nightly-news/51435096/#51435096. Accessed 26th Aug 2014
2. Ring T, Morbin T (2013) Cyber attacks hit UK banks, SC magazine: for IT security professionals, 29 Nov 2013http://www.scmagazineuk.com/cyber-attacks-hit-uk-banks/article/323030/ Accessed 26 Aug 2014
3. Wikipedia, Computer Security. http://en.wikipedia.org/wiki/Cyber_security. Accessed 26 Aug 2014
4. Wikipedia, The Internet of Things. http://en.wikipedia.org/wiki/Internet_of_Things. Accessed 26 Aug 2014
5. Ashton K (2009) That 'Internet of Things' Thing, in the real world things matters more than ideas. RFID J (22 June 2009)